Cardiorenal Syndromes in Critical Care

The present book has been published to offer a scientific guide to the participants of the 19th International Vicenza Course on Critical Care Nephrology and in conjunction with the birth of the

International Renal Research Institute of Vicenza

Contributions to Nephrology

Vol. 165

Series Editor

Claudio Ronco Vicenza

Cardiorenal Syndromes in Critical Care

Volume Editors

Claudio Ronco Vicenza
Rinaldo Bellomo Melbourne, Vic.
Peter A. McCullough Royal Oak, Mich.

44 figures, 1 in color, and 28 tables, 2010

Basel · Freiburg · Paris · London · New York · Bangalore · Bangkok · Shanghai · Singapore · Tokyo · Sydney

Contributions to Nephrology
(Founded 1975 by Geoffrey M. Berlyne)

Claudio Ronco
Department of Nephrology,
Dialysis & Transplantation
International Renal Research Institute
San Bortolo Hospital
Viale Rodolfi 37
IT-36100 Vicenza (Italy)

Rinaldo Bellomo
Department of Intensive Care
Austin Hospital
Melbourne, Vic. 3084 (Australia)

Peter A. McCullough
Division of Nutrition and
Preventive Medicine
William Beaumont Hospital
4949 Coolidge Highway
Royal Oak, MI 48073 (USA)

Library of Congress Cataloging-in-Publication Data

Cardiorenal syndromes in critical care / volume editors, Claudio Ronco,
Rinaldo Bellomo, Peter A. McCullough.
p. ; cm. -- (Contributions to nephrology, ISSN 0302-5144 ; vol. 165)
Includes bibliographical references and indexes.
ISBN 978-3-8055-9472-1 (hard cover : alk. paper)
1. Renal intensive care. 2. Kidneys--Diseases. 3. Heart--Diseases. I.
Ronco, C. (Claudio), 1951- II. Bellomo, R. (Rinaldo), 1956- III. McCullough,
Peter A. IV. Series: Contributions to nephrology, v. 165. 0302-5144 ;
[DNLM: 1. Kidney Failure, Acute--complications. 2. Critical Care. 3.
Heart Failure--complications. 4. Kidney--injuries. 5. Kidney Failure,
Acute--physiopathology. W1 CO778UN v.165 2010 / WJ 342 C2695 2010]
RC903.C27 2010
616.6′14028--dc22 2010009321

Bibliographic Indices. This publication is listed in bibliographic services, including Current Contents® and Index Medicus.

Disclaimer. The statements, opinions and data contained in this publication are solely those of the individual authors and contributors and not of the publisher and the editor(s). The appearance of advertisements in the book is not a warranty, endorsement, or approval of the products or services advertised or of their effectiveness, quality or safety. The publisher and the editor(s) disclaim responsibility for any injury to persons or property resulting from any ideas, methods, instructions or products referred to in the content or advertisements.

Drug Dosage. The authors and the publisher have exerted every effort to ensure that drug selection and dosage set forth in this text are in accord with current recommendations and practice at the time of publication. However, in view of ongoing research, changes in government regulations, and the constant flow of information relating to drug therapy and drug reactions, the reader is urged to check the package insert for each drug for any change in indications and dosage and for added warnings and precautions. This is particularly important when the recommended agent is a new and/or infrequently employed drug.

www.karger.com
Printed in Switzerland on acid-free and non-aging paper (ISO 9706) by Reinhardt Druck, Basel
ISSN 0302–5144
ISBN 978–3–8055–9472–1
e-ISBN 978–3–8055–9473–8

Contents

Preface

Critical care nephrology is an emerging multidisciplinary field in which the competences of different specialists merge into a unified diagnostic and therapeutic approach to the critically ill patient. In the past editions of the International Vicenza Course, attention was placed on multiple organ failure, sepsis and renal replacement therapy in the intensive care unit setting. This volume and the relevant course place great emphasis on cardiorenal syndromes and the multidisciplinary collaboration between cardiology and nephrology. A series of important contributions describe the cardiorenal syndrome in its different varieties and subtypes, report the results from the most recent Acute Dialysis Quality Initiative consensus conference, and propose new diagnostic approaches based on early biomarkers of acute kidney injury. The traditional discussion on the most recent and updated technology for renal replacement therapy and multiple organ support therapy is part of the program and some contributions reflect such important advances. Special emphasis is placed on the potential role of extracorporeal therapies in patients affected by H1N1 influenza. A summary of the most recent trials in the field is also included in the book. The 19th International Vicenza Course on Critical Care Nephrology is highlighted by the timely publication of this volume. The book contains the contributions of all members of the faculty allowing the participants to retain a great part of the information delivered at the course but also allowing the readers of the book that could not attend to obtain comprehensive information about the content of the course. The Vicenza Course has yet another exceptional group of speakers and we are excited that this volume of *Contributions to Nephrology* is ready to guide attendees through the event and serve as a valuable tool for the future. This result has been made possible thanks to the invaluable effort of the team of the Department of Nephrology at San Bortolo Hospital and the International Renal Research Institute of Vicenza. We are also grateful for the important editorial effort undertaken by Professor McCullough and Professor Bellomo to make all chapters consistent and well balanced. We truly appreciate the help of the members of the faculty who have made this possible by submitting their manuscripts in advance. Our sincere thanks go to Anna Saccardo, Ilaria Balbo and Marta Scabardi who worked in Vicenza on the final arrangement of the course. Special thanks also go to our team in Vicenza and to Karger Publishers for the usual outstanding quality of the publication.

We hope that the readers will enjoy the book and will consider it as a companion for daily clinical practice as well as for future research.

C. Ronco, Vicenza
R. Bellomo, Melbourne, Vic.
P.A. McCullough, Royal Oak, Mich.

Ronco C, Bellomo R, McCullough PA (eds): Cardiorenal Syndromes in Critical Care.
Contrib Nephrol. Basel, Karger, 2010, vol 165, pp 1–8

Epidemiology of Acute Kidney Injury

Eric A.J. Hoste[a,b] · John A. Kellum[b] · Nevin M. Katz[c] · Mitchell H. Rosner[d] · Michael Haase[e] · Claudio Ronco[f]

[a]Intensive Care Unit, Ghent University Hospital, Ghent University, Gent, Belgium; [b]The CRISMA (Clinical Research, Investigation, and Systems Modeling of Acute Illness) Laboratory, Department of Critical Care Medicine, University of Pittsburgh School of Medicine, Pittsburgh, Pa., [c]The George Washington University Medical Center, Washington, D.C., [d]University of Virginia Health System, Division of Nephrology, Charlottesville, Va., USA; [e]Department of Nephrology and Intensive Care Medicine, Charité University Medicine, Berlin, Germany, and [f]Department of Nephrology, Dialysis & Transplantation, International Renal Research Institute, San Bortolo Hospital, Vincenza, Italy

Abstract

Different definitions for acute kidney injury (AKI) once posed an important impediment to research. The RIFLE consensus classification was the first universally accepted definition for AKI, and has facilitated a much better understanding of the epidemiology of this condition. The RIFLE classification was adapted by a broad platform of world societies, the Acute Kidney Injury Network group, as the preferred AKI diagnostic and staging system. RIFLE defines three increasing severity stages of AKI. One- to two-thirds of intensive care unit (ICU) patients develop AKI according to these criteria which is associated with worse outcomes such as increased length of ICU stay, costs, and mortality. Over the last decade the incidence of AKI has increased, probably as a consequence that baseline characteristics of ICU patients have changed. Another factor that may explain this is that more patients are treated in clinical settings that are associated with high risk for development of AKI. In addition, there may be genetically predetermined risk profiles for development of AKI such homozygotes for the low activity form of the COMT gene. Mortality of AKI patients has decreased over the last few decades, especially when underlying severity of illness is considered. An important consequence of this is the increasing number of surviving AKI patients who develop chronic kidney disease and end-stage kidney disease. In the specific setting of cardiac surgery, AKI occurs in 19–45% of patients. Renal replacement therapy is necessary in approximately 2% of this cohort. AKI that occurs within a 7-day period after cardiac surgery is related to perioperative risk factors, such as preexisting chronic kidney disease, acute ischemia, aorta cross-clamping, or use of cardiopulmonary bypass. AKI that occurs after the first week is mostly a consequence of sepsis or heart failure.

Over the past 50–60 years the concept of acute kidney injury (AKI) has evolved enormously. In the 1950s, AKI was only considered when kidney function was absent, i.e. acute renal failure. It is from these days that the first patients were successfully treated with primitive dialysis machines designed by the Dutch pioneer William Kolff. Since then, the emphasis gradually moved to recognition of the importance of less severe kidney injury. During this process, over 35 different definitions for acute renal failure were used in medical literature. These definitions used different biomarkers such as blood urea nitrogen, urine output, and serum creatinine concentration or a combination of these, and described different severity grades of AKI. The Acute Dialysis Quality Initiative (ADQI) published in 2004 a consensus definition, the RIFLE classification, that was rapidly accepted by the medical community [1]. This definition was later accepted in a slightly modified form by the Acute Kidney Injury Network (AKIN), which is a consortium uniting representatives from all major international nephrology and critical care societies (fig. 1) [2]. Later, ADQI issued specific guidelines for the use of this definition in the setting of cardiovascular surgery [3]. The introduction of these definitions has been a hallmark in critical care nephrology. Less than 6 years ago, almost every study on AKI used a different definition, which made comparison of study results almost impossible. Today, almost all studies use the RIFLE classification in its original or modified form allowing comparison of epidemiologic results, and interventional studies with external validity.

Conceptual Model for Acute Kidney Injury

The current concept of AKI is that it is in fact a continuum of different severity grades of kidney dysfunction, ranging from decreased glomerular filtration without cellular damage to severe AKI with anuria. As most intensive care unit (ICU) patients in the developed world are older than 50, and suffering from chronic co-morbid diseases such as hypertension and diabetes, many patients will already have chronic kidney disease (CKD). Secondary to another disease, such as infection, the kidney may experience damage without reduction in glomerular filtration rate (GFR). When the initiating nephrotoxic event persists, or another is added (e.g. aminoglycosides, or administration of iodine radiocontrast for a diagnostic CT scan), the damage may evolve to decreased GFR, and eventually even failure, when there is absence of glomerular filtration, as in oligoanuria. Conversely, GFR may be decreased, for example in prerenal states, before damage occurs. In both circumstances, it seems logical that therapeutic interventions for AKI should be administered as early as possible before damage is irreversible, or when GFR is only moderately decreased.

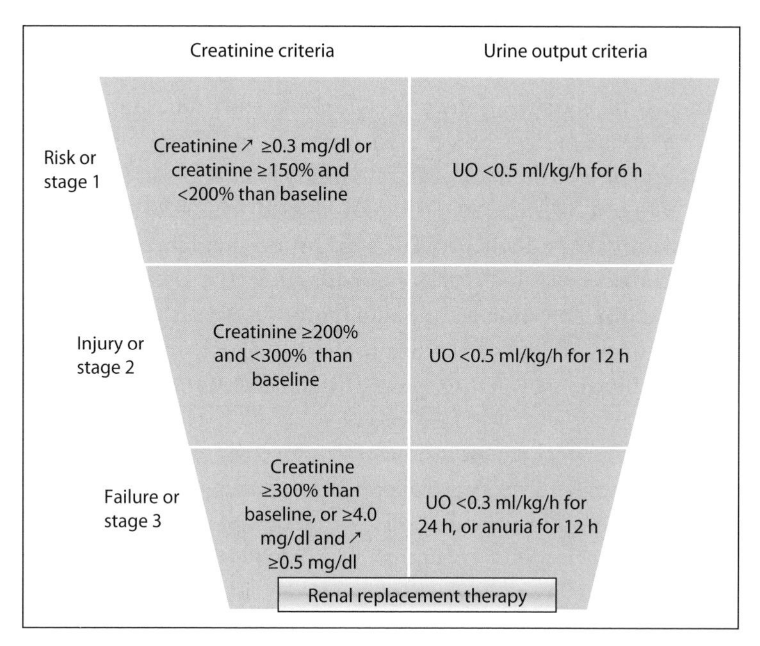

Fig. 1. Classification of AKI: patients are classified according to the worst of creatinine or urine output criteria. Patients do not need to fulfill both creatinine and urine output criteria. Diagnosis of AKI on the basis of creatinine criteria is fulfilled when patients experience an increase of serum creatinine ≥0.3 mg/dl or >150% within a 48-hour period. Figure is adapted to Bellomo et al. [1].

Classification of Acute Kidney Injury

The RIFLE classification defines three increasing severity classes of AKI (Risk, Injury and Failure) on the basis of either a relative increase of serum creatinine or an episode of oliguria (fig. 1). The two outcome classes (Loss and End-Stage Kidney Disease) are defined on the basis of the duration of renal replacement therapy (RRT). The modifications that were proposed upon acceptance of these criteria by AKIN can be summarized as follows: (a) broadening of the 'Risk' category of RIFLE to include an increase in serum creatinine of at least 0.3 mg/dl even if this does not reach the 50% threshold; (b) setting a 48-hour window on the first documentation of any criteria; (c) categorizing patients as 'Failure' if they are treated with RRT regardless of what their serum creatinine or urine output is at the point of initiation, and (d) AKIN also proposed that stages 1, 2 and 3 be used instead of R, I and F. These modifications indeed classify more patients with AKI on the basis of the absolute increase of serum creatinine criterion, however many patients are also not identified as a result of the 48-hour time window [4]. At present, RIFLE has been validated in more than 550,000 patients around the world.

Epidemiology and Outcomes of AKI in Hospital and ICU

AKI defined by the RIFLE classification varies widely across studies due to differences in the populations being studied. AKI was present in 18% of patients admitted to the emergency room in an Australian hospital, and varies in general ICU patients between 30 and 70% [5]. AKI treated with RRT occurs in approximately 4–5% of ICU patients [6]. The population incidence was 2,147 patients per million inhabitants in a region in Scotland. In the USA, Xue et al. [7] have demonstrated that over the last decade there was a yearly increase of patients with the diagnosis of AKI of 11%. The change in patient demographics in ICUs in the Western world is probably the most important factor for this.

Length of Stay

Increasing RIFLE class is associated with increasing length of stay in the hospital and ICU, and therefore, also with greater use of hospital resources and costs [5].

End-Stage Kidney Disease (ESKD)

An increasing proportion of patients with AKI has preexisting CKD. These patients are especially at risk for developing ESKD. Another clinical scenario is that there is incomplete recovery of kidney function, leaving the patients with even more pronounced CKD after hospital discharge. These patients have lower long-term survival (even at 5 years), and are at greater risk for developing ESKD in the years following ICU survival [8].

Mortality

It is clear that severe oliguric AKI leads to death within days unless treated with RRT, but also less severe AKI is independently associated with increased mortality. This was first reported in patients with less severe AKI, such as in contrast nephropathy. Interestingly, the RIFLE classification allowed to establish that there is a strong relationship between severity of AKI and mortality across the entire severity spectrum. Several studies demonstrated that after correction for various covariates, increasing RIFLE class is associated with increasing risk for in-hospital mortality.

The mortality of AKI treated with RRT as reported in the medical literature has remained more or less stable between 50 and 60% over the last 40 years. However, the patient profile changed dramatically over this period. In the period 1960–1980, we treated predominantly younger patients with less co-morbidities, and less severe organ dysfunction. Older and more severely ill patients were often denied RRT because of anticipated futility. In recent years the average ICU patient treated with RRT is typically older than 55 years age, with two or more associated organ dysfunctions [6]. Despite this, longitudinal observations in the USA and elsewhere show that mortality is decreasing [5, 7].

Incidence and Outcomes of Acute Kidney Injury in Cardiac Surgery

AKI has long been recognized as a relatively common complication after cardiac surgery and one that carries significant morbidity and mortality. AKI is reported in 19–45% of cardiac surgery patients [9]. RRT is required in approximately 2% of cardiac surgical patients, and if it is needed, mortality increases to 50–60% [10].

Multivariate analyses have identified independent risk factors for cardiac surgery-associated acute kidney injury (CSA-AKI). They include age, gender, preoperative renal disease, low ejection fraction, diabetes, use of an intra-aortic balloon pump, emergency surgery, mitral valve surgery, reoperation, and duration of cardiopulmonary bypass. Coronary bypass surgery performed off-pump may be associated with a lower risk of AKI compared to surgery performed on-pump. The occurrence of AKI within 7 days of cardiac surgery is usually due to preoperative and operative factors. Preoperative factors are often related to acute ischemia or myocardial infarction. Operative factors include if and how the aorta is cross-clamped, and how anticoagulation and cardiopulmonary bypass are managed. AKI that occurs later, between 7 and 30 days of surgery, often can be related to postoperative factors such as low cardiac output or sepsis.

Although the definition of CSA-AKI has traditionally included the need for dialysis, it is now recognized that classification systems that include various levels of renal dysfunction provide valuable information [3]. In this regard, the RIFLE criteria have importantly refined the knowledge base of CSA-AKI and facilitate early management.

Risk Factors for AKI

Given that AKI is associated with significant morbidity and mortality, it is critically important to prevent its development. The first step in a preventative strategy is to identify those risk factors that are associated with AKI and that can be intervened upon to lower the risk. Broadly speaking, risk factors for AKI can be divided into those that are specific to clinical settings and those that are patient-specific. Clinical setting-associated risk factors include: sepsis, need for ICU care, multiorgan failure, certain postoperative states (such as cardiac or vascular surgery), trauma, burns, human immunodeficiency virus infection, liver disease, non-renal solid organ transplantation and bone marrow transplantation. Patients who enter these clinical settings require vigilance for evidence of the development of AKI as well as measures to prevent it (such as optimization of hemodynamics and avoidance of nephrotoxic insults). It is also important to realize that nearly two-thirds of patients who develop AKI suffer from more than one insult. Patient-specific risk factors include: advanced age, impaired renal function CKD, diabetes mellitus, impaired cardiac function, volume depletion/hypotension, and use of nephrotoxic medications and radiocontrast agents. Common nephrotoxic medications include: non-steroidal anti-inflammatory agents, drugs that block the renin-angiotensin system, aminoglycoside antibiotics, amphotericin B, calcineurin inhibitors, chemotherapeutic agents (cisplatin, ifosfamide), herbal remedies and toxic ingestions (such as ethylene glycol). In some cases, recognition of the nephrotoxic potential of these drug exposures allows for a risk reduction strategy, for instance substituting liposomal formulations of amphotericin B or using once-daily aminoglycoside dosing.

Identification of risk factors has been used to develop risk stratification tools that identify the subset of patients at highest risk for AKI and thus would benefit from close observation and renal protective strategies. These tools are most useful when the renal insult occurs at a predefined time (such as surgery or radiocontrast agent administration) thus allowing implementation of a risk reduction strategy before or at the time of the insult [11]. Several risk stratification scoring systems exist and it is important to realize that these scoring systems have good negative predictive power but poor positive predictive power. Furthermore, many of the more common causes of AKI, such as sepsis, cannot be easily predicted and most patients in these settings often present to the hospital with significant renal damage. Thus, in these cases, risk stratification becomes less important.

Role of Genetic Factors in AKI: Catechol-*O*-Methyltransferase

Circulating catecholamines are primarily catabolized through enzymatic pathways involving the enzyme catechol-*O*-methyltransferase (COMT). A G-to-A polymorphism in the fourth exon of the COMT gene results in a valine-to-methionine amino acid substitution at codon 158, which leads to thermolability and low ('L'), as opposed to high ('H'), enzymatic activity. Genetically determined COMT activity is important to regulation of affective mood and pain perception and also influences outcomes in patients with ischemic heart disease. In the kidney, COMT is essential for catecholamine degradation along the distal parts of proximal tubules and thick ascending limb of loop of Henle.

In a cohort of 260 post-bypass surgery patients, 64 (24.6%) were homozygous (LL), 123 (47.3%) heterozygous (HL), and 73 (28.1%) homozygous (HH) for COMT [12]. Patients who were homozygous for the low activity *COMT*

L allele (LL patients) presented with increased concentrations of plasma catecholamine and MAO-dependent catecholamine degradation products underlining genotype-phenotype correlations. These patients more commonly developed prolonged vasodilatory shock, AKI, more severe AKI requiring RRT, and prolonged hospital stay. After adjustment for important covariates, the LL allele was independently associated with vasodilatory shock, AKI, and prolonged ICU and hospital stay, and we observed a trend for higher mortality. Nevertheless, preoperatively, LL patients could not be distinguished from HL or HH patients in terms of any other known risk factors. Cross-validation analysis revealed a similar graded relationship of adverse outcomes by genotype. Therefore, COMT LL homozygosity is an independent risk factor for shock, AKI, and hospital stay after cardiac surgery. This therefore provides another piece of evidence for a genetic base in the development of shock and AKI.

In summary, AKI is by consensus defined by the RIFLE classification. This classification includes less severe forms of AKI, and classifies into three severity grades. AKI defined by these sensitive criteria occurs in 20% of hospitalized patients and in 30–70% of ICU patients. It is associated with increased length of stay and mortality. Risk factors for AKI include patient-specific risk factors such as advanced age, co-morbid diseases and possibly certain gene polymorphisms, but also clinical risk factors such as severity of acute illness. In cardiac surgery, AKI defined by the RIFLE classification is a frequent complication and is associated with worse outcomes.

References

1 Bellomo R, Ronco C, Kellum JA, Mehta RL, Palevsky P, and the ADQI Workgroup: Acute renal failure – definition, outcome measures, animal models, fluid therapy and information technology needs. The Second International Consensus Conference of the Acute Dialysis Quality Initiative (ADQI) Group. Crit Care 2004;8: R204–R212.

2 Mehta RL, Kellum JA, Shah SV, et al: Acute Kidney Injury Network (AKIN): report of an initiative to improve outcomes in acute kidney injury. Crit Care 2007; 11:R31.

3 Hoste EAJ, Cruz DN, Davenport A, et al: The epidemiology of cardiac surgery-associated acute kidney injury. Int J Artif Organs 2008;31:158–165.

4 Joannidis M, Metnitz B, Bauer P, et al: Acute kidney injury in critically ill patients classified by AKIN versus RIFLE using the SAPS-3 database. Intensive Care Med 2009;35:1692–1702.

5 Hoste EAJ, Schurgers M: Epidemiology of AKI: how big is the problem? Crit Care Med 2008;36:S1–S4.

6 Uchino S, Kellum JA, Bellomo R, et al: Acute renal failure in critically ill patients: a multinational, multicenter study. JAMA 2005;294:813–818.

7 Xue JL, Daniels F, Star RA, et al: Incidence and mortality of acute renal failure in Medicare beneficiaries, 1992–2001. J Am Soc Nephrol 2006;17:1135–1142.

8 Ishani A, Xue JL, Himmelfarb J, et al: Acute kidney injury increases risk of ESRD among elderly. J Am Soc Nephrol 2009;20:223–228.

9 Haase M, Bellomo R, Matalanis G, Calzavacca P, Dragun D, Haase-Fielitz A: A comparison of the RIFLE and Acute Kidney Injury Network classifications for cardiac surgery-associated acute kidney injury: a prospective cohort study. J Thorac Cardiovasc Surg 2009;138:1370–1376.
10 Thakar CV, Worley S, Arrigain S, Yared JP, Paganini EP: Improved survival in acute kidney injury after cardiac surgery. Am J Kidney Dis 2007;50:703–711.
11 Bartholomew BA, Harjai KJ, Dukkipati S, et al: Impact of nephropathy after percutaneous coronary intervention and a method for risk stratification. Am J Cardiol 2004;93:1515–1519.
12 Haase-Fielitz A, Haase M, Bellomo R, et al: Decreased catecholamine degradation associates with shock and kidney injury after cardiac surgery. J Am Soc Nephrol 2009;20:1393–1403.

Eric A.J. Hoste, MD, PhD
ICU, 2K12-C, Ghent University Hospital, De Pintelaan 185
BE–9000 Gent (Belgium)
Tel. +32 9 332 27 75, Fax +32 9 332 49 95
E-Mail Eric.Hoste@UGent.be

Ronco C, Bellomo R, McCullough PA (eds): Cardiorenal Syndromes in Critical Care.
Contrib Nephrol. Basel, Karger, 2010, vol 165, pp 9–17

Pathophysiology of AKI: Injury and Normal and Abnormal Repair

Joseph V. Bonventre

Renal Division, Brigham and Women's Hospital and Department of Medicine, Harvard Medical School, and the Harvard-Massachusetts Institute of Technology, Division of Health Sciences and Technology, Boston, Mass., USA

Abstract

The pathophysiology of acute kidney injury involves a complex interplay among vascular, tubular, and inflammatory factors followed by a repair process that can either restore epithelial differentiation and function to normal or result in progressive fibrotic chronic kidney disease. Innate and acquired immunity play an important role in the injury phase, in the regulation of the inflammatory response, and in processes related to repair of the epithelial layer. Recent data change the direction of focus of the role of the epithelium in fibrosis and attributes myelofibroblast production to perivascular and interstitial fibroblasts. The epithelium plays an important role in abnormal repair through a recently defined link between cell cycle arrest of the epithelial cell and profibrogenic cytokine production.

Our view of the processes of injury and repair to the kidney epithelium, as expressed a number of years ago, is depicted schematically in figure 1. Whether acute kidney injury (AKI) is associated with ischemia reperfusion injury, sepsis or toxins, there is a rapid loss of proximal tubular cell cytoskeletal integrity and cell polarity. There is shedding of the proximal tubule brush border, loss of polarity with mislocalization of adhesion molecules and other membrane proteins such as the Na^+/K^+-ATPase and β-integrins [1], as well as apoptosis and necrosis [2]. Normal cell-cell interactions are disrupted with injury. With severe injury, viable and non-viable cells are desquamated leaving regions where the basement membrane remains as the only barrier between the filtrate and the peritubular interstitium. This allows for backleak of the filtrate, especially under circumstances where the pressure in the tubule is increased due to intratubular obstruction resulting from cellular debris in the lumen interacting with proteins

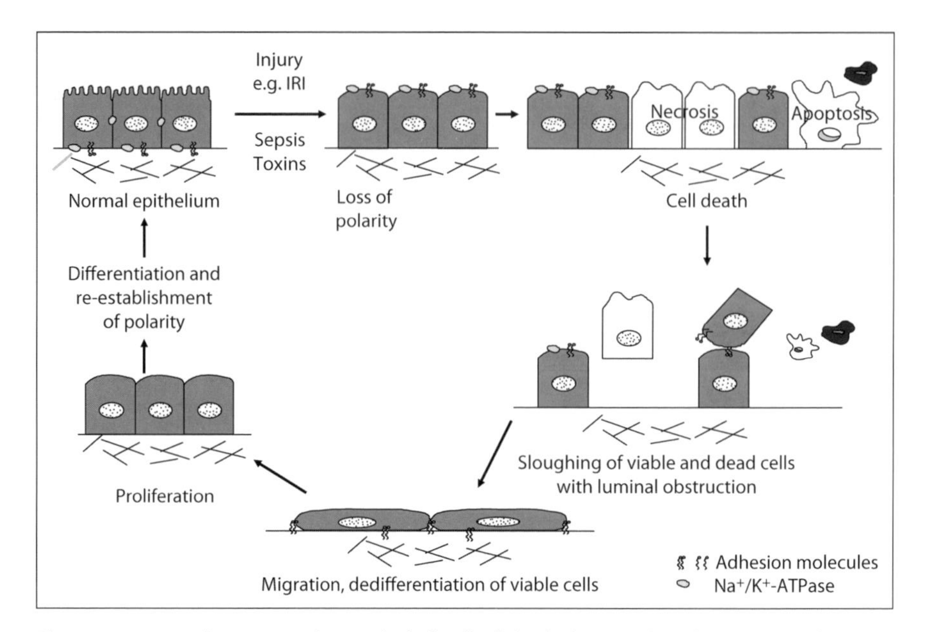

Fig. 1. Injury and repair to the epithelial cell of the kidney with ischemia/reperfusion – the 'classic' model. As a very early response of the epithelium to injury caused by a variety of influences, the normally highly polar epithelial cell loses its polarity. This can be demonstrated by alterations in the location of adhesion molecules and Na^+/K^+-ATPase. In addition, early after ischemia, there is a loss of the brush border of the proximal epithelial cell. With increasing time of ischemia, there is cell death by either necrosis or apoptosis. Some of the necrotic debris is then released into the lumen, where it can ultimately result in obstruction, since it interacts with lumenal proteins. In addition, because of the mislocation of adhesion molecules, viable epithelial cells lift off the basement membrane and are found in the urine. The kidney responds to the injury by initiating a repair process, if there are sufficient nutrients and sufficient oxygen delivery subsequent to the ischemia period. Viable epithelial cells migrate and cover denuded areas of the basement membrane. These cells express proteins that are not normally expressed in an adult mature epithelial cell. The cells then undergo division and replace lost cells. Ultimately, the cells go on to differentiate and re-establish the normal polarity of the epithelium.

such as fibronectin which enter the lumen [3]. This injury to the epithelium results in the generation by this epithelium of inflammatory and vasoactive mediators, which can feedback on the vasculature to worsen the vasoconstriction and inflammation. The pathophysiology of injury represents a complex interplay between the vasculature, the tubules, inflammation and the innate and acquired immune response. The above model has been challenged by those who have suggested that progenitor cells from the bone marrow or some other non-epithelial intra-kidney compartment participate in the repair of the damaged epithelium. Others have also proposed that when injury is severe or prolonged then the epithelial cell may transform to a fibroblast and in this way contribute to the development of fibrosis which is a hallmark of chronic kidney disease.

While there is some controversy as to the relative extent of proximal versus distal tubule injury in humans [4], many studies using biomarkers of proximal tubule injury have confirmed there is a markedly increased amount of kidney injury molecule-1 ectodomain in the urine, reflecting significant proximal tubule injury in the human, similar to what is found in injury models in rodents [5, 6].

While the kidney has the capability to recover from injury, this recovery process can often be suboptimal in humans. Recent studies in animals have provided insight into the mechanisms responsible for normal and abnormal repair and draw an important link between injury, abnormal repair and the development of fibrosis. This provides a link to the well-appreciated fact that AKI in humans frequently accelerates the progression of chronic kidney disease. In this short review I will discuss first the main contributors to the injury phase and then discuss normal and abnormal repair of the kidney after injury.

Vasculature Reactivity

The vasculature plays a critical role in the pathophysiology of AKI with decreased generalized or localized blood flow related to systemic factors or intrarenal factors. The small constrictive vessels respond more vigorously to increased tissue levels of vasoconstrictive agents such as endothelin-1, angiotensin II, thromboxane A_2, prostaglandin H_2, leukotrienes C_4, D_4, adenosine and sympathetic nerve stimulation [7]. At the same time there is decreased responsiveness to vasodilators such as acetylcholine, bradykinin and nitric oxide, and lower production levels of some of the vasodilators [8]. These effects can be mediated by an interaction with the smooth muscle or as a consequence of alterations in the endothelium. There is endothelial injury and enhanced leukocyte-endothelial adhesion particularly in the post-capillary venules. The latter effects can result in small vessel occlusion and activation of the leukocytes with resultant inflammation providing a positive feedback network. The inflammation will result in increased levels of mediators which increase the interactions between leukocytes and endothelial cells, and activate the coagulation pathways. The resultant effects on oxygen and nutrient delivery to the epithelial cells result in damage to those cells.

Endothelium

The endothelial cell contributes to the pathobiology of ischemic AKI in many ways. A normal arteriole can respond to local high concentrations of vasoconstrictive agents associated with injury. This response can be amplified, however, when the endothelium is damaged since these damaged endothelial cells produce less vasodilatory substances such as nitric oxide [9]. Damage results in increased permeability of the endothelium which then results in loss of fluid into the interstitium which is particularly problematic in the outer medulla where the anatomy of the vessels makes them very vulnerable to compromise of blood flow. There is marked congestion of the outer medulla and reduction in blood flow to this region of the kidney after ischemia [10].

The number of microvessels in the inner stripe of the outer medulla declines, potentially facilitated by the downregulation of angiogenic factors such as the VEGF pathway [11] and upregulation of inhibitors of angiogenesis. A reduced number of vessels are associated with chronic hypoxia which can lead to increased tubulointerstitial fibrosis which can be reinforcing and progressive since increased fibrosis will decrease the ability of oxygen and nutrients to get to the tubules and hence enhance tubular stress and epithelial cell injury and death leading to further fibrosis. There are also other functional consequences of vessel dropout including the development of salt-sensitive hypertension and altered concentrating ability. Remaining vessels may have blood flow compromised by endothelial cell swelling. Components of the inflammatory response, including monocytes and T lymphocytes, promote neovascularization through production of proangiogenic factors although the effects of inflammatory cells may be complex since T-regulatory cells have been found to inhibit angiogenesis in a hindlimb ischemia model [12].

Inflammation

The Immune Response. Both innate and adaptive components are important contributors to the pathobiology of AKI. Toll-like receptors (TLRs) recognize host material released during injury. When TLRs interact with their ligands the receptors are activated and initiate a proinflammatory response releasing cytokines/chemokines and attracting inflammatory cells. The role of TLRs has been studied in TLR2–/– and TLR4–/– mice [13, 14]. The absence of TLR2 or TLR4 had anti-inflammatory effects post-injury in models of ischemia/reperfusion injury.

Leukocyte-Endothelial Interactions. With many etiologies of AKI, endothelial cells upregulate integrins, selectins, and members of the immunoglobulin superfamily, including intercellular adhesion molecule-1 (ICAM-1) and vascular cell adhesion molecule (VCAM). The administration of anti-ICAM-1 antibodies either before or 2 h following renal ischemia/reperfusion or the genetic deletion of ICAM-1 in mice resulted in protection of the kidney from injury [15, 16]. We proposed that the upregulation of ICAM-1 was due to the proinflammatory cytokines TNF-α and IL-1 which we measured to be increased by ischemia/reperfusion. The protection using antibodies or the ICAM-1–/– mouse was associated with decreased tissue myeloperoxidase levels reflective of decreased interstitial leukocytes. A number of vasoactive compounds which have an effect on local hemodynamics may also affect blood flow and inflammation by affecting leukocyte-endothelial interactions. For example, endothelin antagonists, in addition to their effects on the vasculature, may also have an effect on neutrophil adhesion or other aspects of leukocyte-endothelial interactions with implications for inflammation [17]. Vasodilators, such as endothelium-derived nitric oxide, in addition to their effects to counteract vasoconstrictor agents as described above, also can have effects to decrease inflammation. NO inhibits adhesion of neutrophils to endothelial cells, stimulated by TNF-α, which would also be protective [18].

Inflammatory Effector Cells. There are many additional mechanisms by which immune/inflammatory cells can potentiate renal injury. Inflammatory mediators, including cytokines, chemokines, eicosanoids, and reactive oxygen species, recruit leukocytes, and upregulate adhesion molecules that engage counter-receptors on the activated endothelium. The activated leukocytes generate proinflammatory cytokines interleukin (IL)-1 and tumor necrosis factor-α (TNF-α). TNF-α, IL-1, and IFN-γ, produce a number of injurious changes in proximal tubular epithelial cells. These cytokines also disrupt cell-matrix adhesion dependent on β_1-integrin, inducing cell shedding into the lumen. Macrophages and dendritic cells also likely play an important role in the inflammatory response to ischemia/reperfusion.

We showed that neutrophils are seen in the interstitium early after ischemic injury in the mouse [16]. If neutrophil accumulation is prevented, tissue injury is ameliorated [16]. It is possible that neutrophil depletion models, however, may not adequately differentiate involvement of neutrophils from T lymphocytes and macrophages. A recent report indicates that IL-17 produced by neutrophils is upstream of IL-12 and regulates IFN-γ-mediated neutrophil migration into the mouse kidney after ischemia/reperfusion [19].

Later phases of AKI are characterized by infiltration of macrophages and T lymphocytes which may predominate over neutrophils at that time. Mice lacking CD4 and CD8, cell adhesion receptors on T lymphocytes, are protected from ischemia-reperfusion injury [20], suggesting a causal role for T lymphocytes in mediating injury. The role of the T cell, however, is complex. Mice deficient in recombination-activating gene (RAG)-1 lack T and B cells and are not protected from AKI induced by ischemia. Tubular necrosis and neutrophil infiltration are present to a degree comparable to that seen in wild-type mice [21, 22]. T cells are also likely to play a role in the repair phase of AKI. TCR-β+ CD4+ CD25+Foxp3+ regulatory T cells infiltrate the kidney after 3 or 10 days in the mouse model of ischemia [23]. Regulatory T-cell depletion resulted in worse functional injury and increased mortality.

Macrophages represent a heterogeneous population of monocytic cells that contribute to kidney fibrosis but also to beneficial effects on repair and matrix remodeling [24]. There is an increase in F4/80 cells in the kidney at 1 h after reperfusion, peaking at 24 h and persisting for 7 days [25]. F4/80^{high} cells were interpreted to be resident dendritic cells and F4/80^{low} cells were considered to be derived from the inflamed monocyte pool. Macrophages are an important source of the C3 component of complement.

Dendritic Cells. Dendritic cells are major effector cells linking injury to immune response in the kidney [26]. During acute tubular injury, these resident myeloid cells are activated and additional blood monocytes are recruited to the tissue. Dendritic cells can activate naive T cells, thus linking the innate immune response to adaptive immunity [27].

Complement. The complement system is an important component of the innate and adaptive immune response. Complement may potentiate leukocyte-endothelial interactions. In a number of different tissues exposed to ischemia-reperfusion, complement-dependent upregulation of endothelial cell adhesion molecules, with resulting neutrophil accumulation in the vasculature, has been implicated as a mechanism for complement-mediated injury [28].

Normal Repair of the Epithelium

Under normal circumstances, human proximal tubule cells divide at a low rate. When the kidney recovers from acute injury it relies on a sequence of events that include epithelial cell spreading and migration to cover the exposed areas of the basement membrane, and rapid proliferation to restore cell number, followed by differentiation ultimately resulting in restoration of the functional integrity of the nephron. We and others have concluded that the bone marrow does not contribute directly by replacing dead cells but bone-marrow-derived cells may have paracrine effects that may facilitate repair potentially by reducing inflammation [29].

Using genetic fate-mapping techniques in transgenic mice, 94–95% of metanephric mesenchyme-derived tubular epithelial cells, but not interstitial or vascular cells, were labeled with either β-galactosidase (lacZ) or red fluorescent protein. Two days after ischemia-reperfusion injury, 50.5% of epithelial cells coexpress Ki-67 and red fluorescent protein, indicating that epithelial cells that survived injury undergo proliferative expansion in response to the injury. After repair was complete, 66.9% of epithelial cells had incorporated BrdU, compared to only 3.5% of cells in the uninjured kidney. Despite this extensive cell proliferation, no dilution of either cell-fate marker was observed after repair. These results indicate that non-labeled cells did not participate in the replacement of lost tubular epithelial cells. Hence the predominant mechanism of normal repair after ischemic tubular injury in the adult mammalian kidney involves regeneration by surviving tubular epithelial cells [29].

Abnormal Repair of the Epithelium

Repair can either leave no lasting evidence of damage or can result in fibrotic lesions which can result in progressive organ dysfunction. AKI can result in incomplete repair and persistent tubulointerstitial inflammation, with proliferation of fibroblasts and excessive deposition of extracellular matrix being a common feature of many different kinds of kidney diseases and a primary determinant of progression to end-stage renal failure [30, 31]. While much has been written about the potential role of epithelial to mesenchymal transition in this process of fibrosis, Humphreys et al. [29, 32], based on lineage tracing, have established that epithelial cells do not contribute to the generation of myelofibroblasts directly by transdifferentiation in vivo in injury models in the mouse. Rather the myelofibroblasts are generated from fibroblasts, many of which are perivascular fibroblasts, or pericytes.

Although the epithelial cell does not transdifferentiate into a myofibroblast, we have recently found that the epithelial cell can nevertheless play a critical role in the proliferation of interstitial cells through the generation of profibrogenic cytokines. Using various models of AKI, we have mapped the cell cycle progression in vivo and found large differences among injury models resulting in 'normal' non-fibrotic repair and those resulting in 'fibrotic' repair [33]. We have found a causal association between epithelial cell cycle arrest G2 and the development of fibrosis. This introduces a new dimension to our understanding of the development of fibrosis after injury and insight into our clinical recognition that AKI is an important potentiating factor in the progression of chronic kidney disease [34].

Conclusions

The pathophysiology of AKI involves a complex interplay among vascular, tubular, and inflammatory factors followed by a repair process that can either restore epithelial differentiation and function to normal or result in fibrotic chronic kidney disease. We have learned a great deal recently about the roles of innate and acquired immunity in the injury phase and processes related to repair of the epithelial layer. Recent data changes the direction of focus of the role of the epithelium in fibrosis and explains myelofibroblast production to perivascular and interstitial fibroblasts. The epithelium plays an important role in abnormal repair through a recently established link between cell cycle arrest of the epithelium and profibrogenic cytokine production.

Acknowledgment

This work was supported by the National Institutes of Health (grants DK 39773, DK72381, DK54741).

References

1 Zuk A, Bonventre JV, Brown D, Matlin KS: Polarity, integrin, and extracellular matrix dynamics in the postischemic rat kidney. Am J Physiol 1998;275:C711–C731.

2 Thadhani R, Pascual M, Bonventre JV: Acute renal failure. N Engl J Med 1996;334:1448–1460.

3 Zuk A, Bonventre JV, Matlin KS: Expression of fibronectin splice variants in the postischemic rat kidney. Am J Physiol 2001;280:F1037–F1053.

4 Heyman SN, Rosenberger C, Rosen S: Experimental ischemia-reperfusion: biases and myths – the proximal vs. distal hypoxic tubular injury debate revisited. Kidney Int 2009;77:9–16.

5 Han WK, Bailly V, Abichandani R, Thadhani R, Bonventre JV: Kidney injury molecule-1: a novel biomarker for human renal proximal tubule injury. Kidney Int 2002;62:237–244.

6 Vaidya VS, Waikar SS, Ferguson MA, Collings FB, Sunderland K, Gioules C, Bradwin G, Matsouaka R, Betensky RA, Curhan GC, Bonventre JV: Urinary biomarkers for sensitive and specific detection of acute kidney injury in humans. Clin Transl Sci 2008;1:200–208.

7 Conger J: Hemodynamic factors in acute renal failure. Adv Ren Replace Ther 1997;4 (suppl 1):25–37.

8 Conger JD: Vascular abnormalities in the maintenance of acute renal failure. Circ Shock 1983;11:235–244.

9 Kwon O, Hong SM, Ramesh G: Diminished NO generation by injured endothelium and loss of macula densa nNOS may contribute to sustained acute kidney injury after ischemia-reperfusion. Am J Physiol 2009; 296:F25–F33.

10 Park KM, Chen A, Bonventre JV: Prevention of kidney ischemia/reperfusion-induced functional injury and JNK, p38, and MAPK kinase activation by remote ischemic pretreatment. J Biol Chem 2001;276:11870–11876.

11 Basile DP, Fredrich K, Chelladurai B, Leonard EC, Parrish AR: Renal ischemia reperfusion inhibits VEGF expression and induces ADAMTS-1, a novel VEGF inhibitor. Am J Physiol 2008;294:F928–F936.

12 Zouggari Y, Ait-Oufella H, Waeckel L, Vilar J, Loinard C, Cochain C, Recalde A, Duriez M, Levy BI, Lutgens E, Mallat Z, Silvestre JS: Regulatory T cells modulate postischemic neovascularization. Circulation 2009;120:1415–1425.

13 Leemans JC, Stokman G, Claessen N, Rouschop KM, Teske GJ, Kirschning CJ, Akira S, van der Poll T, Weening JJ, Florquin S: Renal-associated TLR2 mediates ischemia/reperfusion injury in the kidney. J Clin Invest 2005;115:2894–2903.

14 Pulskens WP, Teske GJ, Butter LM, Roelofs JJ, van der Poll T, Florquin S, Leemans JC: Toll-like receptor-4 coordinates the innate immune response of the kidney to renal ischemia/reperfusion injury. PloS One 2008; 3:e3596.

15 Kelly KJ, Williams WW, Colvin RB, Bonventre JV: Antibody to intercellular adhesion molecule-1 protects the kidney against ischemic injury. Proc Natl Acad Sci USA 1994;91:812–816.

16 Kelly KJ, Williams WW, Colvin RB, Meehan SM, Springer TA, Gutierrez-Ramos JC, Bonventre JV: Intercellular adhesion molecule-1-deficient mice are protected against renal ischemia. J Clin Invest 1996;97:1056–1063.

17 Sanz MJ, Johnston B, Issekutz A, Kubes P: Endothelin causes P-selectin-dependent leukocyte rolling and adhesion within rat mesenteric microvessels. Am J Physiol 1999;277: H1823–H30.

18 Linas S, Whittenburg D, Repine JE: Nitric oxide prevents neutrophil-mediated acute renal failure. Am J Physiol 1997;272:F48–F54.

19 Li L, Huang L, Vergis AL, Ye H, Bajwa A, Narayan V, Strieter RM, Rosin DL, Okusa MD: IL-17 produced by neutrophils regulates IFN-γ-mediated neutrophil migration in mouse kidney ischemia-reperfusion injury. J Clin Invest 2010;120:331–342.

20 Rabb H, Daniels F, O'Donnell M, Haq M, Saba SR, Keane W, Tang WW: Pathophysiological role of T lymphocytes in renal ischemia-reperfusion injury in mice. Am J Physiol 2000;279:F525–F531.

21 Park P, Haas M, Cunningham PN, Bao L, Alexander JJ, Quigg RJ: Injury in renal ischemia-reperfusion is independent from immunoglobulins and T lymphocytes. Am J Physiol 2002;282:F352–F357.
22 Burne-Taney MJ, Yokota-Ikeda N, Rabb H: Effects of combined T- and B-cell deficiency on murine ischemia reperfusion injury. Am J Transplant 2005;5:1186–1193.
23 Gandolfo MT, Jang HR, Bagnasco SM, Ko GJ, Agreda P, Satpute SR, Crow MT, King LS, Rabb H: Foxp3+ regulatory T cells participate in repair of ischemic acute kidney injury. Kidney Int 2009;76:717–729.
24 Ricardo SD, van Goor H, Eddy AA: Macrophage diversity in renal injury and repair. J Clin Invest 2008;118:3522–3530.
25 Li L, Huang L, Sung SS, Vergis AL, Rosin DL, Rose CE Jr, Lobo PI, Okusa MD: The chemokine receptors CCR2 and CX3CR1 mediate monocyte/macrophage trafficking in kidney ischemia-reperfusion injury. Kidney Int 2008;74:1526–1537.
26 John R, Nelson PJ: Dendritic cells in the kidney. J Am Soc Nephrol 2007;18:2628–2635.
27 Reis e Sousa C: Dendritic cells in a mature age. Nat Rev Immunol 2006;6:476–483.
28 Homeister JW, Lucchesi BR: Complement activation and inhibition in myocardial ischemia and reperfusion injury. Annu Rev Pharmacol Toxicol 1994;34:17–40.
29 Humphreys BD, Valerius MT, Kobayashi A, Mugford JW, Soeung S, Duffield JS, McMahon AP, Bonventre JV: Intrinsic epithelial cells repair the kidney after injury. Cell Stem Cell 2008;2:284–291.
30 Forbes JM, Hewitson TD, Becker GJ, Jones CL: Ischemic acute renal failure: long-term histology of cell and matrix changes in the rat. Kidney Int 2000;57:2375–2385.
31 Macedo E, Bouchard J, Mehta RL: Renal recovery following acute kidney injury. Curr Opin Crit Care 2008;14:660–665.
32 Humphreys BD, Lin SL, Kobayashi A, Hudson TE, Nowlin BT, Bonventre JV, Valerius MT, McMahon AP, Duffield JS: Fate tracing reveals the pericyte and not epithelial origin of myofibroblasts in kidney fibrosis. Am J Pathol 2010;176:85–97.
33 Yang L, Besschetnova TY, Brooks CR, Shah JV, Bonventre JV: Epithelial cell cycle arrest in G2/M mediates kidney fibrosis after injury. Nat Med 2010 (in press).
34 Ishani A, Xue JL, Himmelfarb J, Eggers PW, Kimmel PL, Molitoris BA, Collins AJ: Acute kidney injury increases risk of ESRD among elderly. J Am Soc Nephrol 2009;20:223–228.

Joseph V. Bonventre, MD, PhD
Brigham and Women's Hospital, Renal Division, MRB-4
75 Shattuck Street, Boston, MA 02115 (USA)
E-Mail joseph_bonventre@hms.harvard.edu

Ronco C, Bellomo R, McCullough PA (eds): Cardiorenal Syndromes in Critical Care.
Contrib Nephrol. Basel, Karger, 2010, vol 165, pp 18–27

Pathophysiology of Septic Acute Kidney Injury: A Different View of Tubular Injury

Ken Ishikawa[b] · Clive N. May[b] · Glenda Gobe[c] · Christoph Langenberg[b] · Rinaldo Bellomo[a]

[a]Department of Intensive Care, Austin Health, Heidelberg, Melbourne, [b]Howard Florey Institute, Melbourne University, Parkville, Melbourne, and [c]Apoptosis Laboratory, Princess Alexandra Hospital, Brisbane, Vic., Australia

Abstract

Septic acute kidney injury (AKI) is the most common form of AKI seen in critically ill patients in developed countries. Its pathogenesis has been traditionally attributed to ischemia secondary to decreased cardiac output and hypotension, which trigger sustained renal vasoconstriction and in turn exacerbate and sustain the ischemia. This paradigm is supported by the fact that many patients who develop AKI do so in the setting of hemodynamic instability and also by evidence that renal blood flow is decreased and renal vascular resistance increased when they are measured in patients with AKI. However, recent evidence shows that renal blood flow may vary from increased in some animal models to normal in some patients and to decreased in other patients. Furthermore, the induction of prolonged severe subtotal ischemia by acute occlusion of the renal artery does not seem to trigger subsequent renal vasoconstriction and, finally, experimental studies suggest that immune-mediated injury may be a more likely cause of tubular cell dysfunction than ischemia. These lines of evidence suggest that the pathogenesis of AKI is complex, does not simply involve ischemia, and may differ according to the etiological trigger.

Acute kidney injury (AKI) of the pre-renal type is the most common type of AKI seen in intensive care units around the world. It is a major complication of many serious conditions from severe sepsis to open heart surgery, from trauma to liver failure, from hemorrhagic shock to cardiogenic shock, and from major surgery to drug toxicity. As such, AKI seems to represent the final common pathway of different types of injury suggesting that the mechanisms involved are essentially the same in almost all cases: decreased renal blood flow, associated renal vasoconstriction in response to decreased perfusion, tubular cell hypoxia, bioenergetic failure and cell death (acute tubular necrosis (ATN)). Septic AKI, however, seems unlikely to follow this pattern because, under most circumstances, glomerular filtration rate (GFR) decreases rapidly, despite increased cardiac output and an adequate mean blood pressure, which is supported by vasopressor drugs. These observations have led to recent challenges to the concept that ischemic tubular ischemia is responsible for septic AKI. In this article, we will review recent evidence that challenges the traditional view of septic AKI as a condition mostly, if not wholly, secondary to ischemia.

The Ischemia Paradigm

The main cause of septic AKI has been assumed to be ATN caused by hemodynamic instability and subsequent renal vascular vasoconstriction and kidney ischemia [1]. Renal vascular vasoconstriction triggered by hemodynamic instability then exacerbates and sustains renal ischemia and eventually causes established ATN. In response to this pathophysiologic paradigm, over the last 40 years or more, all efforts to prevent AKI or reverse it have been based on a logically related therapeutic paradigm. This therapeutic paradigm states that preservation of 'adequate' perfusion or rapid correction of hypoperfusion by fluid resuscitation and/or vasoactive drugs, and/or the correction of renal vasoconstriction by the administration of renal vasodilators like low-dose dopamine, is the only logical approach to the prevention and treatment of AKI. Unfortunately, all of these strategies, although widely applied, have so far failed to achieve any demonstrable success. This failure may, of course, reflect problems other than an incorrect understanding of the pathogenesis of AKI. For example, failure may have been due to the use of drugs that are not sufficiently effective, the application of interventions too late in the course of AKI to be able to reverse those biological processes responsible for cell injury, the insufficient correction of hypoperfusion, or any combination of the above factors. Another possible explanation for our therapeutic failures is that the renal ischemia paradigm is only partly correct and can only explain a small part of what might be happening during AKI, especially in sepsis. When considering a challenge to a 40- to 50-year-old paradigm, one needs to consider the available evidence as objectively as possible. One initial step in direction of challenging the ischemia-AKI paradigm might come from a discussion of what evidence exists that ATN is the histopathological substrate of septic AKI.

Histopathological Changes in Septic AKI

Light Microscopy in Humans

Recently, Langenberg et al. [2] performed a systematic review of the histopathology of septic AKI. They found that only 22% of all 184 patients with AKI, whose histopathology had been reported in the literature, actually showed signs of ATN on biopsy or post-mortem. There are two additional detailed human studies (beyond those considered by Langenberg) which seem to have examined the histopatology of septic AKI. First, Brun and Munck [3] examined histopathological changes of 33 patients with acute renal failure (ARF) following shock including sepsis by biopsy or necropsy material. ATN findings (predominantly in distal tubules) were observed in only 5 of the 33 patients. So the authors concluded the most striking of feature of AKI was that moderate structural changes contrasted sharply with complete functional breakdown. Second, Diaz de Leon et al. [4] reported that ATN findings were observed in only 20 of 40 patients with septic AKI. Unfortunately, the details of histopathology were not described in this article. Even if these articles are added to Langenberg's systematic review, only 25% of all patients with septic AKI reported in the literature show ATN findings.

More recently, Lerolle et al. [5] reported the histopathological findings in 19 patients who died from severe sepsis shock in the intensive care unit. They found that all patients had changes consistent with ATN. They also found that apoptosis was a common finding with approximately 3% of cells examined showing evidence of apoptotic changes. There is no information on the presence of apoptosis in normal human kidneys. Even when these highly biased cases (the most severe end of the spectrum with sustained septic shock leading to death as well as AKI) are included in our analysis, only a minority of patients with septic AKI appear to have ATN as the histological substrate for the their functional loss. If the paradigm is that ischemia leads to ATN, which then leads to AKI, then this paradigm is clearly not adequately supported, because about 70% of patients with sepsis as AKI appear not to have ATN. Thus, in the severest cases, this paradigm may have some credence, but in most patients with septic AKI, it does not.

In animal experiments, which model sepsis to study its renal effects, the findings are similar. Langenberg et al. [2] also performed a systematic review of animal model of septic AKI using all articles that make specific mention of histopathology. The authors reported that only 23% in all studies showed evidence of ATN, a finding which is remarkably consistent with the human data.

Thus in most cases of human and experimental septic AKI there appears to be a degree of dissociation between the marked loss of GFR and urine output, which often decreases to almost zero, and the histopathologic changes which are surprisingly limited in most patients. This dissociation remains unexplained. Equally important, the major functional event of septic AKI (loss of GFR) would logically lead to a focus on the glomerulus as a potential site of dysfunction and yet essentially all literature has focused on the tubules instead. Despite these concerns, it is possible that injury occurs to both tubules and glomeruli and we simply cannot see it with light microscopy. Electron microscopy might shed some light on this subject.

Electron Microscopy

There are no reports or case series of patients with septic AKI to describe the electron microscopy findings of this condition. Thus, there is no knowledge of whether tubular injury occurs at a level that is potentially clinically significant but not detectable with light microscopy. However, experimentally, cytokines (TNF-α, IL-1α, IFN-γ) can induce shedding of viable, apoptotic and necrotic proximal tubule epithelial cells. This process appears dependent on nitric oxide and not on ischemia. Cytokine administration can also change the morphologic features of proximal tubule epithelial cells. Finally, plasma of burns septic AKI patients can induce alteration in the distribution of cytoskeleton actin fibers in tubular cells [6]. These observations are important because they suggest that tubular injury in septic or inflammatory states may *be mediated by immunological injury*, not ischemia. Knowledge that this can happen experimentally, and our inability to exclude immunological injury in humans, raise the issue of whether septic AKI is a form of organ injury triggered by the innate immune response to infection. Moreover, given this experimental evidence that cytokines can induce tubular cell injury, given our knowledge that sepsis is characterized by a cytokine storm, given that there is no reliable evidence of ischemic ATN in the majority of animal experiments of sepsis or in humans with sepsis, given that most septic patients have a hyperdynamic circulation with a high cardiac output, and given that recent animal models of such hyperdynamic sepsis show *increased* rather than decreased renal blood flow, why do we think that septic AKI is due to ischemia instead of immune injury? How can we exclude immune-mediated injury as a cause of septic AKI? One related consideration is that, under many circumstances, immune-mediated cell injury takes the form of apoptosis. This invites interest in considering what role apoptosis may play in the development of septic AKI.

Apoptosis of Kidney Tubules in Septic AKI

In one report of early autopsy findings organs were examined for apoptosis in 20 septic patients [7]. In this study, despite the high prevalence of clinical renal dysfunction (65%) in patients with sepsis, only 1 septic patient had evidence of kidney necrosis. No renal tubular apoptosis was seen in any septic patient by light microscopy. The authors relied heavily on use of fluorescent terminal deoxynucleotidyl transferase dUTP nick end-labeling (TUNEL) labeling. TUNEL does not always label apoptotic cells and there is often staining of nuclei that show no morphological evidence of apoptosis [8] and thus, it remains unclear to what extent apoptosis may or may not play a role in septic AKI in humans.

The authors of the above study once again described that renal histology did not reflect the severity of renal injury indicated by the decrease in kidney function. The TUNEL method could not discriminate between septic patients and control patients. Importantly, it was shown that with the TUNEL method, a delay in tissue fixation caused a marked increase in the number of apoptotic-positive cells at 3 and 6 h compared with immediate fixation in almost all organs. More recently, another case series examined the post-mortem findings in a cohort of patients who died of septic shock [5]. In this study the major finding was that of ATN in all cases. However, apoptosis was also seen as prominent with 3% of all cells showing evidence of apoptotic injury. The cause of such injury remains unclear and it is also uncertain whether such degree of apoptosis has pathogenetic significance or simply represents an epiphenomenon. Similarly, although it is known that plasma from the septic burns patients with septic AKI induced a pro-apoptotic effect in tubular cells in vitro [6], the clinical significance of apoptosis as a contributor to loss of GFR in AKI remains unknown.

In a model of sepsis, Messaris et al. [9] followed the time course of apoptosis using cecal ligation and puncture (CLP) in rats. The time distribution of all types of cell death was increased significantly 6 h after the induction of sepsis, and declined subsequently. The cells initiating apoptosis were significantly more common at 6 than at 48 h post-CLP. On the other hand, other studies report no apoptotic cells detected in the kidney after CL. Hotchkiss et al. [7] also examined whether apoptosis occurs systemically in lymphoid and parenchymal cells in CLP of mice. They found that the extent of apoptosis was less in kidney compared with other organs such as lymph nodes and spleen. In sepsis models using lipopolysaccharide (LPS) injection, the time course of apoptosis peaks at 8–12 h and is still apparent after 48 h. In some studies, the amount of apoptosis in individual animals correlated with the extent of renal functional impairment, suggesting that apoptosis may be intimately involved in LPS-induced ARF or be a good marker of other mechanisms responsible for loss of GFR. On the other hand, several other studies reported no apoptotic cell 16 h after LPS injection.

Caspases are essential proteases for initiation and execution of apoptosis and for the processing and maturation of the inflammatory cytokines IL-1β and IL-18, and plasma of burns patients with septic AKI increases the activity of caspases-3, -8 and -9 in tubular cells [6]. Broad-spectrum caspase inhibitors such as Z-VAD-FMK can prevent lymphocyte apoptosis in sepsis, and in turn, improve septic animal survival by 40–45%. However, at high doses, caspase inhibitors exacerbated TNF toxicity by enhancing oxidative stress and mitochondrial damage.

It must be noted that in the normal mature adult, there is a balance between normal levels of mitosis and apoptosis, so that individual organs remain the same size. In the conditions like sepsis, however, toxic cytokines may tip the balance

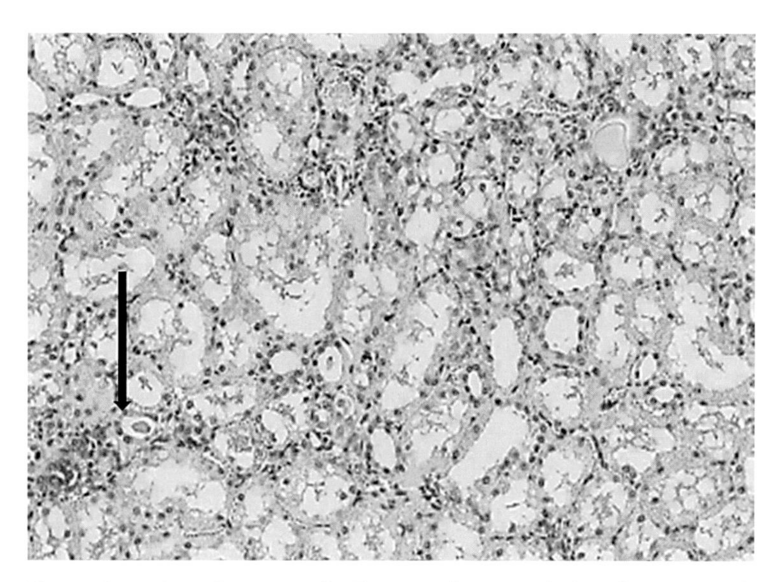

Fig. 1. Detection of apoptosis by TUNEL technique in kidney from septic sheep following infusion of live *E. coli*. The dark staining nuclei indicate the presence of apoptosis (as shown by arrow).

towards more apoptosis than mitosis. Even when the number of apoptotic cells detected in any experiment is not great, this needs to be seen in the light of the rapidity of the apoptotic process, where cells may die, be phagocytosed and the cellular content broken down via lysosomal enzymes and recycled in just a few hours. Even if accumulating apoptotic cell death resulted in cellular loss of up to 50%, the percentage of dying cells identifiable at a single time point could easily be less than a few percent of the entire tissue mass, especially when this atrophic process occurs over a span of weeks [10]. Thus, finding that only a small percentage of tubular cells are apoptotic at the time of sampling cannot exclude a very significant destructive process. In our experimental animals with sepsis induced by *Escherichia coli* infusion, apoptosis is relatively common (fig. 1, 2).

Further evidence against major histopathological changes in sepsis is the ability of the kidney to recover function rapidly after experimental sepsis [11]. This suggests that there is not any major damage to the glomeruli or tubules due to cell death from either necrosis or apoptosis in sepsis.

Role of Toll-Like Receptors

Innate immunity is first line of host defenses for the pathogens. It recognizes pathogens via Toll-like receptors (TLRs), which detect specific molecular

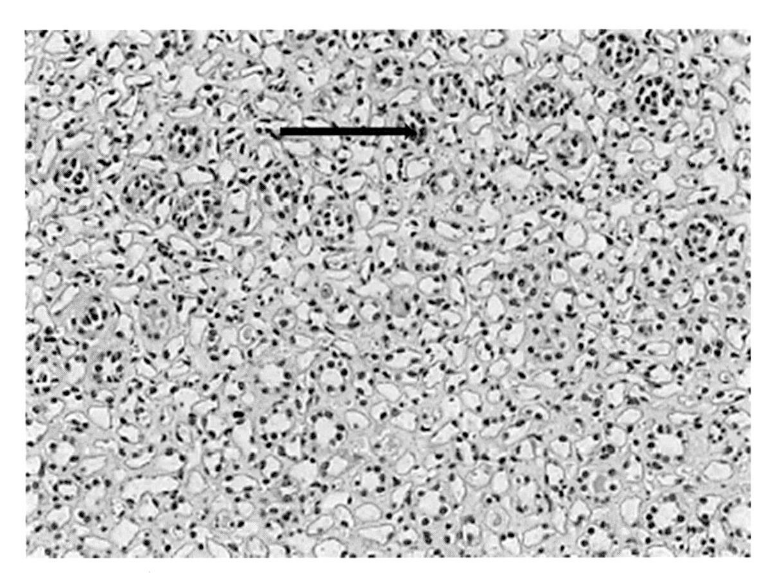

Fig. 2. Detection of apoptosis by TUNEL technique in kidney from septic sheep following infusion of live *E. coli* during the recovery phase. Medullary apoptosis is indicated by the dark staining cells (as shown by arrow).

patterns of pathogens from Gram-negative to Gram-positive organisms to fungi, viruses or parasites. TLRs are thought the main initial modulators of the inflammatory cascade associated with a pathogen attack [12]. So far, at least 11 members of the TLR family have been found in mammals, and TLR2, 4 and 9 appeared particularly important in septic AKI. For example, TLRs 1–10 have all been detected in human kidney cells by several methods including polymerase chain reaction, in situ hybridization, and immunohistochemistry [12]. TLR4 mutations and polymorphisms are reported in human kidney cells.

Using TLR knockout mice, one can understand the possible role of TLRs in the pathophysiology of septic AKI. The C3H/HeJ strain of mice that lack the function of TLR4 are resistant to LPS-induced septic AKI and mortality [13]. These experimental findings are crucial because they further suggest a powerful role by the immune system in the pathogenesis of septic AKI. If this is the case in man, the ischemia-AKI paradigm is seriously challenged.

In a CLP model, monoclonal antibody for TLR4 and myeloid differentiation protein-2 complex improve survival [14]. Myeloid differentiation factor 88 is a main messenger molecule for TLRs, acting as a link between the receptors and downstream kinases such as NF-κB and TNF. Myeloid differentiation factor 88 knockout mice do not develop septic AKI or show the histopathological changes of AKI after CLP. Yasuda et al. [15] recently reported the protective effect of chloroquine, an inhibitor of TLR9, in preventing septic AKI in a CPL model.

After recognition of pathogen by TLRs, pro-inflammatory cytokines (TNF, IL-6, IL-8) are released into the circulation within the first several hours. First, TNF-α and IL-1β are released and, later, IL-6, IL-10 and nitric oxide. Chawal et al. [16] reported that an increased IL-6 level is a significant risk factor for septic AKI, further supporting the notion that inflammation is a significant component of septic AKI. Elevated soluble TNF receptor level is an independent predictor of mortality among patients developing septic ARF [17].

TNF-α directly injures kidney tubules, independent of inducible nitric oxide synthase, hypotension, apoptosis, and morphologic alterations. High-dose TNF-α causes renal tubular necrosis. Low or moderate doses of TNF cause glomerular inflammation, but no histological change in tubules.

TNF receptors (TNFR) are needed to mediate the injurious effects of TNF on kidney cells. TNFR1+/+ kidneys transplanted into TNFR1–/– mice develop severe ARF after LPS injection, but TNFR1–/– kidneys transplanted into TNFR1+/+ mice do not. Therefore, TNF is a key mediator of LPS-induced ARF, acting through its receptor TNFR1 in the kidney.

Mitochondrial Dysfunction and ATP Depletion

ATP depletion can cause either necrosis or apoptosis in mouse proximal tubular cells in vitro. A legitimate question, if ischemia is the cause of septic tubular injury, is to ask whether mitochondrial function is lost in septic AKI, especially because the determination on whether cells die by either necrosis or apoptosis depends on the depletion of ATP. However, continuous infusion of LPS (0.4 μg/kg/h) did not change renal blood flow, renal mitochondrial respiration, and renal lactate/pyruvate ratio [18].

Magnetic resonance imaging (MRI) studies by May et al. [19] demonstrated no change in total ATP or β-ATP/total ATP ratio in the kidney during hyperdynamic sepsis despite profound hypotension and anuria. Furthermore, Dear et al. [20] performed MRI with gadolinium-based G4 dendrimer intravenous contrast in CLP mice. 24 h post-CLP, aged mice had a distinct pattern of renal injury that was different from renal injury induced by either ischemia reperfusion or pre-renal azotemia. Moreover, MRI detected renal dysfunction 6 h post-CLP, a time when serum creatinine was still normal.

Conclusions

The ischemia-ATN paradigm is flawed as an explanation of tubular injury in septic AKI. ATN is not the most common histopathological finding in septic

AKI and only occurs in a third of cases. Apoptosis may be a more important process than previously appreciated. A large body of experimental data supports the notion that the innate immune system is deeply involved in the pathogenesis of septic AKI in a way that is independent of decreased renal perfusion. Furthermore, magnetic resonance spectroscopy shows preserved levels of ATP in severe septic shock and a pattern of injury which is unique in severe sepsis. Our understanding of the pathogenesis of tubular injury in septic AKI is limited. Until it is sufficiently increased, therapeutic strategies will continue to fail.

References

1 Schrier RW, Wang W: Acute renal failure and sepsis. N Engl J Med 2004;351:159–169.
2 Langenberg C, Bagshaw SM, May CN, Bellomo R: The histopathology of septic acute kidney injury: a systematic review. Crit Care 2008;12:R38.
3 Brun C, Munck O: Lesions of the kidney in acute renal failure following shock. Lancet 1957;272:603–607.
4 Diaz de Leon M, Moreno SA, Gonzalez Diaz DJ, Briones GJ: Severe sepsis as a cause of acute renal failure. Nefrologia 2006;26: 439–444.
5 Lerolle N, Nochy D, Guerot E, Bruneval P, Fagon JY, Diehl JL, Hill G: Histopathology of septic shock induced acute kidney injury: apoptosis and leukocytic infiltration. Intensive Care Med 2010;36:471–478.
6 Mariano F, Cantaluppi V, Stella M, Romanazzi GM, Assenzio B, Cairo M, Biancone L, Triolo G, Ranieri VM, Camussi G: Circulating plasma factors induce tubular and glomerular alterations in septic burns patients. Crit Care 2008;12:R42.
7 Hotchkiss RS, Swanson PE, Freeman BD, Tinsley KW, Cobb JP, Matuschak GM, Buchman TG, Karl IE: Apoptotic cell death in patients with sepsis, shock, and multiple organ dysfunction. Crit Care Med 1999;27:1230–1251.
8 Gobe G: Identification of apoptosis in kidney tissue sections. Methods Mol Biol 2009;466:175–192.
9 Messaris E, Memos N, Chatzigianni E, Kataki A, Nikolopoulou M, Manouras A, Albanopoulos K, Konstadoulakis MM, Bramis J: Apoptotic death of renal tubular cells in experimental sepsis. Surg Infect (Larchmt) 2008;9:377–388.
10 Yasuhara S, Asai A, Sahani ND, Martyn JA: Mitochondria, endoplasmic reticulum, and alternative pathways of cell death in critical illness. Crit Care Med 2007;35:S488–S495.
11 Langenberg C, Wan L, Egi M, May CN, Bellomo R: Renal blood flow and function during recovery from experimental septic acute kidney injury. Intensive Care Med 2007;33:1614–1618.
12 Smith KD: Toll-like receptors in kidney disease. Curr Opin Nephrol Hypertens 2009;18:189–196.
13 Poltorak A, He X, Smirnova I, Liu MY, Van Huffel C, Du X, Birdwell D, Alejos E, Silva M, Galanos C, Freudenberg M, Ricciardi-Castagnoli P, Layton B, Beutler B: Defective LPS signaling in C3H/HeJ and C57BL/10ScCr mice: Mutations in TLR4 gene. Science 1998;282:2085–2088.
14 Daubeuf B, Mathison J, Spiller S, Hugues S, Herren S, Ferlin W, Kosco-Vilbois M, Wagner H, Kirschning CJ, Ulevitch R, Elson G: TLR4/MD-2 monoclonal antibody therapy affords protection in experimental models of septic shock. J Immunol 2007;179:6107–6114.
15 Yasuda H, Leelahavanichukul A, Tsunoda S, et al: Chloroquine and inhibition of toll-like receptor 9 protect from sepsis-induced acute kidney injury. Am J Physiol Renal Physiol 2008;294:F1050–F1058.

16 Chawla LS, Seneff MG, Nelson DR, et al: Elevated plasma concentrations of IL-6 and elevated APACHE II scotre predict acute kidney injury in patients with severe sepsis. Clin J Am Soc Nephrol 2007;2:22–30.
17 Iglesias J, Marik PE, Levine JS: Elevated serum levels of the type I and type II receptors for tumor necrosis factor-α as predictive factors for ARF in patients with septic shock. Am J Kidney Dis 2003;41:62–75.
18 Porta F, Takala J, Weikert C, Bracht H, Kolarova A, Lauterburg BH, Borotto E, Jakob SM: Effects of prolonged endotoxemia on liver, skeletal muscle and kidney mitochondrial function. Crit Care 2006;10:R118.
19 May C, Wan L, Williams J, Wellard MR, Pell G, Langenberg C, Jackson G, Bellomo R: A technique for the simultaneous measurement of renal ATP, blood flow and pH in a large animal model of septic shock. Crit Care Resusc 2007;9:30–33.
20 Dear JW, Kobayashi H, Brechbiel MW, Star RA: Imaging acute renal failure with polyamine dendrimer-based MRI contrast agents. Nephron Clin Pract 2006;103:c45–c49.

Prof. Rinaldo Bellomo
Department of Intensive Care, Austin Health
Heidelberg, Melbourne, Vic 3084 (Australia)
Tel. +61 3 9496 5992, Fax +61 3 9496 3932
E-Mail rinaldo.bellomo@austin.org.au

Ronco C, Bellomo R, McCullough PA (eds): Cardiorenal Syndromes in Critical Care.
Contrib Nephrol. Basel, Karger, 2010, vol 165, pp 28–32

Cardiopulmonary Bypass, Hemolysis, Free Iron, Acute Kidney Injury and the Impact of Bicarbonate

Michael Haase[a] · Anja Haase-Fielitz[a] · Rinaldo Bellomo[b]

[a]Department of Nephrology and Intensive Care, Charité – University Medicine, Berlin, Germany, and
[b]Department of Intensive Care, Austin Health, Heidelberg, Melbourne, Vic., Australia

Abstract

Cardiac surgery with cardiopulmonary bypass (CPB) is one of the most common major surgical procedures worldwide often associated with acute kidney injury (AKI) – a frequent and serious complication of cardiac surgery and affects up to 50% of patients. The major determinants of AKI after CPB are hemodynamic and inflammatory factors and the release of heme and labile iron contributing to oxidation from reactive oxygen species. The generation of reactive oxygen species is catalyzed by free labile iron which is most active at acid pH. To date, no simple, safe, and effective intervention to prevent CPB-associated AKI in a broad patient population has been found. However, there is recent evidence from experimental and clinical studies that sodium bicarbonate protects from oxidant renal injury in different settings, at least in part, by scavenging of hydroxyl radicals, peroxynitrite and other reactive species.

With over 1 million operations a year, cardiac surgery with cardiopulmonary bypass (CPB) appears to be the second most common major contributor to acute kidney injury (AKI) in developed countries [1]. AKI is a common and serious postoperative complication of CPB in adults and children [2, 3]. AKI carries significant costs and is independently associated with increased morbidity such as a significantly higher risk for renal replacement therapy and mortality.

The pathophysiology of AKI is complex and not yet completely understood. Multiple causes of AKI after CPB have been proposed and include ischemia/reperfusion, generation of reactive oxygen species (ROS), hemolysis, and activation of inflammatory pathways [4]. Figure 1 gives a schematic overview. During CPB there is injury to red cells and release of free hemoglobin with red blood cell fragmentation resulting in altered rheological properties. Increased levels

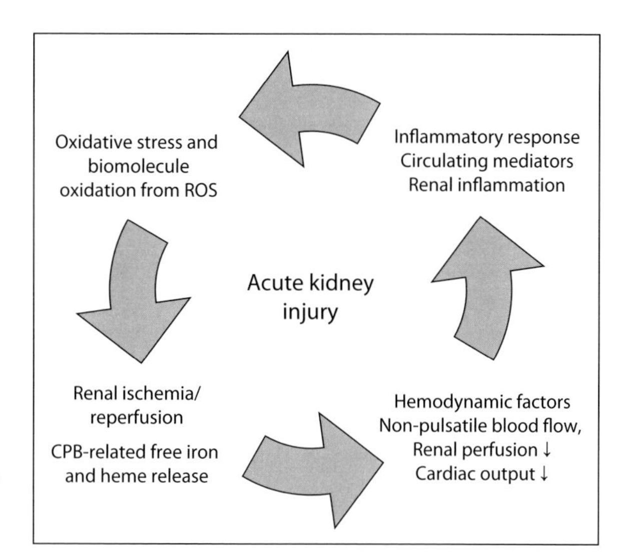

Fig. 1. Overview of the roles of ischemia/reperfusion during CPB, and of ROS, poorly liganded iron and iron metabolism regulators in affecting renal injury.

of free red blood cell constituents result in renal tubular damage and increased mortality [5]. Such injury raises concerns that CPB-associated AKI may be caused by free or inappropriately liganded iron-related toxicity.

Free labile iron is capable of inducing multiple changes in renal tubular epithelial function, including impaired proliferation [6] and the induction of free radical injuries, such as lipid peroxidation and protein oxidation. The generation of hydroxyl radicals is catalyzed by free iron ions released during CPB and most active at acid pH. Urinary acidity may enhance the toxicity of ROS. Furthermore, aciduria converts hemoglobin to methemoglobin, which precipitates, forms distal casts, and induces AKI [7]. Experimental data shows that red blood cell hemolysate is a potent mitogen for renal tubular epithelial cells [8].

Such a relationship was confirmed in a recent observational study where 35 patients undergoing on-pump surgical repair of aortic aneurysms were enrolled of whom 19 developed AKI. Plasma-free hemoglobin was independently correlated with the tubular injury marker urine N-acetyl-β-D-glucosaminidase (NAG), which, in turn, was an independent risk factor for postoperative AKI [9].

The beneficial effect of increasing tubular pH by urinary alkalinization, achieved for example with the use of sodium bicarbonate infusion, was protective in a rat model of acute renal failure [10].

Experimental studies found that sodium bicarbonate protects from oxidant injury by slowing pH-dependent Haber-Weiss free radical production [11], and by scavenging of hydroxyl radicals, peroxynitrite and other reactive species generated from nitric oxide [12].

Under aerobic conditions, ferrous ions will react with oxygen to produce ferric ions. At neutral or alkaline pH, free ferric ions precipitate as insoluble ferric hydroxide, which is excreted as inert complex in the urine. More alkaline urine

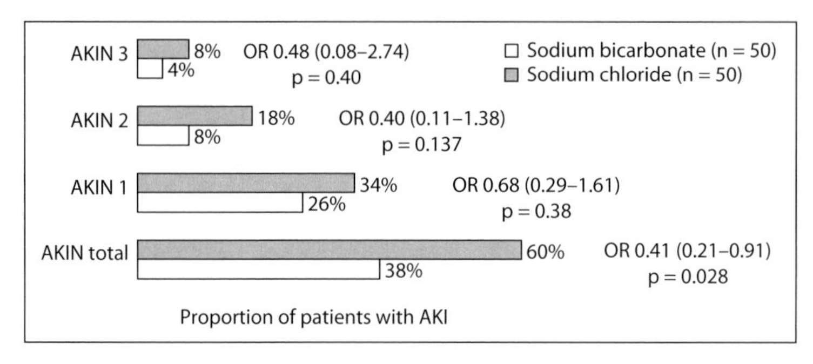

Fig. 2. Sodium bicarbonate reduces the incidence of AKI after cardiac surgery. Number of patients receiving sodium bicarbonate (white bars) developing acute kidney injury after cardiac surgery compared to patients receiving sodium chloride (grey bars). Acute Kidney Injury Network (AKIN) stage 1, 2, 3 is based on the definition of acute kidney injury by the AKI Network. OR = Odds ratio.

reduces the generation of injurious hydroxyl radicals. Bicarbonate directly scavenges hydroxyl ions and – as a not well absorbable anion compared to chloride – causes more rapid volume excretion and thereby might reduce the contact time between injurious radicals and renal tubules.

Urinary alkalinization may thus protect from renal injury induced by iron-mediated free radical pathways, and tubular hemoglobin cast formation.

To date, no simple, safe, and effective intervention to prevent CPB-associated AKI in a broad patient population has been found. Accordingly, Haase et al. [13] hypothesized that urinary alkalinization might protect kidney function and conducted a pilot double-blind, randomized controlled clinical trial to investigate whether sodium bicarbonate infusion with preoperative intravenous loading to achieve urinary alkalinization could reduce the incidence of AKI associated with CPB in cardiac surgical patients at increased renal risk.

In a cohort of 100 cardiac surgical patients [13], sodium bicarbonate treatment successfully alkalinized urine. There were no significant differences between the groups in baseline characteristics including duration of CPB and in hemodynamic and fluid management during and after cardiac surgery, nor in plasma creatinine, plasma urea or urinary neutrophil gelatinase-associated lipocalin (NGAL). The mean dose of sodium bicarbonate was 307 ± 57 mmol and the mean dose of sodium chloride was 309 ± 68 mmol (p = 0.89). Fewer patients in the sodium bicarbonate group developed AKI compared to the control group (OR 0.43; 95% CI 0.19–0.98) (fig. 2). Also, sodium bicarbonate infusion was associated with a significant attenuation in the postoperative increase of plasma urea, urinary NGAL and urinary NGAL/urinary creatinine ratio.

In previous studies, the appearance of NGAL in the urine was related to the dose and duration of renal injury and precedes the appearance of other urinary markers. Also, its expression is induced by hydrogen peroxide making it a useful biomarker of oxidative stress [14]. NGAL is a siderophore-binding lipocalin involved in ischemic renal injury and repair processes. Siderophores are small iron-containing molecules produced from bacteria and plants that, through iron transport and supply, are involved in cellular growth and survival. Siderophores can solubilize and sequester iron (mainly ferric iron) such that it can be internalized via suitable transporter molecules within the plasma membrane. In this regard, the study by Haase et al. [13] suggests a novel utility of urinary NGAL measurements as potential monitoring tool of the efficacy of therapy for AKI and point towards the importance of the role of labile iron in acute renal damage. As the study was not powered to detect group differences in the need for renal replacement therapy and mortality, and because there is justification for larger (phase III) trials, currently several international studies are under way.

In turn, biomarkers of tubular injury might have pathophysiological implications. In this regard, several markers of renal free-iron toxicity have been recently described [15].

Of note, by analogy, increasing urinary pH has also been reported to attenuate renal injury in patients undergoing contrast media infusion. Several years ago, Merten et al. [16] showed that replacing chloride ion with bicarbonate as the anion in sodium-containing hydration fluids significantly reduced nephropathy following radiographic contrast injection. Meanwhile, a growing body of evidence including several meta-analyses favors bicarbonate for nephroprotection in this setting [17, 18].

Every year, millions of percutaneous coronary interventions are performed with postinterventional contrast medium-induced nephropathy affecting up to 15% of patients associated with increased mortality [19]. After administration of contrast media, increased tubular viscosity and direct toxic effects due to the release of free radicals are discussed [20]. The apparent success of sodium bicarbonate in reducing contrast-induced nephropathy is consistent with the hypothesis that contrast injury is from free radicals generated within the acid environment of the renal medulla. In addition, contrast-induced nephropathy appears to be caused by the hyperosmolar and hyperviscid nature of most contrast agents [20, 21]. Hyperosmolar stress triggers prompt cellular generation of ROS and increased urine viscosity might obstruct renal tubules. Effects from hyperosmolar stress and viscosity might be compounded in the renal medulla, which is normally deficient in oxygen, with a PaO_2 of 10–20 mm Hg [22].

Taken together, there is evidence from physicochemical considerations, animal experiments, renal biomarker research and several randomized controlled trials that bicarbonate might exhibit positive renal effects in various settings of AKI potentially improving patients' outcome. The latter however needs to be proven in specifically designed randomized trials. Two multicentre randomized double-blind placebo-controlled trials are under way and will provide more information about the efficacy of bicarbonate infusion as a nephroprotective intervention in patients receiving CPB.

References

1 Uchino S, Kellum JA, Bellomo R, et al: Acute renal failure in critically ill patients: a multinational, multicenter study. JAMA 2005;294:813–818.
2 Skippen PW, Krahn GE: Acute renal failure in children undergoing cardiopulmonary bypass. Cit Care Resusc 2005;7:180–187.
3 Chertow GM: Independent associations between acute renal failure and mortality following cardiac surgery. Am J Med 1998;104:343–348.
4 Rosner MH, Okusa MD: Acute kidney injury associated with cardiac surgery. Clin J Am Soc Nephrol 2006;1:19–32.
5 Vercaemst L: Hemolysis in cardiac surgery patients undergoing cardiopulmonary bypass: a review in search of a treatment algorithm. J Extra Corpor Technol 2008;40:257–267.
6 Sponsel HT, Alfrey AC, Hammond WS, et al: Effect of iron on renal tubular epithelial cells. Kidney Int 1996;50:436–444.
7 Zager RA, Gamelin LM: Pathogenetic mechanisms in experimental hemoglobinuric acute renal failure. Am J Physiol 1989;256:F446–F455.
8 Anderson RJ, Ray CJ, Burke TJ: Human red blood cell hemolysate is a potent mitogen for renal tubular epithelial cells. Ren Fail 2000;22:267–281.
9 Vermeulen Windsant IC, Snoeijs MG, Hanssen SJ, et al: Hemolysis is associated with acute kidney injury during major aortic surgery. Kidney Int 2010 Feb 24 [Epub ahead of print].
10 Atkins JL: Effect of sodium bicarbonate preloading on ischemic renal failure. Nephron 1986;44:70–74.
11 Halliwell B, Gutteridge JM: Role of free radicals and catalytic metal ions in human disease: an overview. Methods Enzymol 1990;186:1–85.
12 Caulfield JL, Singh SP, Wishnok JS, et al: Sodium bicarbonate inhibits *N*-nitrosation in oxygenated nitric oxide solutions. J Biol Chem 1996;271:25859–25863.
13 Haase M, Haase-Fielitz A, Bellomo R, et al: Sodium bicarbonate to prevent increases in serum creatinine after cardiac surgery: a pilot double-blind, randomized controlled trial. Crit Care Med 2009;37:39–47.
14 Roudkenar MH, Kuwahara Y, Baba T, et al: Oxidative stress-induced lipocalin 2 gene expression: addressing its expression under the harmful conditions. J Radiat Res (Tokyo) 2007;48:39–44.
15 Haase M, Bellomo R, Haase-Fielitz A: Novel biomarkers, oxidative stress, and the role of labile iron toxicity in cardiopulmonary bypass-associated acute kidney injury. J Am Coll Cardiol 2010 (in press).
16 Merten GJ, Burgess WP, Gray LV, et al: Prevention of contrast-induced nephropathy with sodium bicarbonate: a randomized controlled trial. JAMA 2004;291:2328–2334.
17 Navaneethan SD, Singh S, Appasamy S, et al: Sodium bicarbonate therapy for prevention of contrast-induced nephropathy: a systematic review and meta-analysis. Am J Kidney Dis 2009;53:617–627.
18 Kanbay M, Covic A, Coca SG, et al: Sodium bicarbonate for the prevention of contrast-induced nephropathy: a meta-analysis of 17 randomized trials. Int Urol Nephrol 2009;41:617–627.
19 McCullough PA, Adam A, Becker CR, et al: Epidemiology and prognostic implications of contrast-induced nephropathy. Am J Cardiol 2006;98:5K–13K.
20 Seeliger E, Flemming B, Wronski T, et al: Viscosity of contrast media perturbs renal hemodynamics. J Am Soc Nephrol 2007;18:2912–2920.
21 Aspelin P, Aubry P, Fransson S-G, et al: Nephrotoxic effects in high-risk patients undergoing angiography. N Engl J Med 2003;348:491–499.
22 Brezis M, Rosen S: Hypoxia of the renal medulla: its implications for disease. N Engl J Med 1995;332:647–655.

Prof. Rinaldo Bellomo
Department of Intensive Care, Austin Health
Heidelberg, Melbourne, Vic 3084 (Australia)
Tel. +61 3 9496 5992, Fax +61 3 9496 3932
E-Mail rinaldo.bellomo@austin.org.au

Ronco C, Bellomo R, McCullough PA (eds): Cardiorenal Syndromes in Critical Care.
Contrib Nephrol. Basel, Karger, 2010, vol 165, pp 33–38

Acute Kidney Injury and Cardiopulmonary Bypass: Special Situation or Same Old Problem?

Michael Haase[a] · Andrew Shaw[b]

[a]Department of Nephrology and Intensive Care, Charité – University Medicine Berlin, Germany, and
[b]Department of Anesthesiology, Duke University Medical Center, Durham, N.C., USA

Abstract

Acute kidney injury (AKI) occurring after cardiac surgery is common and extends intensive care unit and hospital length of stay, as well as increases mortality rates. Its causes are multifactorial, and are not limited to ischemia and nephrotoxin administration. One particular problem is that both anemia and blood transfusion are risk factors for AKI, and neither are generally avoidable in high-risk heart surgery. Certain etiological factors are peculiar to the cardiac surgery setting, such as routine use of cardiopulmonary bypass, and various degrees of hypothermia, which is probably nephroprotective from an ischemia point of view, but which also worsens hemolysis from the pump and thus may actually exacerbate the problem. In this paper we review the place of antifibrinolytic therapy, hemodynamic control on bypass, and the correct level of oxygen delivery in the development of AKI after open heart surgery.

Acute kidney injury (AKI) is a frequent and serious complication of surgery, and is independently associated with increased morbidity and risk of death [1–3]. Cardiac surgery appears to be the second most common cause of AKI in critically ill patients after sepsis [4]. AKI occurs particularly commonly after open-heart surgery involving a period of cardiopulmonary bypass (CPB) or deep hypothermic circulatory arrest. There are many reasons why AKI might develop after a surgical intervention, including hypovolemia, systemic inflammation, anemia, hypotension, infection, administration of nephrotoxic drugs and hypoxemia as well as direct ischemia in the case of operations involving vascular interruption. However, there are certain characteristics of cardiac surgery that are unique, of which the most celebrated is use of the CPB machine, and in this review we focus on the pathophysiological mechanisms most relevant for cardiac surgery-associated acute kidney injury (CSA-AKI).

General Pathophysiology

In the setting of cardiac surgery, AKI has features in common with other types of AKI, and also peculiar to itself. In considering the factors that may give rise to an episode of AKI after heart surgery, we endorse the temporal approach developed by Bellomo et al. [5], reported in 2007. These authors considered a number of different pathophysiological mechanisms occurring in a sequence of insults to the kidney that together lead to an overall reduction in function that itself manifests as a rise in serum creatinine. These mechanisms include exogenous and endogenous toxins, metabolic factors, ischemia/reperfusion, neurohormonal activation, inflammation and oxidative stress.

Each of these mechanisms is likely relevant for CSA-AKI, but they are probably not all significant (or at play) at the same time. For example, antifibrinolytic therapy is clearly important in the intraoperative phase but rarely in the pre- and postoperative phase. Likewise, preexisting chronic kidney disease is important preoperatively since its severity has been shown to define the background on which further renal insults exert their influence on postoperative renal function. There has recently been considerable interest in the role of different antifibrinolytic drugs in heart surgery, and we review this again here, because of its relevance for this particular situation.

Antifibrinolytics

Antifibrinolytic drugs are used in the course of cardiac surgery to reduce bleeding, avoid blood transfusion and prevent reoperation ('take-back') procedures, since these are all known to increase mortality. Exposure of the circulation to the CPB machine leads to widespread intravascular thrombin generation (even in the presence of systemic heparin anticoagulation) and this in turn leads to increased fibrinolysis once the patient is weaned from CPB and the heparin anticoagulation is reversed with protamine. Increased fibrinolysis may lead to 'medical bleeding', particularly in the setting of complex operations (e.g. combined CABG/valve or revision procedures) and is associated with higher chest tube output, more blood transfusion and a higher reoperation rate. Antifibrinolytic drugs such as aprotinin, aminocaproic acid and tranexamic acid have traditionally been given in order to reduce the frequency and severity of this complication.

Aminocaproic acid and tranexamic acid are both known to be effective antifibrinolytic agents, and have a good safety record. Importantly, they block reuptake of small filtered proteins in the proximal tubule, resulting in 'tubular proteinuria' that resolves as soon as they are withdrawn [6]. For many years the greater blood-sparing effect of aprotinin led people to believe that aprotinin was superior to both of these agents, and that there were no specific safety issues relating to its use. However, in 2006, Mangano et al. [7] published a report that questioned both the efficacy and safety of this drug. This report generated much debate and was followed by a series of reports of the safety and efficacy of aprotinin from both muticenter [8] and single-center groups [9]. The US FDA subsequently removed this drug from the market in that country after a large randomized controlled clinical trial was stopped early because of excess mortality in the aprotinin arm [10].

Hemolysis

Of particular importance to cardiac surgery involving CPB is the role of pump-induced hemolysis. On the face of it, this may not seem to be a serious problem, since the urine of patients who have been exposed to CPB is often pink, yet these patients do not always develop AKI. Recently, there has been an investigation of the effects of pump-induced hemolysis on renal function after heart surgery, and of the extent of the injury (cf. decline in function) induced by high levels of free hemoglobin in the setting of cardiac surgery [11]. This working group found a convincing relationship between the magnitude of CPB-induced hemolysis, tubular injury and postoperative AKI. It is distinctly possible that amelioration of the renal manifestations of increased plasma-free hemoglobin is the reason for the apparent success of infusions of sodium bicarbonate in reducing the magnitude of the rise in serum creatinine when compared with an infusion of sodium chloride of similar 'tonic' load [12]. Alkalinization of the urine may in fact be a better therapeutic target than administration of a set dose of buffer base to at-risk patients, since it allows both personalization (e.g. acetazolamide for patients with systemic alkalemia, in whom base infusions would be difficult to justify) and titration (e.g. achieve a urine pH of 7) of the treatment. In fact, a very reasonable trial indeed would be to randomize any CPB patient being positive for any novel renal biomarker indicating structural tubular injury or those with pink urine to urinary alkalinization or control as soon as the urine discoloration appeared.

Blood Pressure

Although the pathophysiology of AKI after cardiac surgery is complex, only some contributing factors appear to be modifiable by those caring for these patients in the perioperative period. These include hemodynamic instability and systemic oxygen content. Systemic arterial hypotension of varying duration and degree is frequent during CPB. Mean arterial pressure (MAP) values may fall below the limit of renal autoregulation and thus initiate or augment ischemic kidney damage. Despite targeting systemic MAP values during CPB, MAP frequently and sometimes for prolonged periods, falls below values required to sustain renal autoregulation. Also, hemodilution is known to decrease the oxygen carrying capacity of blood and thus may contribute to a reduction in systemic oxygen content, in part attenuated by increased local renal capillary plasma flow. However, at which level of hemodilution the oxygen transport decreases below a critical level, i.e. tissue hypoxia occurs, depends on blood flow and the rate of metabolism, which is affected by many factors, such as body temperature and depth and type of anesthesia. Hemodilution is likely to dilute hormones acting distantly from the synthesis site more than hormones acting near to the synthesis site, for example diminishing the effect of systemic hormones such as angiotensin II or epinephrine [13].

The potential association of intraoperative hemodynamic instability with postoperative renal function has been previously investigated in patients receiving open-heart surgery. None of these studies were able to demonstrate low perfusion pressure to be an independent risk factor for AKI. Furthermore, in a small interventional study [14] comparing a higher blood pressure group (range 67–77 mm Hg) with a lower blood pressure group (range 39–70 mm Hg), there were no deleterious effects of a low intraoperative perfusion pressure on postoperative renal function or postoperative mortality.

Too much CPB-associated hemodilution is a major contributing factor to renal damage as it notably reduces oxygen delivery by reducing the oxygen transport capacity of the blood as well as affecting the microcirculation. However, blood transfusion has also been associated with an increased incidence of postoperative AKI [15, 16].

Due to technological limitations, most studies reporting on the possible association of MAP and postoperative AKI used manually recorded and potentially incomplete and biased MAP values, thus they might lack accuracy and had limited statistical power. Therefore, the potential importance of MAP as an independent predictor of postoperative AKI remains uncertain. The MAP values applied during CPB are often near the minimum levels that support normal renal function, and any further disturbance may lead to ischemia and cellular damage [17]. Thus, the investigation of the possible role of intraoperative hypotension as an independent contributor to the development of AKI remains physiologically logical and clinically important.

We conducted an observational study [18] of 920 patients receiving CPB to detect whether modifiable factors such as intraoperative electronically stored MAP indices (measured every 20 s), or oxygenation indices obtained by 30-min arterial blood gas monitoring are independent predictors of subsequent AKI. AKI was defined as an increase in serum creatinine >50%. We linked such data with development of postoperative AKI. Patients who developed postoperative AKI had worse clinical outcomes including a need for initiation of renal replacement therapy and an increased length of stay in the intensive care unit and in hospital.

In this study, univariate analysis showed that patients who subsequently developed AKI had more prolonged duration and greater AUC of intraoperative MAP values <60 and <70 mm Hg. However, we found no difference in the median intraoperative MAP or MAP variability according to differential creatinine increase. Also, low intraoperative systemic MAP – in the setting of a targeted MAP >60, or >70 mm Hg in patients at renal risk – was not independently associated with postoperative kidney injury in this setting. On the other hand, systemic oxygen content was lower in patients developing AKI compared to those without AKI. In multivariate regression analyses, duration of CPB, emergency cardiac surgery, atrial fibrillation and AKICS score [3] were independent risk factors for AKI in this cohort. Sensitivity analysis showed that the study results remained essentially unchanged when the AKI definition was varied.

Further studies are needed to investigate if modified management of systemic arterial oxygen content might help to reduce the incidence of AKI. Indeed, prospective randomized trials are warranted to identify the optimal CPB hematocrit and to test the efficacy of nephroprotective measurements such as preoperative erythropoietin treatment, the use of miniaturized CPB circuit designs or a restrictive threshold for intraoperative crystalloid infusion or blood transfusion.

Conclusions

It is clear that AKI occurring in the setting of cardiac surgery has additional (i.e. different) etiological factors when compared with other causes of acute reductions in kidney function. We believe that intraoperative hemodynamic and hemolysis control provide potential areas for further study, and further that the field of CSA-AKI is an attractive domain for pathophysiological and mechanistic studies owing to its temporal predictability and relatively homogeneous injury type.

References

1 Chertow GM, Burdick E, Honour M, et al: Acute kidney injury, mortality, length of stay, and costs in hospitalized patients. J Am Soc Nephrol 2005;16:3365–3370.

2 Lassnigg A, Schmidlin D, Mouhieddine M, et al: Minimal changes of serum creatinine predict prognosis in patients after cardiothoracic surgery: a prospective cohort study. J Am Soc Nephrol 2004;15:1597–1605.

3 Palomba H, de Castro I, Neto AL, et al: Acute kidney injury prediction following elective cardiac surgery: AKICS Score. Kidney Int 2007;72:624–631.

4 Uchino S, Kellum JA, Bellomo R, et al: Acute renal failure in critically ill patients: a multinational, multicenter study. JAMA 2005; 294:813–818.

5 Bellomo R, Auriemma S, Fabbri A, et al: The pathophysiology of cardiac surgery-associated acute kidney injury (CSA-AKI). Int J Artif Organs 2008;31:166–178.
6 Stafford Smith M: Antifibrinolytic agents make α_1- and β_2-microglobulinuria poor markers of post-cardiac surgery renal dysfunction. Anesthesiology 1999;90:928–929.
7 Mangano DT, Tudor IC, Dietzel C: The risk associated with aprotinin in cardiac surgery. N Engl J Med 2006;354:353–365.
8 Schneeweiss S, Seeger JD, Landon J, et al: Aprotinin during coronary-artery bypass grafting and risk of death. N Engl J Med 2008;358:771–783.
9 Shaw AD, Stafford-Smith M, White WD, et al: The effect of aprotinin on outcome after coronary-artery bypass grafting. N Engl J Med 2008;358:784–793.
10 Fergusson DA, Hebert PC, Mazer CD, et al: A comparison of aprotinin and lysine analogues in high-risk cardiac surgery. N Engl J Med 2008;358:2319–2331.
11 Vermeulen Windsant IC, Snoeijs MG, Hanssen SJ, et al: Hemolysis is associated with acute kidney injury during major aortic surgery. Kidney Int 2010; Febr 24 [Epub ahead of print].
12 Haase M, Haase-Fielitz A, Bellomo R, et al: Sodium bicarbonate to prevent increases in serum creatinine after cardiac surgery: a pilot double-blind, randomized controlled trial. Crit Care Med 2009;37:39–47.
13 Vermeer H, Teerenstra S, de Sevaux RG, et al: The effect of hemodilution during normothermic cardiac surgery on renal physiology and function: a review. Perfusion 2008; 23:329–338.
14 Urzua J, Troncoso S, Bugedo G, et al: Renal function and cardiopulmonary bypass: effect of perfusion pressure. J Cardiothorac Vasc Anesth 1993;6:299–303.
15 Habib RH, Zacharias A, Schwann TA, et al: Role of hemodilutional anemia and transfusion during cardiopulmonary bypass in renal injury after coronary revascularization: implications on operative outcome. Crit Care Med 2005;33:1749–1756.
16 Karkouti K, Beattie WS, Wijeysundera DN, et al: Hemodilution during cardiopulmonary bypass is an independent risk factor for acute renal failure in adult cardiac surgery. J Thorac Cardiovasc Surg 2005;129:391–400.
17 Rosner MH, Okusa MD: Acute kidney injury associated with cardiac surgery. Clin J Am Soc Nephrol 2006;1:19–32.
18 Haase M, Haase-Fielitz A, Story D: Low systemic mean arterial pressure during cardiopulmonary bypass is not independently associated with postoperative acute kidney injury in cardiac surgery patients. J Am Soc Nephrol 2007;18:3206A.

Dr. Michael Haase
Department of Nephrology and Intensive Case, Charité, University of Medicine Berlin
Augustenburger Platz 1, DE–13353 Berlin (Germany)
Tel. +49 30 450 553132, Fax +49 30 450 553909
E-Mail michael.haase@charite.de

Ronco C, Bellomo R, McCullough PA (eds): Cardiorenal Syndromes in Critical Care.
Contrib Nephrol. Basel, Karger, 2010, vol 165, pp 39–45

Pathophysiology of Acute Kidney Injury: A New Perspective

Xiaoyan Wen · Raghavan Murugan · Zhiyong Peng · John A. Kellum

Clinical Research, Investigation, and Systems Modeling of Acute illness (CRISMA) Laboratory, Department of Critical Care Medicine, University of Pittsburgh, Pittsburgh, Pa., USA

Abstract

Acute kidney injury (AKI) in critically ill patients is a devastating illness associated with prolonged hospital stay and high mortality. Limited progress has been made in the field of AKI, and its treatment using renal replacement therapy, at best, only provides partial renal support. Ischemia-reperfusion rodent AKI models do not resemble human renal injury and the absence of renal biopsy data limits our understanding of the pathophysiology of human AKI. However, laboratory and clinical evidence suggests that the inflammatory milieu leads to dysfunction of renal cells and this may be the key factor leading to AKI. Cells in injured tissues release immunological danger signals or danger-associated molecular pattern molecules which communicate with remote organs including the kidney, where they activate dendritic cells and T cells and thus initiate inflammation. Once the initial insult has passed, tubular epithelial cells undergo dedifferentiation, reacquire progenitorial ability to proliferate, migrate, and redifferentiate into mature intrinsic cells. Dissonance of mediator secretion and cell responses may lead to persistent injury and de novo chronic kidney disease. A number of soluble mediators including transforming growth factor-β (TGF-β) initiate a variety of pathophysiological processes at the beginning of kidney injury. TGF-β also plays a fundamental role in cell proliferation and interstitial fibrosis in later phases. The renin-angiotensin-aldosterone system, especially angiotensin II, contributes to kidney injury through the angiotensin II type 1 receptor, TGF-β receptor Smad and epidermal growth factor receptor by affecting general angiostasis and vascular remodeling, indirectly modulating inflammation and cell reactions. We review the pathophysiology of AKI in light of new information regarding renal injury and repair.

Acute kidney injury (AKI) occurs in up to 70% of critically ill patients. Overall 4–5% of patients admitted to intensive care units (ICU) receive renal replacement therapy. Development of AKI results in considerable morbidity, high mortality, and significant healthcare resource utilization. Despite improvements in dialysis technology and supportive care, the mortality associated with severe AKI has only mildly ameliorated over the past 2 decades – though typical patients are older and sicker than before. AKI in the ICU rarely occurs in isolation and has a profound influence on the function and outcomes of non-renal organ systems. A systematic review of 24 AKI studies [1], using kidney injury defined according to the Risk-Injury-Failure-Loss-Endstage (RIFLE) criteria, showed an increase in the relative risk of death with the severity of AKI. Even in the mildest kidney dysfunction risk category, AKI was associated with a 2.4-fold increase in the risk of death. Data from a multicenter ICU study found that sepsis accounted for 32.4% of all patients with AKI, and was associated with greater severity of AKI and higher mortality [2]. Hence understanding the pathophysiology of this set of AKI, including developing tools for early diagnosis is pivotal to improve outcomes.

Although many efforts have been made to understand the pathophysiology of septic and inflammatory AKI, the precise mechanisms responsible for functional deterioration is still limited and therefore therapies are mainly supportive. However, recent studies suggest that leukocyte activation, upregulation of soluble mediators such as cytokines and chemokines, as well as their interactions are key factors that significantly influence the occurrence of AKI, its progression, and the development of complications. In a prospective multicenter study of 1,836 patients with community-acquired pneumonia, Murugan et al. [3] reported that 34% of patients with pneumonia developed AKI. The occurrence, severity, and outcomes of AKI were significantly associated with high concentrations of plasma cytokines, in particular, interleukin-6 (IL-6). Moreover, the majority of patients with sepsis-induced AKI are not admitted to ICU and are not in shock. As such, animal models of AKI that rely on ischemia-reperfusion bear little resemblance to the clinical condition and produce histopathologic changes not seen in human AKI [4]. However, numerous laboratory results support the view that the inflammatory milieu leads to derangements in the structure and function of renal cells including epithelial cells [5] and this may be a key factor leading to kidney and multiple organ dysfunction. The mechanisms responsible for cell dysfunction associated with acute inflammation are elusive, many of which are the focus of intensive research.

Renal Response Following Injury

AKI occurs in the setting of a variety of insults and the form of injury may cause very different forms of this disease. Indeed, there is evidence that sepsis, while being the most common cause of AKI in humans, may be a special form of the disease, typified by a systemic inflammatory state in which the 'primary disease' is usually not in the kidney. Although renal biopsies are seldom performed in cases of septic AKI, available data from autopsies and animal studies suggest that sepsis-induced AKI is largely an acellular process with little evidence of tubular necrosis [6]. However, a strong association with proinflammatory cytokine expression and the development of AKI [3, 7] suggests an important role of systemic inflammatory mediators in the process of septic AKI.

By contrast pure ischemic AKI may be relatively uncommon clinically, but it is certainly the best studied form of AKI in terms of mechanisms and histopathology [4]. While a unified understanding of AKI pathophysiology is still lacking, it appears that some features may well be consistent across different forms of AKI. For example, regardless of the type of injury, sub-lethally injured cells release immunological danger signals or danger-associated molecular pattern molecules (DAMPs) such as the chromatin-associated protein high-mobility group box 1, DNA, ATP, adenosine and uric acid.

These molecules activate immune effector cells, such as dendritic cells, which are constantly in communication with other cells capable of priming and activating the immune system for response against various insults. Several important inflammatory molecules and pathways can become involved including Toll-like receptors and the nuclear factor-κB system. The complement cascade may also become activated as part of the DAMP signaling.

Experimental data show that intrinsic renal cells can undergo phenotypic transformation to progenitor cells known as 'dedifferentiation', influenced by certain soluble mediators such as inflammatory cytokines and growth factors. The dedifferentiation from renal tubular epithelial cells to mesenchymatous cells is the best documented cell transition in the kidney, and has been termed the epithelial-mesangial trans-differentiation process. Dedifferentiated epithelial cells lose cell polarity and cell–cell junction, detach from the basement membrane, and lose their normal phenotype and function. They acquire new characters of both muscle cells and fibroblasts called myofibroblasts. These activated fibroblasts are capable of proliferation, migration and synthesis of extracellular matrix and can play integral roles in both repair and in fibrosis. Further studies also show tubular epithelial cells can dedifferentiate into other cell types such as phagocytes, which are also characterized by the expression of proteins such as kidney injury molecule-1. These cells acquire the ability to internalize adjacent apoptotic bodies and necrotic cells, and remove irreversibly damaged tubular cells in absence of infiltrated macrophages [8]. During the process of renal injury and subsequent recovery, phenotypically transformed cells produce extracellular matrix components that are deposited in the extracellular compartment. Normal kidney tissue architecture is maintained by the dynamic balance between cellular matrix production and enzyme matrix degradation. Extracellular matrix turnover imbalance leads to deposition of abnormal constituents such as interstitial collagen I and III (fig. 1).

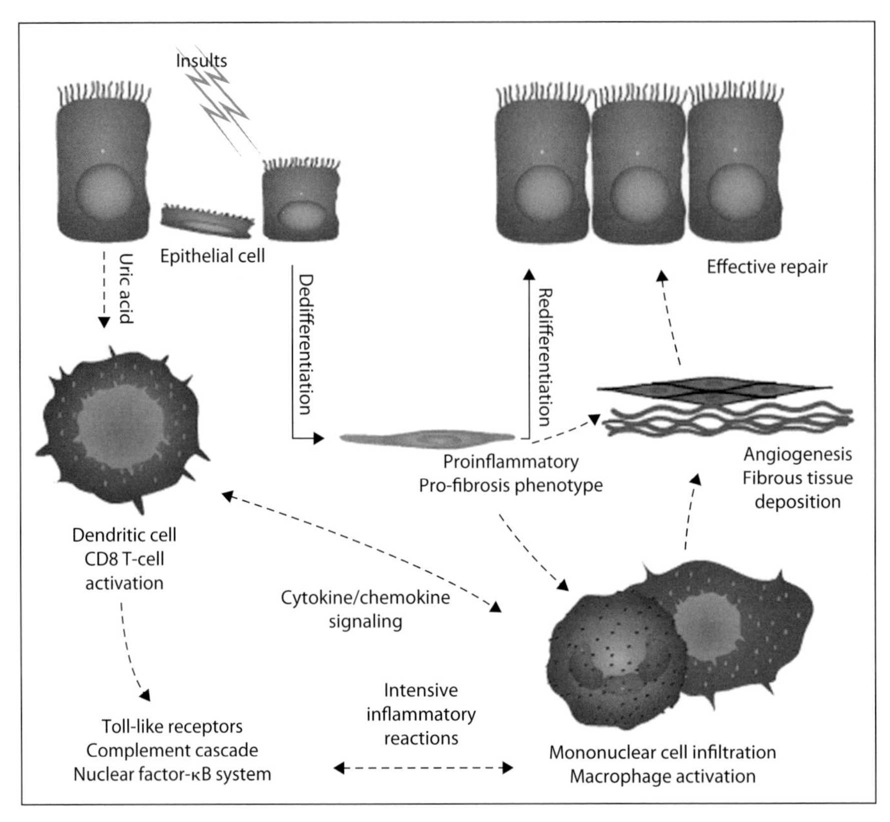

Fig. 1. In response to kidney injury, sub-lethally injured cells release uric acid as the immunologic danger signal, which subsequently activates dendritic cells and T cells. Released cytokine/chemokines promote intense inflammation in locally injured regions through activated Toll-like receptors, complement cascade, and nuclear factor-κ-B. The initial injury and proinflammatory state may lead to mononuclear cell infiltration and macrophage activation, and promote angiogenesis and fibrous tissue deposition in cases of AKI non-recovery. Surviving intrinsic cells dedifferentiate to ancestor phenotype due to an imbalance in TGFβ1 and hepatocyte growth factor. They acquire the ability to migrate, engulf, proliferate, secrete cytokine/chemokines and synthesize matrix in the process of repair, and redifferentiate into normal cells when inflammation has subsided. A well-orchestrated cellular response leads to effective tissue repair.

Renal Tubular Inflammation and Repair

Renal tubular epithelial cells are the most fragile yet actively regenerating population of cells following renal injury. In response to fluctuation in soluble mediators, cells undergo dedifferentiation, migration, proliferation, and redifferentiation leading to complete recovery. Tubular epithelial cells acquire the ability to migrate and proliferate only after they have dedifferentiated into progenitor phenotype. Research has revealed that the timing and direction of cell's transition is mainly determined by the counterbalance of several pairs of growth factors, adhesive molecules, and cytokines such as transforming growth factor-β1 (TGFβ1) and hepatocyte growth factor (HGF). Through binding with its cell membrane receptors, Smad, TGFβ1 transfers into nucleolus and modulates multiple gene transcriptional factors. TGFβ1 is therefore capable of activating mitogen-activated protein kinase, phosphoinositide-3-kinase and Wnt/β-catenin-signaling pathways widely involved in cell differentiation, migration and extracellular matrix deposition. HGF antagonizes TGFβ1 and prevents TGFβ1 gene transcription thereby inhibiting protein synthesis. HGF also directly promotes cell regeneration and tissue repair against TGFβ1. Genetic fate-mapping techniques show that the majority of regenerated tubule epithelial cells come from injured intrinsic tubular cells [9]. Transformed mesangial cells and endothelial cells eventually undergo apoptosis. Disturbances involving anatomic/functional changes or concomitant diseases interfering with soluble mediator secretion, cell–cell communication and cellular phenotype transition will lead to inappropriate inflammation and extensive apoptosis. Intrinsic cell losses, glomerulosclerosis, and progressive interstitial fibrosis will progress to incomplete renal recovery and ultimate organ death.

Important Elements that Mediate Injury Responses and Recovery

TGFβ

The TGFβ is a member of the transforming growth factor superfamily that suppresses cell proliferation, promotes cell differentiation, and plays fundamental roles in cell proliferation and response in embryologic development, oncogenesis, and tissue repair. It initiates a variety of pathophysiological processes at the beginning of kidney injury, including tubular epithelial cell apoptosis, intrinsic cell dedifferentiation and extracellular matrix deposition, and is correlated closely with acute deterioration in renal function and renal fibrosis. Several cells such as macrophages, tubular epithelial cells and myofibroblasts are all capable of secreting TGFβ at different stages in the process of renal fibrosis. The fact that macrophage ablation markedly attenuated renal fibrosis in various etiologies of renal diseases suggests that macrophages might be the main source of TGFβ. Activated TGFβ first binds with its receptor 'Smad' on the surface of the cell membrane, and is then endocytosed into the cell. After a series recruiting and binding reactions, part of TGF-Smad forms a complex that gets into the nucleus thereby regulating target gene transcription. Another part of TGFβ-Smad is rapidly digested and turned over. The way in which TGF-Smad is processed and which kind of formation is dominant are determined by local signals. Recent research shows that IL-6 enhances TGFβ signaling by reinforcing sustained endocytosis [10], which may help explain the negative association between plasma IL-6 levels and the adverse outcome we observe in humans with kidney injury [3].

Renin-Angiotensin-Aldosterone System
The renin-angiotensin-aldosterone system has emerged as an important contributor to chronic fibrotic kidney disease. Angiotensin II, the main physiological effector molecule of the renin-angiotensin-aldosterone system, via binding with the angiotensin II type 1 receptor, mediates cell cytoskeleton rearrangements leading to a morphological change in cells. This classical ligand-receptor combination in the kidney affects general angiotasis, and its oversecretion causes hypertension which further causes renal injury. Recently, it has been reported that high level angiotensin II activity contributes to long-term renal fibrosis through conjugation with the TGFβ receptor Smad, facilitating tubular epithelial cell dedifferentiation in the absence of TGFβ [11]. When angiotensin II binds with epidermal growth factor receptor mediated by TGFα, it modulates kidney vascular remodeling, playing an important role in the outcome of kidney injury. Angiotensin II promotes T-cell proliferation and activation, increases expression of the proinflammatory cytokine tumor necrosis factor and profibrotic cytokine TGFβ, thus indirectly modulating inflammation, cell proliferation, cell transdifferentiation and fibrosis.

Potential Immune Modulation Therapies for AKI

Increasing knowledge of AKI pathophysiology could potentially result in the development of novel therapeutic interventions. Based on the conclusions drawn from experimental and human studies, several treatments have been confirmed showing efficiency in preliminary clinical trials. The most acceptant strategy is to modulate immunological states and hence alleviate ultimate outcome of AKI using mesenchymal stem cell injection or corporate circulation within the synthetic membrane device consisting of cytopheretic membranes and/or renal progenitor/stem cells. Another way is to adopt biological inhibitors or chemical reagents, such as monoclonal antibody to TGFβ, the angiotensin-converting enzyme inhibitor, the adenosine 2A receptor agonist, and Toll-like receptor-9 inhibitor, -chloroquine, to inactivate the inflammatory cascade, and ameliorate renal damage.

Conclusion

AKI has a high incidence rate in critically ill patients and inflammation plays an important pathophysiological role in injury and renal repair. A significant percentage of AKI may progress to chronic renal dysfunction due to dissonant mediator secretion and cell responses. Although much of the mechanisms are still unknown, better understanding of the pathophysiology is important for early diagnosis and to design interventions.

References

1 Ricci Z, Cruz D, Ronco C: The RIFLE criteria and mortality in acute kidney injury: a systematic review. Kidney Int 2008;73:538–546.

2 Bagshaw SM, George C, Dinu I, Bellomo R: A multi-centre evaluation of the RIFLE criteria for early acute kidney injury in critically ill patients. Nephrol Dial Transplant 2008; 23:1203–1210.

3 Murugan R, Karajala-Subramanyam V, Lee M, Yende S, Kong L, Carter M, Angus DC, Kellum JA: Acute kidney injury in non-severe pneumonia is associated with an increased immune response and lower survival. Kidney Int 2009; Dec 23 [Epub ahead of print].

4 Heyman SN, Rosenberger C, Rosen S: Experimental ischemia-reperfusion: biases and myths-the proximal vs. distal hypoxic tubular injury debate revisited. Kidney Int 2010;77:9–16.

5 Fink MP, Delude RL: Epithelial barrier dysfunction: a unifying theme to explain the pathogenesis of multiple organ dysfunction at the cellular level. Crit Care Clin 2005; 21:177–196.

6 Langenberg C, Bagshaw SM, May CN, Bellomo R: The histopathology of septic acute kidney injury: a systematic review. Crit Care 2008;12:R38.

7 Bagshaw SM, Langenberg C, Haase M, Wan L, May CN, Bellomo R: Urinary biomarkers in septic acute kidney injury. Intensive Care Med 2007;33:1285–1296.

8 Ichimura T, Asseldonk EJ, Humphreys BD, Gunaratnam L, Duffield JS, Bonventre JV: Kidney injury molecule-1 is a phosphatidylserine receptor that confers a phagocytic phenotype on epithelial cells. J Clin Invest 2008;118:1657–1668.

9 Humphreys BD, Valerius MT, Kobayashi A, et al: Intrinsic epithelial cells repair the kidney after injury. Cell Stem Cell 2008;2:284–291.

10 Zhang XL, Topley N, Ito T, Phillips A: Interleukin-6 regulation of transforming growth factor (TGF)-beta receptor compartmentalization and turnover enhances TGF-beta1 signaling. J Biol Chem 2005;280:12239–12245.

11 Carvajal G, Rodríguez-Vita J, Rodrigues-Díez R, et.al: Angiotensin II activates the Smad pathway during epithelial mesenchymal transdifferentiation. Kidney Int 2008; 74:585–595.

John A. Kellum, MD
3550 Terrace Street
604 Scaife Hall
Pittsburgh, PA 15261 (USA)
Tel. +1 412 647 7810, Fax +1 412 647 8060, E-Mail kellumja@ccm.upmc.edu

Ronco C, Bellomo R, McCullough PA (eds): Cardiorenal Syndromes in Critical Care.
Contrib Nephrol. Basel, Karger, 2010, vol 165, pp 46–53

Multiphoton Imaging Techniques in Acute Kidney Injury

Bruce A. Molitoris · Ruben M. Sandoval

Division of Nephrology, Department of Medicine, Indiana Center for Biological Microscopy, Indiana University School of Medicine, Indianapolis, Ind., USA

Abstract

Multiphoton microscopy allows investigators the opportunity to study dynamic events within the functioning kidney and during acute kidney injury. This enables investigators to follow complex multifactorial processes in the kidney with improved spatial and temporal resolution, and sensitivity. Furthermore, the ability to obtain volumetric data (3-D) makes quantitative 4-D (time) analysis possible. Finally, use of up to three fluorophores concurrently in multiphoton microscopy allows for three different or interactive processes to be observed simultaneously. Therefore, this approach compliments existing molecular, biochemical and pharmacologic techniques by allowing for direct visualization at the cellular and subcellular levels for molecules without the requirement for fixation. Its use in acute kidney injury is in its infancy but offers much promise for unraveling the complex interdependent processes known to contribute to cell injury and organ failure.

Developments in kidney multiphoton microscopy over the last decade have provided researchers new and tremendously in-depth insights into complex yet highly interdependent cellular and subcellular processes. A significant advance in these developments has been the emerging technology of studying cells within their natural living environment, rather than in isolated ex-vivo controlled settings. This gives the unique opportunity of analyzing cellular pathophysiology of acute kidney injury (AKI) and the effects of potential treatments in the context of an intact functioning organ. It also offers hope for additional diagnostic tools for diagnosis, stratification and prognostic purposes. The purpose of this review is to emphasize the use of multiphoton microscopy in understanding the pathophysiology and therapy of AKI.

Multiphoton Microscopy

The potential to image deeper, with far less phototoxicity, into biological tissue was accomplished by utilizing multiphoton microscopy where increased penetration (up to 150 μm in the kidney) occurs [1–3]. Multiphoton microscopy allows for utilization of multiple fluorescent probes simultaneously, enabling labeling of different physiological compartments. Distinguishing these fluorescent emissions from endogenous or autofluorescence is also easier and enhanced in multiphoton fluorescence microscopy, as compared to confocal microscopy, as the fluorescence excitation occurs only at the focal point of the excitation beam. Therefore, out-of-focus fluorescent excitation is eliminated [2]. The longer wavelength light used for multiphoton microscopy also penetrates deeper into the kidney.

Table 1 lists the possible types of data that can be acquired using multiphoton imaging of the kidneys. Migration of white blood cells (WBCs) out of the microvasculature has not been seen up to 24 h post injury.

Yu et al. [4] have developed a quantitative ratiometric approach using a generalized polarity (GP) concept that was implemented to analyze the multidimensional data obtained from multidextran infusion experiments, the concept being the comparison of relative intensities of two fluorescent dyes. Ratiometric imaging methods are relatively independent of the amount of fluorescent probes injected, the excitation power and the depth of field being imaged. These properties are particularly advantageous for the quantitative imaging of animals where there is variability in the quantity of dyes injected, the appropriate levels of laser power used, and imaging depth. In addition, using ratiometric techniques also minimizes spatial variations of the fluorescence signals across the field of view due to detector/sample non-uniformity. Finally, these investigators were able to quantify GFR based upon the disappearance (filtration) of a small molecular weight dextran, compared to a non-filtered fluoroscopy dextran, over time.

Intracellular Endocytosis, Trafficking, Transcytosis

Intracellular uptake, compartmentation and metabolism can be studied and quantified once the fluorescent probe has entered the cell. Using multiphoton microscopy it is now possible to observe and quantify endocytosis occurring across the apical membrane of the proximal tubule cells [1, 5]. Furthermore, it is possible to follow the intracellular accumulation and subcellular distribution over time in the same animal, and to undertake repeated observations in the same animal at varying intervals over days to weeks. Such experiments are particularly useful in understanding drug delivery for AKI states. We have utilized this approach to observe and quantify fluorescent siRNA

Table 1. Uses for multiphoton microscopy in the kidney

Glomerular
– Size/volume/cellularity
– Permeability/filtration
– Fibrosis/sclerosis
– Glomerular sieving coefficient determination
Microvasculature
– RBC flow rate
– Endothelial permeability
– WBC adherence/rolling
– Vascular diameter
– Clot formation
Cellular uptake
– Cell type-specific uptake
– Site – apical vs. basolateral membrane
– Mechanism – endocytosis vs. carrier/transporter-mediated
Cellular trafficking
– Intracellular organelle distribution
– Cytosol, nuclear localization
– Transcytosis
Cellular metabolism
– Fluorescence decay over time
Cell toxicity
– Cell injury with necrosis, apoptosis
– Surface membrane/blebbing
– Mitochondrial function
– Actin cytoskeletal dystruption, aggregation

uptake and metabolism differentiating endosomal and cytosolic concentrations [6].

Intracellular organelles such as mitochondria and lysosomes can be studied in acute injury states by specific labeling of these organelles and quantifying individual number and fluorescence potential of respective organelles. DNA fluorescent markers can help identify specific cell types based on their nuclear morphology (e.g. nuclei of podocytes are characteristically bean-shaped, while endothelial cells have characteristic flattened elongated morphology). It also permits evaluation of intranuclear uptake of other fluorescent compounds in disease and therapeutic states, and analysis of necrosis and apoptosis [7].

Glomerular Permeability

Glomerular permeability and filtration of different sized compounds across glomerular capillaries can be quantified and visualized using Munich-Wistar rats with surface glomeruli (100 μm in diameter). Areas of interest can be defined and isolated and fluorescence intensities measured to obtain quantitative date. By using the change in GP or ratiometric methods it is possible to quantify glomerular permeability by measuring relative change intensities of dyes in the bloodstream or within the Bowman's space [4]. Filtration and clearance of smaller size molecules is typically faster than larger size molecules. By using two different sized fluorescent dextrans, it is thus possible to measure glomerular filtration fraction by measuring the changes in concentration in the bloodstream over time of each dextran. In figure 1 we show glomerular filtration of both a truly filtered small MWT red fluorescent dextran and a large 500-kDa green fluorescent dextran in the same glomerulus following a 45-min renal pedicle clamp injury. Note the small dextran is seen throughout Bowman's space. The large dextran is also filtered in a portion of the glomerulus suggesting a worked increase in the glomerular sieving coefficient post ischemic injury, consistent with known changes in podocyte structure following ischemic injury [8]. A large clot fills the lower large capillary on the right.

Proximal Tubule Reabsorption

Since many AKI states affect the proximal tubules (PT), it is very important to be able to study the reabsorptive profile of the tubular epithelium and its recovery with or without therapeutic intervention. It is possible to evaluate the reabsorptive properties of the PT by (a) direct observation of the epithelial cells and lumen of the PT, (b) luminal GP changes in the PT, and (c) comparing the

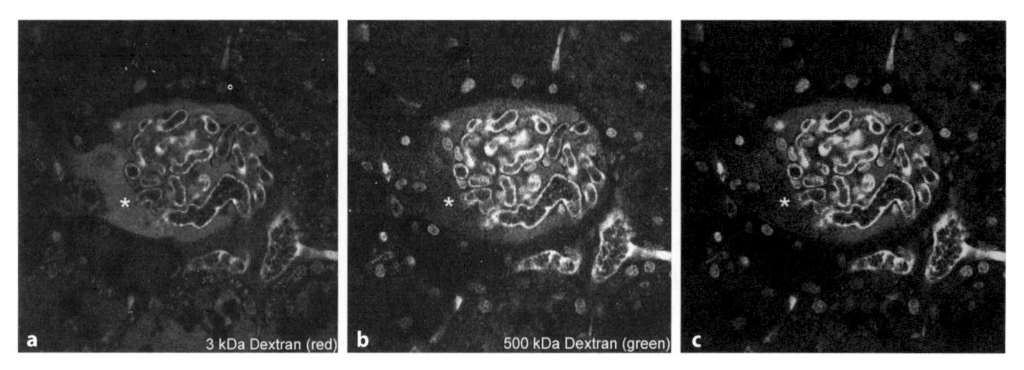

Fig. 1. Intravital multiphoton kidney imaging showing cortical tissue following ischemic injury. A rat following 45 min of renal pedicle clamping was administered intravenously a 500-kDa fluorescein-conjugated dextran (green) to label the vasculature, Hoechst 33342 (cyan) to label nuclei systemically, and a small filterable 3-kDa rhodamine-conjugated dextran (red). Circulating RBCs within the microvasculature appear as black streaks as the 500-kDa dextran (green) fills the plasma space but remains excluded from the RBCs. There is a noticeable flux of the 500-kDa dextran (green) into Bowman's space, except for the area on the left (*), implying injury to part of the glomerular capillaries resulting in an abnormally high glomerular sieving coefficient (bar = 20 μm). **a** 3-kDa dextran channel with uniform filtration into Bowman's space. **b** 500-kDa dextran channel. **c** Overlap of the two channels showing differential filtration of the large 500-kDa green dextran.

change in GP values from within PT lumen with those from within Bowman's capsule [1–4].

Red Blood Cell Flow Rates

Multiphoton microscopy can also be utilized to study and quantify red blood cell (RBC) flow rates in the microvasculature. It is possible to image and differentiate the motion of RBCs in the renal cortex microvasculature. RBCs exclude the non-filterable fluorescent dye used to label the circulating plasma and consequently appear as dark, non-fluorescent objects on the images. The acquisition of repetitive scans along the central axis of a capillary (line-scan method) gives the flow of RBCs, which leaves dark striped bands in the images where the slope of the bands is inversely proportional to the velocity; the more shallow the slope, the faster the flow rate [9]. Using standardized vessel length, diameter and angles, it is possible to calculate the RBC flow rates in different states, e.g. ischemia, sepsis, and response to specific treatments [10–14]. We have recently quantified the effect of soluble thrombomodulin on endothelial protection from ischemic injury and the resulting improvement in RBC velocity and reduced WBC adherence [15].

Inflammation and Leukocytes

Inflammation is being increasingly recognized as an important and central process in the initiation, maintenance and progression of AKI and chronic kidney injury. Hence understanding the dynamic roles and functions of different leukocytes and their interactions with soluble and endothelial cell factors is key to develop and test preventive and therapeutic strategies. Multiphoton microscopy provides the opportunity to observe these crucial processes in vivo. The most effective method to image leukocytes in the kidney involves direct fluorescent labeling of specific WBC lineages. Fluorescent agents such as rhodamine 6G, acridine red or orange are preferentially concentrated in WBCs and allow intravascular detection. The DNA dye (Hoechst) used to label all nuclei, also permits detection of WBCs and differentiation of RBCs in the vascular space. Various markers are available for B lymphocytes, CD8+ T lymphoblasts, macrophages, naive T cells, etc. [12]

Vascular Pathology

AKI resulting from ischemia is associated with microvascular permeability defects and this is eventually linked with microvascular dropout [14]. Various other pathological conditions affect the vascular space, especially the glomerular vasculature. Apart from studying flow rates as mentioned above, using mixtures of different sized fluorescent-labeled dextrans, we can now examine the effect of injury on the microvasculature by observing and measuring the extravasation of these dextrans into the interstitium as elegantly demonstrated by Sutton et al. [16]. Furthermore, using multiple fluorescent labeling there is also the opportunity to study the effect of injury on the intricate and dynamic interaction of the endothelium with the matrix proteins such as metalloproteinases [17]. It also allows for correlation and further quantification of the relationship between the endothelial permeability and alterations in blood flow rates.

The regulation of glomerular hemodynamics is dependent on the interplay between the microvasculature, tubule and the juxtaglomerular apparatus. Peti-Peterdi and colleagues [18] have now extended their studies into actually being able to quantify the renin content of the juxtaglomerular apparatus in vivo in anesthetized Munich-Wistar rats using quinacrine tagging of the renin-secreting granules. They have also been able to measure the diameters of the afferent and efferent arterioles as well as glomerular volume.

Fluorescence microscopy application also extends its application to gene transfer into specific targets, e.g. tubular or endothelial cells. Green fluorescent protein linked with the protein of interest, e.g. actin, that is cloned into an adenovirus vector, can be delivered into the urinary Bowman's space, proximal tubule lumen or superficial efferent arteriole by micropuncture techniques. The expression of these fluorescent-labeled proteins is an invaluable tool in studying dynamic changes in the cell in various disease models [19, 20].

Conclusion

In summary, recent advances in fluorescent imaging techniques, especially in intravital multiphoton microscopy, have now enabled investigators to employ and develop unique, tailor-made techniques to visualize the functioning kidney and analyze, in a dynamic fashion, the cellular and subcellular processes that occur in various AKI states and the response to therapy. Application of some of these techniques to patients is close at hand and will add to the diagnostic repertoire of the clinician.

References

1 Molitoris BA, Sandoval RM: Pharmacophotonics: utilizing multi-photon microscopy to quantify drug delivery and intracellular trafficking in the kidney. Adv Drug Deliv Rev 2006;58:809–823.
2 Dunn KW, Sandoval RM, Molitoris BA: Intravital imaging of the kidney using multiparameter multiphoton microscopy. Nephron Exp Nephrol 2003;94:e7–e11.
3 Dunn KW, Sandoval RM, Kelly KJ, Dagher PC, Tanner GA, Atkinson SJ, Bacallao RL, Molitoris BA: Functional studies of the kidney of living animals using multicolor two-photon microscopy. Am J Physiol 2002; 283:C905–C916.
4 Yu W, Sandoval RM, Molitoris BA: Quantitative intravital microscopy using a generalized polarity concept for kidney studies. Am J Physiol 2005;289:C1197–C1208.
5 Sandoval RM, Molitoris BA, Sandoval RM, Molitoris BA: Quantifying endocytosis in vivo using intravital two-photon microscopy. Methods Mol Biol 2008;440:389–402.
6 Molitoris BA, Dagher PC, Sandoval RM, Campos SB, Ashush H, Fridman E, Brafman A, Faerman A, Atkinson SJ, Thompson JD, Kalinski H, Skaliter R, Erlich S, Feinstein E, Molitoris BA, Dagher PC, Sandoval RM, Campos SB, Ashush H, Fridman E, Brafman A, Faerman A, Atkinson SJ, Thompson JD, Kalinski H, Skaliter R, Erlich S, Feinstein E: siRNA targeted to p53 attenuates ischemic and cisplatin-induced acute kidney injury. J Am Soc Nephrol 2009;20:1754–1764.
7 Kelly KJ, Plotkin Z, Vulgamott SL, Dagher PC: p53 mediates the apoptotic response to GTP depletion after renal ischemia-reperfusion: protective role of a p53 inhibitor. J Am Soc Nephrol 2003;4:128–138.
8 Wagner MC, Rhodes G, Wang E, Pruthi V, Arif E, Saleem MA, Wean SE, Garg P, Verma R, Holzman LB, Gattone V, Molitoris BA, Nihalani D, Wagner MC, Rhodes G, Wang E, Pruthi V, Arif E, Saleem MA, Wean SE, Garg P, Verma R, Holzman LB, Gattone V, Molitoris BA, Nihalani D: Ischemic injury to kidney induces glomerular podocyte effacement and dissociation of slit diaphragm proteins Neph1 and ZO-1. J Biol Chem 2008; 283:35579–35589.
9 Kleinfeld D, Mitra PP, Helmchen F, Denk W: Fluctuations and stimulus-induced changes in blood flow observed in individual capillaries in layers 2 through 4 of rat neocortex. Proc Natl Acad Sci USA 1998;95:15741–15746.
10 Molitoris BA, Sandoval RM: Intravital multiphoton microscopy of dynamic renal processes. Am J Physiol Renal Physiol 2005; 288:F1084–F1089.
11 Kang JJ, Toma I, Sipos A, McCulloch F, Peti-Peterdi J: Quantitative imaging of basic functions in renal (patho)physiology. Am J Physiol Renal Physiol 2006;291:F495–F502.
12 Atkinson SJ: Functional intravital imaging of leukocytes in animal models of renal injury. Nephron Physiol 2006;103:86–90.

13 Sandoval RM, Molitoris BA: Quantifying endocytosis in vivo using intravital two-photon microscopy. Methods Mol Biol 2008;440:389–402.
14 Sutton TA, Horbelt M, Sandoval RM: Imaging vascular pathology. Nephron Physiol 2006;103:82–85.
15 Sharfuddin AA, Sandoval RM, Berg DT, McDougal GE, Campos SB, Phillips CL, Jones BE, Gupta A, Grinnell BW, Molitoris BA, Sharfuddin AA, Sandoval RM, Berg DT, McDougal GE, Campos SB, Phillips CL, Jones BE, Gupta A, Grinnell BW, Molitoris BA: Soluble thrombomodulin protects ischemic kidneys. J Am Soc Nephrol 2009;20:524–534.
16 Sutton TA, Mang HE, Campos SB, Sandoval RM, Yoder MC, Molitoris BA: Injury of the renal microvascular endothelium alters barrier function after ischemia. Am J Physiol Renal Physiol 2003;285:F191–F198.
17 Sutton TA, Kelly KJ, Mang HE, Plotkin Z, Sandoval RM, Dagher PC: Minocycline reduces renal microvascular leakage in a rat model of ischemic renal injury. Am J Physiol Renal Physiol 2005;288:F91–F97.
18 Kang JJ, Toma I, Sipos A, McCulloch F, Peti-Peterdi J: Imaging the renin-angiotensin system: an important target of anti-hypertensive therapy. Adv Drug Deliv Rev 2006;58:824–833.
19 Ashworth SL, Tanner GA: Fluorescent labeling of renal cells in vivo. Nephron Physiol 2006;103:91–96.
20 Ashworth SL, Sandoval RM, Tanner GA, Molitoris BA: Two-photon microscopy: visualization of kidney dynamics. Kidney Int 2007;72:416–421.

Prof. Bruce A. Molitoris, MD
Division of Nephrology, Indiana University School of Medicine
950 W. Walnut Street, R2-202
Indianapolis, IN 46202 (USA)
Tel. +1 317 274 5287, E-Mail bmolitor@iupui.edu

Ronco C, Bellomo R, McCullough PA (eds): Cardiorenal Syndromes in Critical Care.
Contrib Nephrol. Basel, Karger, 2010, vol 165, pp 54–67

Cardiorenal Syndromes: An Executive Summary from the Consensus Conference of the Acute Dialysis Quality Initiative (ADQI)

Claudio Ronco[a] · Peter A. McCullough[b] · Stefan D. Anker[c] · Inder Anand[d] · Nadia Aspromonte[e] · Sean M. Bagshaw[f] · Rinaldo Bellomo[g] · Tomas Berl[h] · Ilona Bobek[a] · Dinna N. Cruz[a] · Luciano Daliento[i] · Andrew Davenport[j] · Mikko Haapio[k] · Hans Hillege[l] · Andrew House[m] · Nevin M. Katz[n] · Alan Maisel[o] · Sunil Mankad[p] · Pierluigi Zanco[q] · Alexandre Mebazaa[r] · Alberto Palazzuoli[s] · Federico Ronco[i] · Andrew Shaw[t] · Geoff Sheinfeld[u] · Sachin Soni[a,v] · Giorgio Vescovo[w] · Nereo Zamperetti[x] · Piotr Ponikowski[y] for the Acute Dialysis Quality Initiative (ADQI) consensus group*

Abstract

The cardiorenal syndrome (CRS) is a disorder of the heart and kidneys whereby acute or chronic dysfunction in one organ may induce acute or chronic dysfunction of the other. The general definition has been expanded into five subtypes reflecting the primacy of organ dysfunction and the time-frame of the syndrome: CRS type 1 = acute worsening of heart function leading to kidney injury and/or dysfunction; CRS type 2 = chronic abnormalities in heart function leading to kidney injury or dysfunction; CRS type 3 = acute worsening of kidney function leading to heart injury and/or dysfunction; CRS type 4 = chronic kidney disease leading to heart injury, disease and/or dysfunction, and CRS type 5 = systemic conditions leading to simultaneous injury and/or dysfunction of heart and kidney. Different pathophysiological mechanisms are involved in the combined dysfunction of heart and kidney in these five types of the syndrome.

* Author's affiliation at the end of the article.

Combined heart and kidney dysfunction is common [1]. A disorder of one of these two organs often leads to dysfunction or injury to the other [2]. This is the pathophysiological basis for the clinical entity defined cardiorenal syndrome (CRS) [3]. Generally defined as a condition characterized by the initiation and/or progression of renal insufficiency secondary to heart failure [4], the term CRS should also be used to describe conditions of renal dysfunction leading to heart dysfunction (renocardiac syndrome) [5]. The absence of a clear definition contributed in the past to a lack of clarity with regard to diagnosis and management [6]. The common view is that a relatively normal kidney is dysfunctional because of a diseased heart [7, 8]. This concept, however, has been challenged and the most recent definition includes a variety of conditions, either acute or chronic, where the primary failing organ can be either the heart or the kidney (table 1) [9]. Such advances in the definition and classification of CRS enabled the characterization of the complex organ crosstalk and have proposed specific prevention strategies and therapeutic interventions to attenuate end organ injury [5, 6, 10]. A major problem with previous terminology was that it did not allow identification of the pathophysiological interactions occurring in the different types of combined heart/kidney disorder [11]. The subdivision into different subtypes seems to provide a better approach to this syndrome.

Cardiorenal Syndrome Type 1 (Acute Cardiorenal Syndrome)

Type 1 or acute CRS is characterized by an acute heart disorder leading to acute kidney injury (AKI; fig. 1). Several patients are admitted to hospital with either de novo acute heart failure (AHF) or with an acute decompensated heart failure [12]. Among these patients, pre-morbid renal dysfunction is common and predisposes to AKI [13, 14]. In AHF, AKI seems to be more severe in patients with impaired left ventricular ejection fraction (LVEF) compared to those with preserved LVEF [15, 16]. Furthermore, impaired renal function is consistently found as an independent risk factor for 1-year mortality in AHF patients including patients with ST-elevation myocardial infarction [16, 17]. This independent effect might be due to an associated acceleration in cardiovascular pathobiology due to kidney dysfunction through the activation of inflammatory pathways [8, 18]. AKI induced by primary cardiac dysfunction implies inadequate renal perfusion until proven otherwise. This should prompt clinicians to consider the diagnosis of a low cardiac output state and/or marked increase in venous pressure leading to kidney congestion. In this condition diuretic responsiveness may decrease. The physiological phenomena of diuretic breaking [19] and post-diuretic sodium retention [20] may also play a role in this setting. AKI can also be worsened by the administration of diuretics at higher doses. Accordingly, diuretics may best be given in AHF patients with evidence of systemic fluid overload with the goal of achieving gradual diuresis. Furosemide can be titrated according

Table 1. Definitions of cardiorenal syndromes

Cardiorenal syndrome (CRS) general definition: A complex pathophysiological disorder of the heart and kidneys whereby acute or chronic dysfunction in one organ may induce acute or chronic dysfunction in the other organ
CRS type 1 (acute cardiorenal syndrome) Abrupt worsening of cardiac function (e.g. acute cardiogenic shock or acute decompensation of chronic heart failure) leading to kidney injury
CRS type 2 (chronic cardiorenal syndrome) Chronic abnormalities in cardiac function (e.g. chronic heart failure) causing progressive chronic kidney disease
CRS type 3 (acute renocardiac syndrome) Abrupt worsening of renal function (e.g. acute kidney failure or glomerulonephritis) causing acute cardiac disorder (e.g. heart failure, arrhythmia, pulmonary edema)
CRS type 4 (chronic renocardiac syndrome) Chronic kidney disease (e.g. chronic glomerular disease) contributing to decreased cardiac function, cardiac hypertrophy and/or increased risk of adverse cardiovascular events
CRS type 5 (secondary cardiorenal syndrome) Systemic condition (e.g. diabetes mellitus, sepsis) causing both cardiac and renal dysfunction

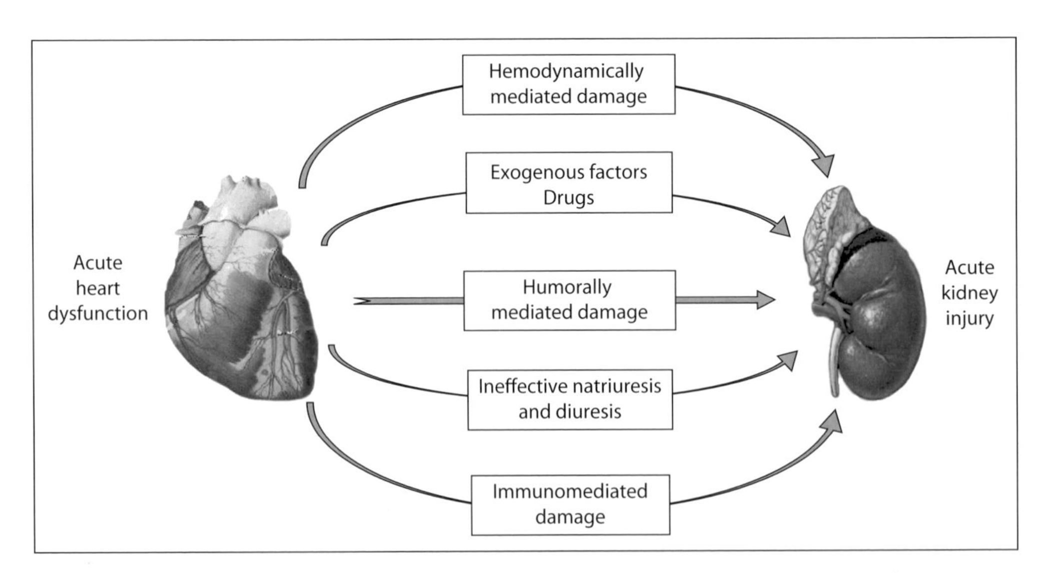

Fig. 1. Diagram illustrating and summarizing the major pathophysiological interactions between the heart and kidney in type 1 cardiorenal syndrome.

to renal function, systolic blood pressure and monitoring of AKI biomarkers. A continuous diuretic infusion might be helpful [21]. In parallel, measurement of cardiac output, venous pressure and bioelectric impedance vector analysis may also help ensure continued and targeted diuretic therapy while preventing unwanted iatrogenic complications [22]. Accurate estimation of cardiac output can now be easily achieved by means of arterial pressure monitoring combined with pulse contour analysis or by Doppler ultrasound [23–25]. Knowledge of cardiac output allows physicians to develop a physiologically safer and more logical approach to the simultaneous treatment of AHF and acute decompensated heart failure and AKI. If diuretic-resistant fluid overload exists despite an optimized cardiac output, removal of isotonic fluid can be achieved by ultrafiltration. This approach can be efficacious and clinically beneficial [26]. The presence of AKI with or without concomitant hyperkalemia may also affect patient outcome by inhibiting the prescription of ACE inhibitors and aldosterone inhibitors [27]. Acute administration of β-blockers in the setting of type 1 CRS is generally not advised. Such therapy should wait until the patient has stabilized physiologically and concerns about a low cardiac output syndrome have been resolved. In some patients, stroke volume cannot be increased and relative or absolute tachycardia sustains the adequacy of cardiac output. Blockade of such compensatory tachycardia and sympathetic system-dependent inotropic compensation can precipitate cardiogenic shock and can be lethal [28]. Particular concern applies to β-blockers excreted by the kidney such as atenolol or sotalol, especially if combined with calcium antagonists [29]. These considerations should not inhibit the slow introduction of treatment with β-blockers once patients are hemodynamically stable [30]. Attention should be paid to preserving renal function, perhaps as much attention as is paid to preserving myocardial muscle. Worsening renal function (WRF) during admission for ST-elevation myocardial infarction is a powerful and independent predictor of in-hospital and 1-year mortality [16, 17]. In a study involving 1,826 patients who received percutaneous coronary intervention, even a transient rise in serum creatinine (>25% compared to baseline) was associated with increased hospital stay and mortality [31]. Similar findings have also been shown among coronary artery bypass graft cohorts [32].

Cardiorenal Syndrome Type 2 (Chronic Cardiorenal Syndrome)

Type 2 or chronic CRS is characterized by chronic abnormalities in cardiac function causing progressive chronic kidney disease (CKD; fig. 2). WRF in the context of heart failure is associated with significantly increased adverse outcomes and prolonged hospitalizations [33]. The prevalence of renal dysfunction in chronic heart failure has been reported to be approximately 25% [33]. Even limited decreases in estimated glomerular filtration rate (GFR) of >9 ml/min appears to confer a significantly increased mortality risk [33]. Some researchers

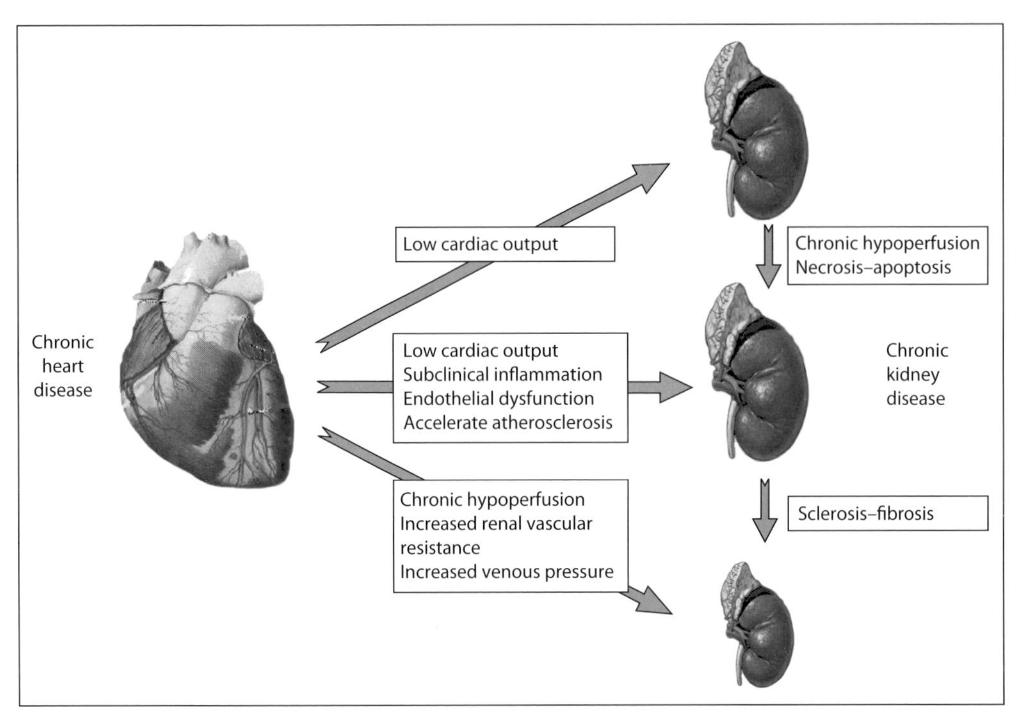

Fig. 2. Diagram illustrating and summarizing the major pathophysiological interactions between the heart and kidney in type 2 cardiorenal syndrome.

have considered WRF a marker of severity of generalized vascular disease [33]. Independent predictors of WRF include: old age; hypertension; diabetes mellitus, and acute coronary syndromes.

Chronic heart failure is characterized by a relatively stable long-term situation of probably reduced renal perfusion, often predisposed by both micro- and macrovascular disease in the context of the same vascular risk factors associated with cardiovascular disease. No evidence of association between LVEF and estimated GFR can be consistently demonstrated [34]. Neurohormonal abnormalities are present with excessive production of vasoconstrictive mediators (epinephrine, angiotensin, endothelin) and altered sensitivity and/or release of endogenous vasodilatory factors (natriuretic peptides, nitric oxide). Pharmacotherapies used in the management of heart failure have been touted as contributing to WRF. Diuresis-associated hypovolemia, early introduction of renin-angiotensin-aldosterone system blockade, and drug-induced hypotension have all been suggested as contributing factors [4]. However, their role remains highly speculative. Recently, there has been increasing interest in the pathogenetic role of erythropoietin deficiency and decrease in vitamin D receptor activation [35–37]. Regardless of the cause, WRF in the context of heart failure is associated with an increased risk for adverse outcomes. The proportion of individuals with WRF or CKD receiving appropriate risk factor modification and/or interventional strategies is lower than the general population [38–42]. Potential reasons for this therapeutic failure include concerns about worsening of residual renal function, and/or therapy-related toxic effects due to low clearance rates [43–45]. However, several studies have shown that when appropriately titrated and monitored, cardiovascular medications used in the general population can be safely administered to those with renal impairment and with similar benefits [42, 44, 46].

Newer approaches to the treatment of cardiac failure such as cardiac resynchronization therapy have not yet been studied in terms of their renal functional effects, although preserved renal function after cardiac resynchronization therapy may predict a more favorable outcome [47]. Vasopressin V2-receptor blockers have been reported to decrease body weight and edema in patients with chronic heart failure [48], but their effects in patients with CRS have not been systematically studied and a recent large randomized controlled trial showed no evidence of a survival benefit with these agents [49].

The difficulty of distinguishing in advanced stages whether patients belong to type 2 or type 4 CRS has been recognized. This however should not be of concern as the classification system describes clearly that patients may move among different CRS subtypes along the natural history of the syndrome.

Cardiorenal Syndrome Type 3 (Acute Renocardiac Syndrome)

Type 3 CRS or acute renocardiac syndrome is characterized by an abrupt and primary worsening of renal function (e.g. AKI, ischemia or glomerulonephritis) which then causes or contributes to acute cardiac dysfunction (e.g. heart failure, arrhythmia, ischemia; fig. 3).

The development of AKI as a primary event leading to cardiac dysfunction has recently been observed with increased frequency especially in critically ill patients. Using the recent RIFLE consensus definitions and its injury and failure categories, AKI has been identified in close to 9% of hospital patients [50] and, in a large ICU database, AKI was observed in more than 35% of critically ill patients [51]. AKI can affect the heart through several pathways whose hierarchy is not yet established. Fluid overload can contribute to the development of pulmonary edema. Hyperkalemia can contribute to arrhythmias and may cause cardiac arrest. Untreated uremia affects myocardial contractility through the accumulation of myocardial-depressant factors [52] and can cause pericarditis [53]. Partially corrected or uncorrected acidemia produces pulmonary vasoconstriction [54] which, in some patients, can significantly contribute to right-sided heart failure. Acidemia appears to have a negative inotropic effect [55] and may, together with electrolyte imbalances, contribute to an increased risk of

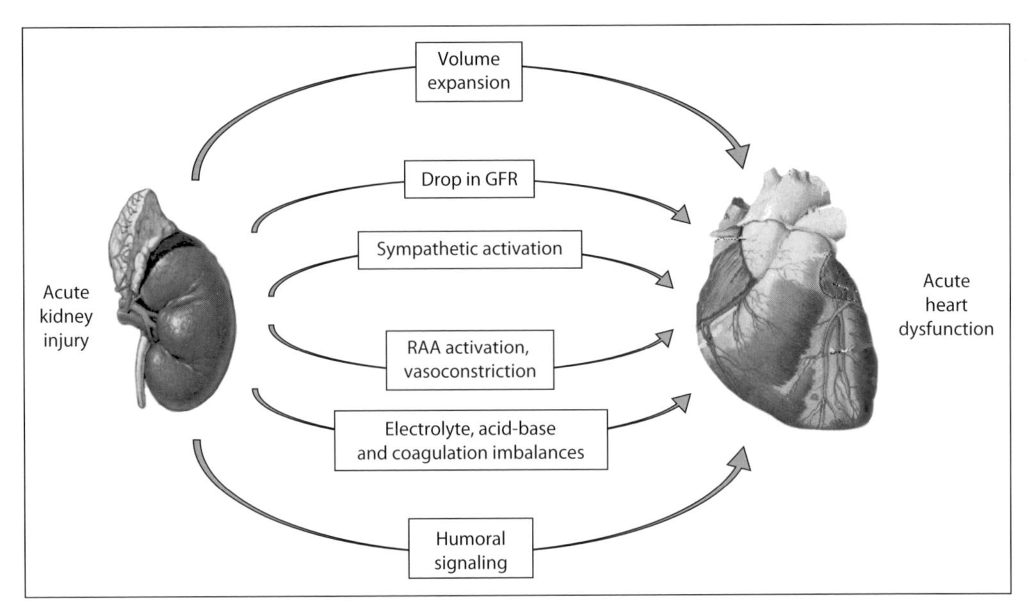

Fig. 3. Diagram illustrating and summarizing major pathophysiological interactions between the heart and kidney in type 3 cardiorenal syndrome. GFR = Glomerular filtration rate; RAA = renin-angiotensin-aldosterone.

arrhythmias [56]. Finally, as discussed above, renal ischemia itself may precipitate activation of inflammation and apoptosis at the cardiac level [8]. Finally, if AKI is severe and renal replacement therapy is necessary, cardiovascular instability generated by rapid fluid and electrolyte shifts secondary to conventional dialysis can induce hypotension, arrhythmias, and myocardial ischemia [57]. Continuous techniques of renal replacement, which minimize such cardiovascular instability appear physiologically safer and more logical in this setting [58].

Cardiorenal Syndrome Type 4 (Chronic Renocardiac Syndrome)

Type 4 CRS or chronic renocardiac syndrome is characterized by a primary CKD condition leading to decreased cardiac function, ventricular hypertrophy, diastolic dysfunction and/or increased risk of adverse cardiovascular events (fig. 4).

CKD is divided into 5 stages based on a combination of the severity of kidney damage and GFR [59]. Individuals with CKD, particularly those receiving renal replacement therapies (CKD stage 5) are at extremely high cardiovascular risk [60]. In CKD stage 5 cohorts, more than 50% of deaths are attributed to cardiovascular disease; namely coronary artery disease and its associated complications

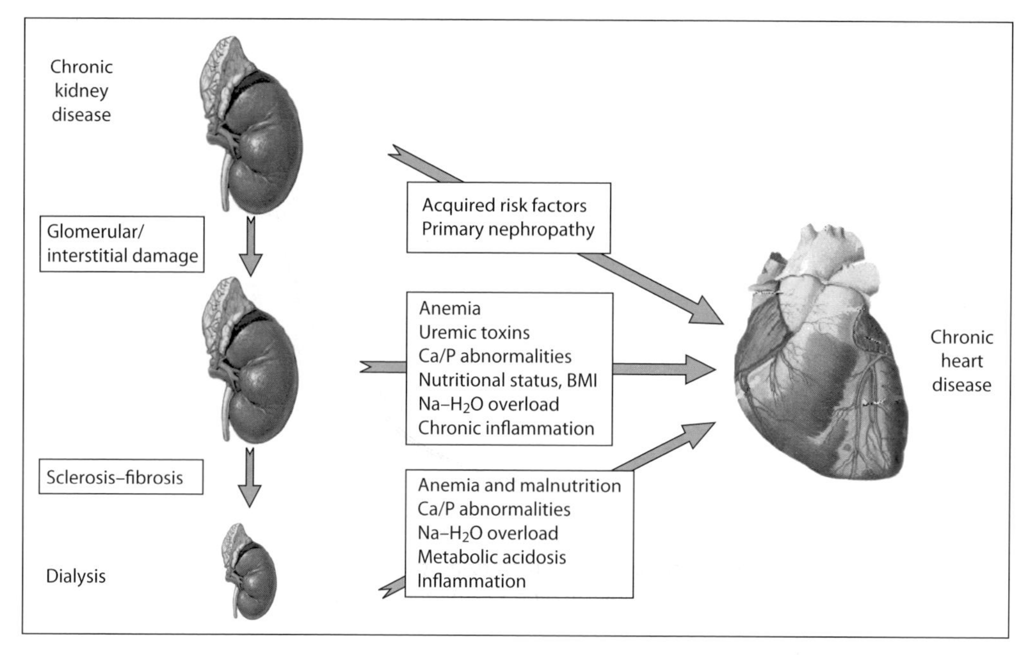

Fig. 4. Diagram illustrating and summarizing the major pathophysiological interactions between the heart and kidney in type 4 cardiorenal syndrome.

[54]. The 2-year mortality rate following myocardial infarction in patients with CKD stage 5 is high and estimated to be 50% [61]. In comparison, for the general population the 10-year mortality rate after myocardial infarction is 25%. A large population of individuals entering the transition phase towards end-stage kidney disease is emerging [62, 63]. The leading cause of death in such patients is cardiovascular with >40% of mortality being cardiovascular event-related. Based on these findings, it is now well established that CKD is a significant risk factor for cardiovascular disease, such that individuals with evidence of CKD have between a 10- to 20-fold increased risk of cardiac death compared to age- and sex-matched controls without CKD [63]. As discussed, part of this problem may be related to the fact that such individuals are also less likely to receive risk-modifying interventions compared to their non-CKD counterparts [61, 64–76].

Renal insufficiency is highly prevalent among patients with heart failure and is an independent prognostic factor in both diastolic and systolic ventricular dysfunction [77]. It is an established negative prognostic indicator in patients with severe heart failure [77, 78]. The logical practical implications of the plethora of data linking CKD with cardiovascular disease is that more attention needs to be paid to reducing risk factors and optimizing medications in these patients and that under-treatment due to concerns about pharmacodynamics in this setting, may have lethal consequences at an individual level and huge

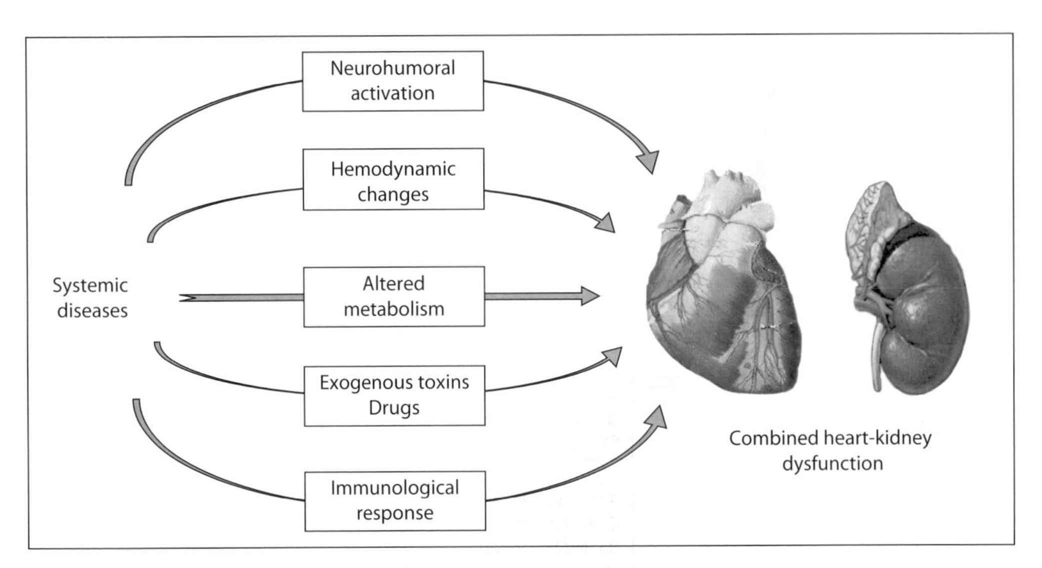

Fig. 5. Diagram illustrating and summarizing the major pathophysiological interactions between the heart and kidney in type 5 cardiorenal syndrome.

potential adverse consequences at a public health level. The presence of concomitant factors such as anemia and calcium-phosphate abnormalities seems to play an important role. Emerging data regarding treatment with erythropoietin and vitamin D receptor activators seem to overcome the expected beneficial effects leading to hypotheses on the important pleiotropic actions of these classes of drugs.

Cardiorenal Syndrome Type 5 (Secondary Cardiorenal Syndrome)

Type 5 or secondary CRS is characterized by the presence of combined cardiac and renal dysfunction due to systemic disorders (fig. 5). Several acute and chronic diseases can affect both organs simultaneously, and the disease induced in one can affect the other and vice versa. Several chronic conditions such as diabetes and hypertension are discussed as part of type 2 and type 4 CRS.

In the acute setting, severe sepsis represents the most common and serious condition which can affect both organs. It can induce AKI while leading to profound myocardial depression. The mechanisms may involve the effect of chemical mediators on both organs [79, 80]. Treatment is directed at the prompt identification, eradication and treatment of the source of infection while supporting organ function with invasively guided fluid resuscitation, and inotropic and vasopressor drug support. Blood purification with extracorporeal therapies may play a role in improving myocardial performance while providing optimal small solute clearance and kidney support [81].

Conclusions

The new definition classification system for CRS seems to represent an important advance and seems to be a valuable tool in the renewed collaboration between nephrology and cardiology for a complete multidisciplinary approach.

References

1 Dar O, Cowie MR: Acute heart failure in the intensive care unit: epidemiology. Crit Care Med 2008;36:(suppl) S3–S8.
2 Schrier RW: Cardiorenal versus renocardiac syndrome: is there a difference? Nat Clin Pract Nephrol 2007;3:637.
3 Ronco C: Cardiorenal and reno-cardiac syndromes: Clinical disorders in search of a systematic definition. Int J Artif Organs 2008; 31:1–2.
4 Liang KV, Williams AW, Greene EL, Redfield MM: Acute decompensated heart failure and the cardio-renal syndrome. Crit Care Med 2008;36(suppl):S75–S88.
5 Ronco C, House AA, Haapio M: Cardiorenal syndrome: refining the definition of a complex symbiosis gone wrong. Intensive Care Med 2008;34:957–962.
6 Ronco C, Haapio M, House AA, Anavekar N, Bellomo R: Cardiorenal syndrome. J Am Coll Cardiol 2008;52:1527–1539.
7 Bongartz LG, Cramer MJ, Doevendans PA, Joles JA, Braam B: The severe cardio-renal syndrome: 'Guyton revisited'. Eur Heart J 2005;26:11–17.
8 Kelly KJ: Acute renal failure: much more than a kidney disease. Semin Nephrol 2006; 26:105–113.
9 Berl T, Henrich W: Kidney-heart interactions: epidemiology, pathogenesis, and treatment. Clin J Am Soc Nephrol 2006;1:8–18.
10 Ronco C, McCullough P, Anker SD, Anand I, Aspromonte N, Bagshaw SM, Bellomo R, Berl T, Bobek I, Cruz DN, Daliento L, Davenport A, Haapio M, Hillege H, House AA, Katz N, Maisel A, Mankad S, Zanco P, Mebazaa A, Palazzuoli A, Ronco F, Shaw A, Sheinfeld G, Soni S, Vescovo G, Zamperetti N, Ponikowski P; for the Acute Dialysis Quality Initiative (ADQI) consensus group: Cardio-renal syndromes:report from the consensus conference of the Acute Dialysis Quality Initiative. Eur Heart J 2009 [Epub ahead of print].
11 Kramer BK, Schweds F, Riegger GA: Diuretic treatment and diuretic resistance in heart failure. Am J Med 1999;106:90–96.
12 Mebazaa A, Gheorghiade M, Pina IL: Practical recommendations for prehospital and early in-hospital management of patients presenting with acute heart failure syndromes. Crit Care Med 2008;36(suppl): S129–S139.
13 Haldeman GA: Hospitalization of patients with heart failure: National Hospital Discharge Survey, 1985 to 1995. Am Heart J 1999;137:352–360.
14 Adams KF, Fonarow GC, Emerman CL: Characteristics and outcome of patients hospitalized for heart failure in the United States: rationale, design, and preliminary observations from the first 100,000 cases in the Acute Decompensated Heart Failure National Registry (ADHERE). Am Heart J 2005;149:209–216.

15 Fonarow GC, Gattis Stough W, Abraham WT, et al: Characteristics, treatments and outcomes of patients with preserved systolic function hospitalized for heart failure. J Am Coll Cardiol 2007;50:768–777.
16 Jose P, Skali H, Anavekar N, et al: Increase in creatinine and cardiovascular risk in patients with systolic dysfunction after myocardial infarction. J Am Soc Nephrol 2006;17:2886–2891.
17 Goldberg A, Hammerman H, Petcherski S, et al: In-hospital and 1-year mortality of patients who develop worsening renal function following acute ST-elevation myocardial infarction. Am Heart J 2005;150:330–337.
18 Tokuyama H, Kelly DJ, Zhang Y, et al: Macrophage infiltration and cellular proliferation in the non-ischemic kidney and heart following prolonged unilateral renal ischemia. Nephron Physiol 2007;106:p54–p62.
19 Ellison DH: Diuretic resistance: physiology and therapeutics. Semin Nephrol 1999;19:581–597.
20 Almeshari K, Ahlstrom NG, Capraro FE, et al: A volume independent component to post diuretic sodium retention in humans. J Am Soc Nephrol 1993;102:450–458.
21 Howard PA, Dunn MI: Aggressive diuresis for severe heart failure in the elderly. Chest 2001;119:807–610.
22 Ronco C, Costanzo MR, Bellomo R, Maisel AS (eds): Fluid Overload: Diagnosis and Management. Contrib Nephrol. Basel, Karger, 2010, vol 164.
23 Mayer J, Boldt J, Wolf MW, Lang J, Sutton S: Cardiac output derived from arterial pressure waveform analysis in patients undergoing cardiac surgery. Validity of a second generation device Anesth Analg 2008;106:867–872.
24 Wan L, Naka T, Uchino S, Bellomo R: A pilot study of pulse contour cardiac output monitoring in patients with septic shock. Crit Care Resusc 2005;7:165–169.
25 Nguyen BH, Losey T, Rasmussen BA, et al: Interrater reliability of cardiac output measurements by transcutaneous Doppler ultrasound: implications for non-invasive hemodynamic monitoring in the ED. Am J Emerg Med 2006;24:828–835.
26 Costanzo MR, Guglin ME, Saltzberg MT, et al: Ultrafiltration versus diuretics for patients hospitalized for acute decompensated heart failure. J Am Coll Cardiol 2007;49:675–683.
27 Verma A, Solomon SD: Optimizing care of heart failure after acute myocardial infarction with an aldosterone receptor antagonist. Curr Heart Fail Rep 2007;4:183–189.
28 Chen ZM, Pan HC, Chen YP, et al: Early intravenous then oral metoprolol in 45,852 patients with acute myocardial infarction: a randomized placebo-controlled trial. Lancet 2005;366:1622–1632.
29 Yorgun H, Deniz A, Aytemir K: Cardiogenic shock secondary to combination of diltiazem and sotalol. Intern Med J 2008;38:221–222.
30 Tessone A, Gottlieb S, Barbash IM, et al: Underuse of standard care and outcome of patients with acute myocardial infarction and chronic renal insufficiency. Cardiology 2007;108:193–199.
31 McCullough PA: Beyond serum creatinine: defining the patient with renal insufficiency and why? Rev Cardiovasc Med 2003;4(suppl 1):S2–S6.
32 Damman K, Navis G, Voors AA, et al: Worsening renal function and prognosis in heart failure: systematic review and meta-analysis. J Card Fail 2007;28:287–285.
33 Hillege HL, Nitsch D, Pfeffer MA, et al: Renal function as a predictor of outcome in a broad spectrum of patients with heart failure. Circulation 2006;113:671–678.
34 Bhatia RS, Tu JV, Lee DS, et al: Outcome of heart failure with preserved ejection fraction in a population-based study. N Engl J Med 2006;355:260–269.
35 Jie KE, Verhaar MC, Cramer MJ, et al: Erythropoietin and the cardio-renal syndrome: cellular mechanisms on the cardiorenal connectors. Am J Physiol Renal Physiol 2006;291:F392–F344.
36 Palazzuoli A, Silverberg DS, Iovine F, et al: Effects of beta-erythropoietin treatment of left ventricular remodelling, systolic function, and B-type natriuretic peptide levels in patients with the cardio-renal anemia syndrome. Am Heart J 2007;154:e9–e15.
37 Butler J, Forman DE, Abraham WT, et al: Relationship between heart failure treatment and development of worsening renal function among hospitalized patients. Am Heart Journal 2004;147:331–339.
38 McCullough PA: Cardiorenal risk: an important clinical intersection. Rev Cardiovasc Med 2002;3:71–76.

39 Wright RS, Reeder GS, Herzog CA, et al: Acute myocardial infarction and renal dysfunction: a high-risk combination. Ann Intern Med 2002;137:563–570.
40 Beattie JN, Soman SS, Sandberg KR, et al: Determinants of mortality after myocardial infarction in patients with advanced renal dysfunction. Am J Kidney Dis 2001;37:1191–1200.
41 Gibson CM, Pinto DS, Murphy SA, et al: Association of creatinine and creatinine clearance on presentation in acute myocardial infarction with subsequent mortality. J Am Coll Cardiol 2003;42:1535–1543.
42 Berger AK, Duval S, Krumholz HM: Aspirin, beta-blocker, and angiotensin-converting enzyme inhibitor therapy in patients with end-stage renal disease and an acute myocardial infarction. J Am Coll Cardiol 2003; 42:201–208.
43 French WJ, Wright RS: Renal insufficiency and worsened prognosis with STEMI: a call for action. J Am Coll Cardiol 2003;42:1544–1546.
44 Levin A, Foley RN: Cardiovascular disease in chronic renal insufficiency. Am J Kidney Dis 2000;36(suppl 3):S24–S30.
45 Keltai M, Tonelli M, Mann JF, et al: Renal function and outcomes in acute coronary syndrome: impact of clopidogrel. Eur J Cardiovasc Pre Rehab 2007;14:312–318.
46 Ruggenenti P, Perna A, Remuzzi G: ACE inhibitors to prevent end-stage renal disease: when to start and why possibly never to stop: a post hoc analysis of the REIN trial results. Ramipril Efficacy in Nephropathy. J Am Soc Nephrol 2001;12:2832–2837.
47 Fung JW, Szeto CC, Cahn JY, et al: Prognostic value of renal function in patients with cardiac resynchronization therapy. Int J Cardiol 2007;122:10–16.
48 Gheorghiade M, Niazi I, Ouyang J, et al: Vasorpessin-V-2 receptor blockade with tolvaptan in patients with chronic heart failure: results of a double-blind randomized trial. Circulation 2003;107:2690–2696.
49 Konstan MA, Gheorghiade M, Burnett JC Jr, et al: Effects of oral tolvaptan in patients hospitalized for worsening heart failure: the EVEREST outcome trial. JAMA 2007; 297:1319–1331.
50 Uchino S, Bellomo R, Goldsmith D, Bates S: An assessment of the RIFLE criteria for ARF in hospitalized patients. Crit Care Med 2006; 34:1913–1917.
51 Bagshaw SM, George C, Dinu I, Bellomo R: A multicenter evaluation of the RIFLE criteria of early acute kidney injury in critically ill patients. Nephrol Dial Transplant 2008; 23:1203–1210.
52 Blake P, Hasegawa Y, Khosla MC, et al: Isolation of myocardial depressant factor(s) from the ultrafiltrate of heart failure patients with acute renal failure. ASAIO J 1996;42: M911–M915.
53 Meyer TW, Hostetter TH: Uremia. N Engl J Med 2007;357:1316–1325.
54 Figueras J, Stein L, Diez V, et al: Relationship between pulmonary hemodynamics and arterial pH and carbon dioxide tension in critically ill patients. Chest 1976;70:466–472.
55 Brady JP, Hasbargen JA: A review of the effects of acidosis on nutrition in dialysis patients. Semin Dial 2000;13:252–255.
56 McCullough PA, Sandberg KR: Chronic kidney disease and sudden death: strategies for prevention. Blood Purif 2004;22:136–142.
57 Selby NM, McIntyre CW: The acute cardiac effects of dialysis. Semin Dial 2007;20:220–228.
58 Ronco C, Bellomo R, Ricci Z: Continuous renal replacement therapy in the critically ill patients. Nephrol Dial Transplant 2001; 16(suppl):67–72.
59 Sarnak MJ, Levey AS, Schoolwerth AC, et al: Kidney disease as a risk factor for development of cardiovascular disease: a statement from the American Heart Association Councils on Kidney in Cardiovascular Disease, High Blood Pressure Research, Clinical Cardiology, and Epidemiology and Prevention. Hypertension 2003;42:1050–1065.
60 Chertow GM, Normand SL, Silva LR, McNeil BJ: Survival after acute myocardial infarction in patients with end-stage renal disease: results from the cooperative cardiovascular project. Am J Kidney Dis 2000;35:1044–1051.
61 Herzog CA: Dismal long-term survival of dialysis patients after acute myocardial infarction: can we alter the outcome? Nephrol Dial Transplant 2002;17:7–10.

62 National Kidney Foundation: Summaries for patients. Diagnosis and evaluation of patients with chronic kidney disease: recommendations from the National Kidney Foundation. Ann Intern Med 2003;139:I36.

63 Johnson DW, Craven AM, Isbel NM: Modification of cardiovascular risk in hemodialysis patients: an evidence-based review. Hemodial Int 2007;11:1–14.

64 Logar CM, Herzog CA, Beddhu S: Diagnosis and therapy of coronary artery disease in renal failure, end-stage renal disease, and renal transplant populations. Am J Med Sci 2003;325:214–227.

65 Collins AJ, Li S, Gilbertson DT, Liu J, Chen SC, Herzog CA: Chronic kidney disease and cardiovascular disease in the Medicare population. Kidney Int 2003;87(suppl):S24–S31.

66 Go AS, Chertow GM, Fan D, et al: Chronic kidney disease and the risks of death, cardiovascular events, and hospitalization. N Engl J Med 2004;351:1296–1305.

67 Coresh J, Astor BC, Greene T, et al: Prevalence of chronic kidney disease and decreased kidney function in the adult US population: Third National Health and Nutrition Examination Survey. Am J Kidney Dis 2003;41:1–12.

68 Garg AX, Clark WF, Haynes RB, House AA: Moderate renal insufficiency and the risk of cardiovascular mortality: results from the NHANES I. Kidney Int 2002;61:1486–1494.

69 Keith DS, Nichol GA, Guillon CM, et al: Longitudinal follow up and outcomes among a population with chronic kidney disease in a large managed care organization. Arch Intern Med 2004;164:659–663.

70 Sarnak MJ, Coronado BE, Greene T, et al: Cardiovascular disease risk factors in chronic renal insufficiency. Clin Nephrol 2002;57: 327–335.

71 al Suwaidi J, Reddan DN, Williams K, et al: Prognostic implications of abnormalities in renal function in patients with acute coronary syndromes. Circulation 2002;106:974–980.

72 Gibson CM, Pinto DS, Murphy SA, et al: Association of creatinine and creatinine clearance on presentation in acute myocardial infarction with subsequent mortality. J Am Coll Cardiol 2003;42:1535–1543.

73 Al Ahmad A, Rand WM, Manjunath G, et al: Reduced kidney function and anemia as risk factors for mortality in patients with left ventricular dysfunction. J Am Coll Cardiol 2001; 38:955–962.

74 Sorensen CR, Brendorp B, Rask-Madsen C, et al: The prognostic importance of creatinine clearance after acute myocardial infarction. Eur Heart J 2002;23:948–952.

75 Tokmakova MP, Skali H, Kenchaiah S, et al: Chronic kidney disease, cardiovascular risk, and response to angiotensin-converting enzyme inhibition after myocardial infarction: the Survival And Ventricular Enlargement (SAVE) study. Circulation 2004; 110:3667–3673.

76 Anavekar NS, McMurray JJ, Velazquez EJ, et al: Relation between renal dysfunction and cardiovascular outcomes after myocardial infarction. N Engl J Med 2004;351:1285–1295.

77 McAlister FA, Ezekowitz J, Tonelli M, Armstrong PW: Renal insufficiency and heart failure: prognostic and therapeutic implications from a prospective cohort study. Circulation 2004;109:1004–1009.

78 Hjalmarson A, Goldstein S, Fegerberg B, et al: Effects of controlled-release metoprolol on total mortality, hospitalization, and well-being in patients with heart failure. JAMA 2000;283:1295–1302.

79 Cunningham PN, Dyanov HM, Park P, et al: Acute renal failure in endotoxemia is caused by TNF acting directly on TNF-1 receptors in kidney. J Immunol 2007;168:5817–5823.

80 Kumar A, Paladugu B, Mensing J, Kumar A, Parrillo JE: Nitric oxide-dependent and independent mechanisms are involved in TNF-alpha induced depression of cardiac myocyte contractility. Am J Physiol Regul Integr Comp Physiol 2007;292:R1900–R1906.

81 Honore PM, Jamez J, Wauthier M, et al: Prospective evaluation of short term high volume hemofiltration on the hemodynamic course and outcome in patients with intractable circulatory failure resulting from septic shock. Crit Care Med 2000;28:3581–3587.

C. Ronco[a] · P.A. McCullough[b] · S.D. Anker[c] · I. Anand[d] · N. Aspromonte[e] · S.M. Bagshaw[f] · R. Bellomo[g] · T. Berl[h] · I. Bobek[a] · D.N. Cruz[a] · L. Daliento[i] · A. Davenport[j] · M. Haapio[k] · H. Hillege[l] · A. House[m] · N.M. Katz[n] · A. Maisel[o] · S. Mankad[p] · P. Zanco[q] · A. Mebazaa[r] · A. Palazzuoli[s] · F. Ronco[i] · A. Shaw[t] · G. Sheinfeld[u] · S. Soni[a,v] · G. Vescovo[w] · N. Zamperetti[x] · P. Ponikowski[y] for the Acute Dialysis Quality Initiative (ADQI) consensus group

[a]Department of Nephrology, Dialysis & Transplantation, International Renal Research Institute, San Bortolo Hospital, Vicenza, Italy; [b]Department of Medicine, Divisions of Cardiology, Nutrition and Preventive Medicine, William Beaumont Hospital, Royal Oak, Mich., USA; [c]Division of Applied Cachexia Research, Department of Cardiology, Charité-Universitätsmedizin, Berlin, Germany; [d]Department of Cardiology, VA Medical Center, Minneapolis, Minn., USA; [e]Department of Cardiology, St Spirito Hospital, Rome, Italy; [f]Division of Critical Care Medicine, University of Alberta Hospital, Edmonton, Alta., Canada; [g]Department of Intensive Care, Austin Hospital, Melbourne, Australia; [h]Department of Nephrology, University of Colorado Health Sciences Center, Denver, Colo., USA; [i]Department of Cardiology, University of Padua, Padua, Italy; [j]UCL Center for Nephrology, Royal Free & University College Medical School, London, UK; [k]Division of Nephrology, Helsinki University Central Hospital, Helsinki, Finland; [l]Trial Coordination Center, Department of Cardiology and Epidemiology, University Medical Center Groningen, Hanzeplein, Netherlands; [m]Division of Nephrology, London Health Sciences Centre, University Hospital, London, Ont., Canada; [n]Department of Surgery, George Washington University, Washington, DC, USA; [o]Department of Medicine and Cardiology, San Diego VA Medical Center and University of California, San Diego, Calif., USA; [p]Department of Cardiology , Mayo Clinic, Rochester, Minn., USA; [q]Department of Nuclear Medicine, San Bortolo Hospital, Vicenza, Italy; [r]Department of Anesthesiology and Critical Care Medicine, Hôpital Lariboisière, University of Paris, Paris Diderot, France; [s]Department of Internal Medicine, University of Siena, Le Scotte Hospital, Siena, Italy; [t]Department of Anesthesiology, Duke University Medical Center, Durham, N.C., USA; [u]Department of Critical Care, University of Maryland, Baltimore, Md., USA; [v]Division of Nephrology, Mediciti Hospitals, Hyderabad, India; [w]Department of Internal Medicine, San Bortolo Hospital, Vicenza, Italy; [x]Department of Intensive Care, San Bortolo Hospital, Vicenza, Italy, and [y]Cardiac Department, Faculty of Public Health, Medical University, Military Hospital, Wroclaw, Poland

Prof. Claudio Ronco
Department of Nephrology, Dialysis & Transplantation, International Renal Research Institute
San Bortolo Hospital
Viale Rodolfi 37, IT–36100 Vicenza (Italy)
Tel. +39 444 753869, Fax +39 444 753949, E-Mail cronco@goldnet.it

Ronco C, Bellomo R, McCullough PA (eds): Cardiorenal Syndromes in Critical Care.
Contrib Nephrol. Basel, Karger, 2010, vol 165, pp 68–82

Epidemiology of Cardiorenal Syndromes

Sean M. Bagshaw[a] · Dinna N. Cruz[b]

[a]Division of Critical Care Medicine, University of Alberta Hospital, Edmonton, Alta., Canada, and [b]Department of Nephrology, Dialysis & Transplantation, International Renal Research Institute, San Bortolo Hospital, Vicenza, Italy

Abstract

Cardiac and kidney disease are common, increasingly encountered and frequently coexist. A consensus classification for the cardiorenal syndrome (CRS) and its specific subtypes has been developed by the Acute Dialysis Quality Initiative (ADQI). The five CRS subtypes have a similar underlying pathophysiology, however they likely have distinguishing features in terms of precipitating events, risk identification, natural history and outcomes. An appreciation for the epidemiology of heart-kidney interface, across the proposed CRS subtypes, is fundamental for understanding the overall burden of disease for each CRS subtype, along with associated morbidity, mortality, and health resource utilization. Likewise, an understanding of the epidemiology of CRS is fundamental for characterizing whether important knowledge gaps exist and for design of further epidemiologic investigations and interventional and/or therapeutic clinical trials. This review will provide a summary of the epidemiology of the CRS and its subtypes.

Cardiac and kidney disease are exceedingly common, are increasing in prevalence and frequently coexist. The cardiorenal syndromes (CRS) are exemplified by significant heart-kidney interactions that share similarities in pathophysiology, however they also have important discriminating features in terms of predisposing events, risk identification, natural history and outcomes. A description of the epidemiology of heart-kidney interaction, as defined by the proposed consensus CRS definitions, is a critical initial step towards understanding not only the overall burden of disease for each of the proposed CRS subtypes, but also their natural history, associated morbidity and mortality and potential health resource implications [1]. Likewise, a surveillance of the available literature on the epidemiology of CRS is a necessary step for detailing whether there are important knowledge gaps and for the design of future epidemiologic investigations and clinical trials.

Acute Cardiorenal Syndrome (Type 1 CRS)

Type 1 CRS is characterized by acute worsening of heart function leading to acute kidney injury (AKI) and/or dysfunction. Acute cardiac events that may contribute to AKI include acute decompensated heart failure (ADHF), acute coronary syndrome (ACS), cardiogenic shock and cardiac surgery-associated low cardiac output syndrome.

Several studies have examined AKI in the context of ADHF and ACS (tables 1, 2) Unfortunately, the majority of these studies are limited in scope due to being performed retrospectively, being secondary and/or post-hoc analyses from large databases [2–6] or secondary analyses of clinical trials of drug therapy [7, 8]. These studies often use the term 'worsening renal failure (WRF)' to describe the acute and/or subacute changes in kidney function that occurs following ADHF or ACS. Incidence estimates of WRF associated with ADHF and ACS have ranged between 24–45 and 9–19%, respectively. This broad incidence is mainly attributed to disparities in the definition used for WRF, differences in the observed time at risk and heterogeneity of selected populations.

The majority of studies have used variable durations of observed time at risk for ascertainment of WRF (or AKI). These variations have the potential to introduce bias and misclassification. The most common definition for WRF has been any time within hospital admission [3, 4]. However, data have shown that the vast majority of WRF occurs early after hospital admission. In a cohort admitted with ADHF, Gottlieb and Abraham [3] showed 47% had WRF within 3 days of hospital admission. Cowie and Komajda [9] found 50% occurred within 4 days, whereas two large observational studies found 70–90% of all WRF had occurred within the first week of hospitalization [4, 9].

In both ADHF and ACS, AKI has consistently shown association with higher morbidity and mortality [4]. In ADHF, AKI is associated with increased risk for both short- and long-term all-cause and cardiovascular mortality [3, 4, 8–11]. In addition, data suggest there is a biological gradient between AKI severity and mortality [12]. Several studies have shown development of AKI in association with ADHF prolongs stay in hospital [3, 4, 9–11]. Selected studies have also shown AKI in ADHF was associated with increased readmission rates [10, 11].

AKI associated with ACS increases the risk of poor outcome [2, 5–7, 13]. Even small acute changes in serum creatinine modify the risk of death [7]. Among those developing AKI, greater risk of cardiovascular events such as CHF, recurrent ACS, and stroke and need for rehospitalization have been shown [7]. Newsome and Warnock [5] reported a greater likelihood of progression to

Table 1. Summary of studies fulfilling criteria for type 1 CRS with a presenting diagnosis of ADHF

Study	Population	Study type (data source)	KI (WRF) definition	Incidence AKI, %	Cardiac disease	Outcome
Krumholz 2000 [4]	n = 1,681 mean age 79 years male 42% DM 38% HTN 60%	retrospective (Medicare)	SCr >26.5 μmol/l	28	hospitalized ADHF LVEF ≤40% 29% prevalence: prior HF 64% prior AMI 37%	in-hospital death (OR) 2.72 increase hospital LOS 2.3 days increase costs USD 1,758 no association with readmission at 30, 60 days
Gottlieb 2002 [3]	n = 1,002 mean age 67 years male 51% DM 40% HTN 70%	retrospective	SCr >26.5 and >44.2 μmol/l	39 & 20	hospitalized ADHF reduced LVEF 73% prevalence: prior HF 63% prior AMI 30%	increase in-hospital death increase hospital LOS >10 days
Smith 2003 [12]	n = 412 mean age 72 years male 51%	prospective	SCr >8.8, >26.5, and >44.2 μmol/l	75, 45 & 24	hospitalized ADHF reduced LVEF 39% prevalence: prior HF 72%	all-cause death at 6 months modified by severity of AKI (HR: SCr >0.8: 0.88, SCr >26.5: 1.61, SCr >44.2: 2.86) no association with readmissions trend in functional decline
Cowie 2006 [9]	n = 299 mean age 68 years male 74% DM 12.7% HTN 47%	prospective (POSH)	SCr >26.5 μmol/l	29	hospitalized ADHF and LVEF ≤40% prevalence: prior AMI 51% NYHA III-IV 87%	increased LOS by 2 days death/readmission rates similar
Nohria 2008 [8]	n = 433 mean age 56 years male 59% DM 34% HTN 47%	retrospective (ESCAPE Trial)	SCr >26.5 μmol/l	29.5	hospitalized ADHF	all-cause death (6 months) (HR) increased for SCr >106.1 AKI (>26.5 μmol/l) not associated with death/readmission

Table 1. Continued

Study	Population	Study type (data source)	KI (WRF) definition	Incidence AKI, %	Cardiac disease	Outcome
Logeart 2008 [10]	n = 416 age 71 years male 59% DM 23% HTN 42%	prospective	SCr >26.5 µmol/l	37	hospitalized ADHF prevalence: LVEF 0.35 LVEF ≤0.45 70% prior HF 45% prior MI 55%	all-cause death (6 months) or readmission (adj. HR) 1.74 increased LOS 3 days risk persisted whether AKI transient or not
Metra 2008 [11]	n = 318 mean age 68 years male 60% DM 34% HTN 53%	prospective	SCr >26.5 µmol/l and ≥25%	34	hospitalized ADHF prevalence: prior MI 51% prior HF 58%	CV death or readmission (adj. HR) 1.47 increased LOS 7 days

Cr = Serum creatinine; CV = cardiovascular; LOS = length of stay.

end-stage kidney disease (ESKD) in those with ACS complicated by AKI. These data would suggest that AKI in association with ADHF or ACS may further exacerbate cardiac injury and/or function and also contribute to exaggerated declines in kidney function. This would imply that the observed heart-kidney interface in Type 1 CRS may synergistically act to further accelerate injury and/or dysfunction.

Chronic Cardiorenal Syndrome (Type 2 CRS)

This syndrome is characterized by chronic abnormalities in cardiac function leading to kidney injury or dysfunction. The term 'chronic cardiac abnormalities' encompasses a number of different conditions and may include chronic left ventricular (LV) dysfunction, atrial fibrillation, congenital heart disease, constrictive pericarditis, and chronic ischemic heart disease (table 3).

Observational data clearly show that chronic heart and kidney disease commonly coexist. This often creates difficulty for determining which disease process was primary versus secondary. This also presents challenges when appraising the literature and attempting to classify patients into the CRS subtype

Table 2. Summary of studies fulfilling criteria for type 1 CRS with a presenting diagnosis of ACS

Study	Population	Study type (data source)	AKI (WRF) definition	Incidence AKI, %	Cardiac disease	Outcome
Goldberg 2005 [2]	n = 1,038 mean age 61 years male 72–79% DM 20–25% HTN 41–52%	retrospective	Scr >44.2 µmol/l	9.6	post-STEMI prevalence: prior AMI 21–25% prior HF 5–9%	in-hospital death (OR): 11.4; 1-year death (HR): 7.2; if both AKI and CKD: in-hospital death (OR): 20.9; 1-year death (HR): 14.4
Jose 2006 [7]	n = 1,854 age 54–61 years male 76–84% DM 20–28% HTN 42–46%	retrospective (SAVE Trial)	SCr >26.5 µmol/l by 2 weeks	12.0	AMI and LVEF ≤40% LVEF 30–31% prevalence: prior AMI 29–38% prior HF 4–6%	all-cause death (HR) 1.46; CV death (HR) 1.62; composite (HR) 1.32; trend for readmission for CHF
Latchamsetty 2007 [13]	n = 1,417 age 63–71 years male 63% DM 31% HTN 72%	prospective	SCr >44.2 µmol/l	any 9.7, transient 6.7, sustained 3.0	ACS prevalence: prior AMI 41% prior HF 18%	all-cause death (6 months) (%): any: 26%, (crude OR) 4.3 transient: 27%, (adj. OR) 2.1 sustained: 23%, (adj. OR) 1.6
Newsome 2008 [5]	n = 87,094 mean age 77 years male 50% DM 28–37% HTN 59–71%	retrospective (CCP)	variable	any 43.2 ΔSCr quartiles: Q1: Δ0.1, 13.1; Q2: Δ0.2, 9.3; Q3: Δ0.3–0.5, 12.3; Q4: Δ0.6–3.0, 8.4	AMI prevalence: prior MI 31–35% prior HF 21–35%	all-cause death (1,000 p-y)/HR): Q1 146/1.1; Q2 157/1.2; Q3 194/1.3; Q4 275/1.4 ESKD (incidence (1,000 p-y)/HR): Q1 2.3/1.5; Q2 3.6/2.0; Q3 6.3/2.4; Q4 20/3.3

Table 2. Continued

Study	Population	Study type (data source)	AKI (WRF) definition	Incidence AKI, %	Cardiac disease	Outcome
Parikh 2008 [6]	n = 147,007 age 76–78 years male 49–50% DM 29–41% HTN 60–69%	retrospective (CCP)	ΔSCr: mild (0.3–0.4); moderate (0.5–0.9); severe (≥1.0)	any 19.4, mild 7.1, moderate 7.1, severe 5.2	AMI prevalence: prior AMI 30–36% prior HF 19–35%	all-cause (10 years) (death (crude %)/adj. HR): none 68/1.00; mild 79/1.15; mod 88/1.23 severe >90/1.33

WRF = Worsening renal function; SCr = serum creatinine; CKD = chronic kidney disease; ESKD = end-stage kidney disease.

definitions. Large database studies often describe patient populations based on the existence of one disease process (i.e. congestive heart failure (CHF)) and subsequently estimate the prevalence/incidence of the other (i.e. kidney dysfunction).

In the ADHERE study, a large dataset of 118,465 individual hospitalizations for ADHF, 27.4, 43.5 and 13.1% of patients were found to have mild, moderate and severe kidney dysfunction at hospital admission [14]. Greater severity of kidney dysfunction correlated with worse clinical outcomes, including need for intensive care unit admission, need for mechanical ventilation, longer hospitalization and higher mortality. Similarly, in the Digitalis Investigation Group trial, Campbell et al. [15] found preexisting chronic kidney disease (CKD) in 45% of chronic CHF patients that correlated with a higher rate of hospitalization and death. This study also found evidence of a biologic gradient between CKD severity and outcome. These two studies underscore the common coexistence of heart and kidney dysfunction and how they interface.

The recent Atherosclerosis Risk in Communities Study (ARIC) and Cardiovascular Health Study (CHS) provide some additional insight into the epidemiology of Type 2 CRS [16]. Patients with baseline CVD comprised 12.9% of the study cohort. At study entry, these patients had a mean SCr of 79.6 μmol/l and estimated GFR of 86.2 ml/min/1.73 m^2. After an average follow-up of 9.3 years, 7.2% of CVD patients had declines in kidney function when defined as an increase in SCr ≥35.4 μmol/l and 34% when defined as a decline in eGFR ≥15 ml/min/1.73 m^2. During follow-up, 5.6% developed new CKD. By multivariable analysis, baseline CVD was independently associated with both decline

Table 3. Summary of studies fulfilling criteria for Type 2 CRS

Study	Population	Study type (data source)	Cardiac disease	CKD	Cardiac-specific outcomes	Outcome
Heywood 2007 [14]	n = 118,465 mean age 61.7–76.3 years male 42–57%	ADHERE registry	ADHF	eGFR 60–89: 27.4%; eGFR 30–59: 43.5%; eGFR 15–29: 13.1%; eGFR <15: 7%	use of cardioprotective meds (ACE-I and ARB) decreased with increasing degree of CKD	OR for in-hospital mortality: eGFR ≥90: 1.0; eGFR 60–89: 2.3; eGFR 30–59: 3.9; eGFR 15–29: 7.6; eGFR <15: 6.5
Elsayed 2007 [16]	n = 13,826 mean age 58 years male 56%	prospective (ARIC and CHS)	Baseline CVD in 12.9%	eGFR decrease of at least 15 ml/min/1.73 m^2 to a final level <60 ml/min/ 1.73 m^2 was seen in 34% of patients with baseline CVD	NA	OR for development of kidney disease 1.54 (CVD vs. non-CVD)
Ahmed 2007 [34]	total = 7,788 mean age 59.9–65.4 years Male 76–81%	retrospective (DIG trial); propensity-matched study	ambulatory patients with CHF	eGFR <60 in 45%	a graded association was found between CKD-related deaths and LVEF	matched HR: (CKD vs. non-CKD) all-cause death 1.71
Campbell 2009 [15]	total = 7,788 mean age 59.9–65.4 years male 76–81%	retrospective (DIG trial); propensity-matched study	ambulatory patients with CHF	eGFR <60 in 45%	matched HR: (CKD vs. non-CKD) CV hospitalization 1.17 HF hospitalization 1.08 CV death 1.24 HF death 1.42	matched HR: (CKD vs. non-CKD) all-cause hospitalization 1.18
Dimopoulos 2008 [35]	n = 1,102 mean age 36 years male 48.5%	retrospective (single-center)	adult congenital heart disease	eGFR 60–89 41% eGFR <60 9%	NA	all-cause death (HR) eGFR ≥90 1.0; eGFR <60 3.25

Table 3. Continued

Study	Population	Study type (data source)	Cardiac disease	CKD	Cardiac-specific outcomes	Outcome
Hillege 2003 [36]	n = 298 mean age 51–67 years male 70% DM 16–27% HTN 6–14%	retrospective (CATS trial)	first anterior wall MI	Change in GFRc placebo: –5.5 ml/min/year captopril: –0.5 ml/min/year	new CHF (RR) GFRc >103: 1.0 GFRc 81–103: 1.23 GFRc <81: 1.55	all-cause death: 1-year 8%

ARIC = Atherosclerosis Risk in Communities Study; ADHF = Acute Decompensated Heart Failure: GFRc = GFR estimated by Cockroft-Gault; CATS = Captopril and Thrombolysis Study; CVD = Cardiovascular Disease; CVS= Cardiovascular Health Study; DIG = Digoxin Investigator Group.

in kidney function and development of new CKD. These data strongly suggest CVD is an important risk for measurable declines in kidney function (OR 1.70, 95% CI 1.36–2.31) and CKD (OR 1.75, 95% CI 1.32–2.32) and empirical proof of the concept of Type 2 CRS.

Acute Reno-Cardiac Syndrome (Type 3 CRS)

Type 3 CRS is characterized by acute worsening of kidney function that leads to acute cardiac injury and/or dysfunction (i.e. ACS, CHF, arrhythmia). Conditions contributing to this syndrome would include contrast-induced AKI (CI-AKI), other drug-induced nephropathies, cardiac surgery-associated AKI, AKI after major non-cardiac surgery, and rhabdomyolysis.

The association of AKI and acute cardiac dysfunction with these conditions likely share similar predisposing risks, pathophysiologic mechanisms and risk factors for development (i.e. volume overload, systemic hypertension, retention of uremic solutes, hyperkalemia). However, the pathophysiologic mechanisms contributing to Type 3 CRS are likely to extend beyond simply retention of uremic solutes and/or volume overload. Defining the precise epidemiology of Type 3 CRS is challenging for several reasons: (1) there is heterogeneity in predisposing conditions causing AKI; (2) there are differing methods that have been used to define AKI; (3) there is a variable baseline risk for acute cardiac dysfunction (i.e. increased susceptibility in individuals with subclinical CVD), and (4) few clinical studies of AKI have reported on the event rates of acute cardiac dysfunction. As such, incidence and associated outcomes of acute cardiac dysfunction with AKI are largely context- and disease-specific.

Table 4. Summary of selected studies fulfilling criteria for Type 4 CRS

Study	Population	Study type (data source)	CKD stage	Cardiac outcomes	Outcome
Herzog 1998 [20]	n = 34,189 age ≥65 years 55% male 56%	Retrospective (USRDS)	ESKD	cardiac death: 1-year 41%; 2-year 52%; 5-year 70.2%; 10-year 83%	all-cause: 1-year 59%; 2-year 73%; 5-year 90%; 10-year 97%
Muntner 2002 [26]	n = 6,534 age 48–63 years male 51–61%	Retrospective (NHANES II)	eGFR <70 75.9%	CV death (rate per 1,000 p-y): eGFR ≥90: 4.1; eGFR 70–89: 8.6; eGFR <70: 20.5	all-cause death (HR): eGFR ≥90: 1.00 eGFR 70–89: 1.64; eGFR <70: 2.00
Go 2004 [19]	n = 1.1 million mean age 52 years male 45%	Retrospective (Kaiser Permanente)	≥CKD stage III or eGFR<60	CV event (rate per 100 p-y/HR): eGFR 45–59: 3.65/1.4; eGFR 30–44: 11.3/2.0; eGFR 15–29: 21.8/2.8; eGFR <15: 36.6/3.4	all-cause mortality (per 100 p-y/HR): eGFR 45–59: 1.1/1.2; eGFR 30–44: 4.8/1.8; eGFR 15–29: 11.4/3.2; eGFR <15: 14.1/5.9
Foley 2005 [27]	n = 1,091,201 age ≥75 years 56.1 male 39%	Retrospective (Medicare/ USRDS)	CKD 3.8% (diagnostic coding)	CV event incidence: AMI 4–7 per 100 p-y; CHF 31–52 per 100 p-y (HR 1.28–1.79)	all-cause death: HR 1.38–1.56
HIllege 2006 [21]	n = 2,680 mean age 65 years male 67%	Retrospective (CHARM)	eGFR <60 36%	CV death/hosp. (HR) eGFR ≥90: 1.0; eGFR 75–89: 1.17; eGFR 60–74: 1.24; eGFR 45–59: 1.54 eGFR <45: 1.86	all-cause death (HR) eGFR ≥90: 1.0; eGFR 75–89: 1.13; eGFR 60–74: 1.14; eGFR 45–59: 1.50 eGFR <45: 1.91
McCullough 2007 [22]	n = 37,153 Mean Age 53 years male 31%	Retrospective (KEEP)	eGFR <60 14.8%	prevalence CVD (OR): eGFR ≥90 1.0; eGFR 60–89 1.1; eGFR 30–59 1.4; eGFR <30 1.3	all-cause death (HR): CKD only 1.98; CVD only 3.02 CKD+CVD 3.80

Table 4. Continued

Study	Population	Study type (data source)	CKD stage	Cardiac outcomes	Outcome
McCullough 2008 [24]	n = 31,417 mean age 45 years male 24.5%	Retrospective (KEEP)	eGFR <60 or ACR ≥30: 20.6%	risk CVD/death (OR): CKD 1.44 no CKD 1.0	worst survival for combined CKD and CVD at time of screening

ESKD = End-stage kidney disease; CKD = chronic kidney disease; DM = diabetes mellitus; HTN = hypertension; CHF = congestive heart failure; CVD = cerebrovascular disease; MA = microalbuminuria; CHD = coronary heart disease; LVH = left ventricular hypertrophy; CV = cardiovascular; eGFR = estimated glomerular filtration rate; ADHF = acute decompensated heart failure.

Chronic Reno-Cardiac Syndrome (Type 4 CRS)

Type 4 CRS is a condition where primary CKD contributes to a reduction in cardiac function (i.e. cardiac remodeling, LV diastolic dysfunction, LV hypertrophy) and/or an increased risk for cardiovascular events (i.e. myocardial infarction, heart failure, stroke). This CRS subtype refers to cardiac dysfunction and/or disease primarily occurring in response to CKD (table 4).

Observational data have evaluated the cardiovascular event rates and outcomes in selected CKD-specific populations [17–24]. Most have been retrospective and/or secondary/post-hoc analyses from large clinical datasets or randomized trials. Despite this, defining the epidemiology of Type 4 CRS is difficult and estimates are variable due to differences in: (1) the populations at risk, (2) the clinical outcomes evaluated, (3) duration of time for ascertainment of study endpoints, and (4) the operational definitions used for defining CKD, cardiac disease and/or mortality (i.e. all-cause or CVD-specific).

For example, the populations at risk in these studies, based on the presence and/or severity of CKD, ranged from near-normal kidney function to ESKD. In a secondary analysis of the multicenter hemodialysis (HEMO) study, Cheung and Sarnak [17] found 80% of ESKD patients had cardiac disease at baseline. Older patients, diabetics, and those having receiving a longer duration of maintenance renal replacement therapy (>3.7 years) had higher prevalence of pre-existing cardiac disease. During follow-up, 39.8% were admitted to hospital for cardiac-related diagnoses. Of these, 42.7% were attributable to ischemic coronary heart disease. Of the 39.4% of cardiac deaths, 61.5% were attributable to ischemic coronary heart disease. Baseline cardiac disease was highly predictive of cardiac-specific death during follow-up. The presence of cardiac disease in

ESKD patients is exceedingly common, and cardiac-specific mortality rates are 10- to 20-fold higher when compared with age- and sex-matched non-CKD populations. Moreover, recent data have emerged to suggest dialysis prescription in selected patients with ESKD receiving chronic maintenance dialysis may precipitate cardiac injury and contribute to accelerated declines in myocardial performance [25].

In CKD patients not receiving maintenance renal replacement therapy, the prevalence of CVD varies considerably with CKD severity (and likely the overall time at risk) [22–24]. The risk of CVD events and death is also likely further modified by older age, additional comorbid illness, and presence of concomitant CHF [20, 23, 24]. In data from the NHANES II study, Muntner and He [26] found a CVD prevalence of 4.5, 7.9 and 12.9% for patients with eGFR ≥90, 70–89 and <70 ml/min/cm^2, respectively. Likewise, in a large population-based cohort, Go and Chertow [19] found similar graded increases in CVD prevalence and HF, along with a higher risk of subsequent cardiac events during follow-up associated with degree of decline in eGFR <60 ml/min/1.73 cm^2. This dose-response gradient in CVD prevalence by severity of CKD was also associated with higher trends in cardiac-specific and all-cause mortality [19, 21, 22, 26].

Observational data have also shown that CKD accelerates the risk for and development of CVD [23, 24, 27]. This accelerated risk for cardiovascular events and disease in CKD may be the consequence of the unique pathophysiology that exists in these patients [28]. Moreover, there may be a unique subset with identifiable clinical features and/or susceptibility to accelerated CVD. Likewise, the exaggerated risk in CKD may also result from additional factors. First, observational data may be prone to bias and may fail to account for residual or uncontrolled confounding. Second, patients with CKD have often been excluded from clinical trials of interventions in CVD [29], and may receive less or suboptimal risk modifying and/or cardioprotective therapies. Finally, the genuine concern for treatment toxicities, intolerance and/or risks in CKD patients or AKI may be such that therapy is not offered due to an unfounded perception of a less favorable risk-benefit ratio. These factors, in part, may provide explanation for the excess of CVD and associated poor outcomes for CKD patients.

Finally, characterization of the epidemiology of Type 4 CRS from the available literature is difficult due, in part, to inability to clearly discriminate between the primary versus secondary disease process and the considerable overlap that likely exists with Type 2 CRS. We also recognize that the current definition for Type 4 CRS explicitly only considers establishing the presence of pathologic heart-kidney interaction and does not identify and/or incorporate provision for subsets of patients where CKD may act to significantly modify the risk of and/or accelerate CVD.

Secondary Cardiorenal Syndromes (Type 5 CRS)

This syndrome is characterized by acute or chronic systemic illnesses that concurrently induce cardiac and kidney injury and/or dysfunction. Limited data is available on the epidemiology of Type 5 CRS due largely to the vast number of contributing acute and/or chronic systemic conditions that may predispose to it. As such, incidence estimates, risk identification and characterization of associated outcomes for Type 5 CRS are either not available or are considered disease- and/or context-specific.

Sepsis represents a prototypical condition that may cause an acute form of Type 5 CRS. Approximately 11–64% of septic patients develop AKI [30], and 46–58% have sepsis as a major contributing factor to development of AKI [31]. Observational data have shown higher morbidity and mortality for those with septic AKI when compared to either sepsis or AKI alone [30, 31]. Similarly, abnormalities in cardiac function are common in septic patients [32]. The incidence of cardiac dysfunction in sepsis is conditional on the population at risk being studied, the definition used for detection of cardiac dysfunction (i.e. troponin, B-type natriuretic peptide, pulmonary artery catheter, echocardiography), severity of illness, resuscitation and duration of illness prior to evaluation. However, observational data have found approximately 30–80% have elevated cardiac-specific troponins that often correlate with reduced cardiac function [33]. Coexisting acute kidney and myocardial dysfunction is, accordingly, common in sepsis, however there is a lack of integrative and epidemiologic studies that have specifically evaluated the pathophysiology, incidence, risk identification, and associated outcomes for septic patients with concomitant AKI and myocardial depression who fulfill criteria for Type 5 CRS.

Conclusions

A large body of accumulated observational and clinical trial data has found that acute/chronic cardiac disease can directly contribute to acute/chronic worsening kidney function and vice versa. CRS subtypes are characterized by important heart-kidney interactions that share some similarities in pathophysiology, however they appear to have important discriminating features in terms of predisposing or precipitating events, risk identification, natural history and outcomes. Type 1 CRS is common, with incidence estimates of AKI in ADHF or ACS between 24–45 and 9–19%, respectively. Type 1 CRS clearly translates into higher morbidity and worse clinical outcome. Chronic heart disease and CKD are increasingly prevalent and frequently coexist. Accordingly, this presents challenges for applying the proposed definitions for Type 2 and 4 CRS 'retrospectively' to the existing literature when the primary versus secondary process cannot be clearly distinguished. Due to heterogeneity, the incidence and outcome estimates associated with Type 3 CRS are largely context- and disease-specific. Limited data is available on the pathophysiology or epidemiology of secondary Type 5 CRS. Accordingly, the epidemiology of Type 5 CRS is also largely disease- and context-specific. In summary, there is a clear need for additional prospective studies to characterize the epidemiology of heart-kidney interactions across the CRS subtypes, not only for a better understanding of the overall burden of disease, but also for risk identification and design of potential targets for intervention.

Acknowledgements

Dr. Bagshaw is supported by a Clinical Investigator Award from the Alberta Heritage Foundation for Medical Research Clinical Fellowship, and Dr. Cruz is supported by a fellowship from the International Society of Nephrology.

References

1 Ronco C, McCullough P, Anker SD, Anand I, Aspromonte N, Bagshaw SM, Bellomo R, Berl T, Bobek I, Cruz DN, et al: Cardio-renal syndromes: report from the consensus conference of the Acute Dialysis Quality Initiative. Eur Heart J 2009; Epub December 25.
2 Goldberg A, Hammerman H: In-hospital and 1-year mortality of patients who develop worsening renal function following acute ST-elevation myocardial infarction. Am Heart J 2005;150:330–337.
3 Gottlieb SS, Abraham W: The prognostic importance of different definitions of worsening renal function in congestive heart failure. J Card Fail 2002;8:136–141.
4 Krumholz HM, Chen Y: Correlates and impact on outcomes of worsening renal function in patients ≥65 years of age with heart failure. Am J Cardiol 2000;85:1110–1113.
5 Newsome BB, Warnock DG: Long-term risk of mortality and end-stage renal disease among elderly after small increases in serum creatinine level during hospitalization for acute myocardial infarction. Arch Intern Med 2008;168:609–616.
6 Parikh CR, Coca SG: Long-term prognosis of acute kidney injury after acute myocardial infarction. Arch Intern Med 2008;168:987–995.
7 Jose P, Skali H: Increase in creatinine and cardiovascular risk in patients with systolic dysfunction after myocardial infarction. J Am Soc Nephrol 2006;17:2886–2891.
8 Nohria A, Hasselblad V, Stebbins A, Pauly DF, Fonarow GC, Shah M, Yancy CW, Califf RM, Stevenson LW, Hill JA: Cardiorenal interactions: insights from the ESCAPE trial. J Am Coll Cardiol 2008;51:1268–1274.
9 Cowie MR, Komajda M: Prevalence and impact of worsening renal function in patients hospitalized with decompensated heart failure: results of the prospective study in heart failure (POSH). Eur Heart J 2006; 27:1216–1222.
10 Logeart D, Tabet J: Transient worsening of renal function during hospitalization for acute heart failure alters outcome. Int J Cardiol 2008;127:228–232.
11 Metra M, Nodari S: Worsening renal function in patients hospitalized for acute heart failure: clinical implications and prognostic significance. Eur J Heart Fail 2008;10:188–195.

12 Smith GL, Vaccarino V: Worsening renal function: what is a clinically meaningful change in creatinine during hospitalization with heart failure? J Card Fail 2003;9:13–25.

13 Latchamsetty R, Fang J: Prognostic value of transient and sustained increase in in-hospital creatinine on outcomes of patients admitted with acute coronary syndromes. Am J Cardiol 2007;99:939–942.

14 Heywood JT, Fonarow GC: High prevalence of renal dysfunction and its impact on outcome in 118,465 patients hospitalized with acute decompensated heart failure: a report from the ADHERE database. J Card Fail 2007;13:422–430.

15 Campbell RC, Sui X, Filippatos G, Love TE, Wahle C, Sanders PW, Ahmed A: Association of chronic kidney disease with outcomes in chronic heart failure: a propensity-matched study. Nephrol Dial Transplant 2009:24:186–193.

16 Elsayed EF, Tighiouart H, Griffith J, Kurth T, Levey AS, Salem D, Sarnak MJ, Weiner DE: Cardiovascular disease and subsequent kidney disease. Arch Intern Med 2007;167:1130–1136.

17 Cheung AK, Sarnak MJ: Cardiac diseases in maintenance hemodialysis patients: results of the HEMO Study. Kidney Int 2004;65:2380–2389.

18 Culleton BF, Larson MG: Cardiovascular disease and mortality in a community-based cohort with mild renal insufficiency. Kidney Int 1999;56:2214–2219.

19 Go AS, Chertow GM: Chronic kidney disease and the risks of death, cardiovascular events, and hospitalization. N Engl J Med 2002;351:1296–1305.

20 Herzog CA, Ma JZ, Collins AJ: Poor long-term survival after acute myocardial infarction among patients on long-term dialysis. N Engl J Med 1998;339:799–805.

21 Hillege HL, Nitsch D: Renal function as a predictor of outcome in a broad spectrum of patients with heart failure. Circulation 2006; 113:671–678.

22 McCullough PA, Jurkovitz CT: Independent components of chronic kidney disease as a cardiovascular risk state. Arch Intern Med 2007;167:1122–1129.

23 McCullough PA, Li S: CKD and cardiovascular disease in screened high-risk volunteer and general populations: The Kidney Early Evaluation Program (KEEP) and the National Health and Nutrition Examination Survey (NHANES) 1999–2004. Am J Kidney Dis 2008;51:S38–S45.

24 McCullough PA, Li S: Chronic kidney disease, prevalence of premature cardiovascular disease, and relationship to short-term mortality. Am Heart J 2008;156:277–283.

25 Burton JO, Jefferies HJ, Selby NM, McIntyre CW: Hemodialysis-induced cardiac injury: determinants and associated outcomes. Clin J Am Soc Nephrol 2009;4:914–920.

26 Muntner P, He J: Renal insufficiency and subsequent death resulting from cardiovascular disease in the United States. J Am Soc Nephrol 2002;13:745–753.

27 Foley RN, Murray AM: Chronic kidney disease and the risk for cardiovascular disease, renal replacement, and death in the United States Medicare population, 1998–1999. J Am Soc Nephrol 2005;16:489–495.

28 Yerkey MW, Kernis SJ: Renal dysfunction and acceleration of coronary disease. Heart 2004;90:961–966.

29 Ronco C, Haapio M, House AA, Anavekar N, Bellomo R: Cardiorenal syndrome. J Am Coll Cardiol 2008;52:1527–1539.

30 Bagshaw SM, Lapinsky S, Dial S, Arabi Y, Dodek P, Wood G, Ellis P, Guzman J, Marshall J, Parrillo JE, et al: Acute kidney injury in septic shock: clinical outcomes and impact of duration of hypotension prior to initiation of antimicrobial therapy. Intensive Care Med 2009;35:871–881.

31 Bagshaw SM, Uchino S, Bellomo R, Morimatsu H, Morgera S, Schetz M, Tan I, Bouman C, Macedo E, Gibney N, et al: Septic acute kidney injury in critically ill patients: clinical characteristics and outcomes. Clin J Am Soc Nephrol 2007;2:431–439.

32 Charpentier J, Luyt CE: Brain natriuretics peptide: a marker of myocardial dysfunction and prognosis during severe sepsis. Crit Care Med 2004;32:660–665.

33 Ammann P, Maggiorinin M, Bertel O: Troponin as a risk factor for mortality in critically ill patients without acute coronary syndromes. J Am Coll Cardiol 2003;41:2004–2009.

34 Ahmed A, Rich MW, Sanders PW, Perry GJ, Bakris GL, Zile MR, Love TE, Aban IB, Shlipak MG: Chronic kidney disease associated mortality in diastolic versus systolic heart failure: a propensity-matched study. Am J Cardiol 2007;99:393–398.
35 Dimopoulos K, Diller G: Prevalence, predictors and prognostic value of renal dysfunction in adults with congenital heart disease. Circulation 2008;117:2320–2328.
36 Hillege HL, van Gilst WH: Accelerated decline and prognostic impact of renal function after myocardial infarction and the benefits of ACE inhibition: the CATS randomized trial. Eur Heart J 2003;24:412–420.

Dr. Sean M. Bagshaw
Division of Critical Care Medicine, University of Alberta Hospital
3C1.12 Walter C. Mackenzie Centre, 8440-122 Street
Edmonton, Alta T6G2B7 (Canada)
Tel. +1 780 407 6755, Fax +1 780 407 1228, E-Mail bagshaw@ualberta.ca

Ronco C, Bellomo R, McCullough PA (eds): Cardiorenal Syndromes in Critical Care.
Contrib Nephrol. Basel, Karger, 2010, vol 165, pp 83–92

Biomarkers of Cardiac and Kidney Dysfunction in Cardiorenal Syndromes

Dinna N. Cruz[a] · Sachin Soni[b] · Leo Slavin[c] · Claudio Ronco[a] · Alan Maisel[d]

[a]Department of Nephrology, Dialysis & Transplantation, International Renal Research Institute, San Bortolo Hospital, Vicenza, Italy; [b]Department of Nephrology, Seth Nandlal Dhoot Hospital, Aurangabad, India; [c]Division of Cardiology, University of California, San Diego, Calif., USA, and [d]Department of Medicine and Cardiology, San Diego VA Medical Center and University of California, San Diego, Calif., USA

Abstract

The role of biomarkers is rapidly emerging as an important tool in the management of the cardiorenal syndromes (CRS). Natriuretic peptides (NPs), due to their low cost and rapid and accurate ability to provide additional information not surmised from clinical evaluation, are the standard bearer for the newer biomarkers. Although the NP-guided therapy has been shown to improve patient outcomes, this has yet to be demonstrated for the novel renal biomarkers. Most of the renal biomarkers studies in CRS have been performed in the setting of cardiac surgery. It will be critical to validate these new biomarkers in multicenter and prospective studies encompassing a broad spectrum of patients. Work with NPs has also shown that novel biomakers are not to be used as 'stand-alone' tests; rather they are best used as adjuncts to everything else the health care provider brings to the table. It is likely that panels of multiple biomarkers will be needed for optimal evaluation, risk stratification, timely treatment initiation and follow-up of patients with CRS.

Kidney and cardiac disease are exceedingly common, are increasing in prevalence and frequently coexist. Observational and clinical trial data have accrued to show that acute/chronic cardiac disease can directly contribute to acute/chronic worsening kidney function and vice versa. This leads to the issue whether assessment of cardiorenal biomarkers will improve the ability to predict: (1) who is at risk for cardiorenal disease, (2) who is at risk over and above the predictive value of established biomarkers/risk factors, and finally (3) who will gain substantial improvement through elimination of these cardiorenal risk indicators.

Morrow and de Lemos proposed three criteria required for a biomarker to be clinically useful [1]. First, the assay should be precise, accurate, and rapidly available to the clinician at a relatively low cost. Second, the biomarker should provide additional information that is not surmised from clinical evaluation. Lastly, the absolute measured value should help in clinical decision-making. Biomarkers that do meet these criteria hail from several domains, reflecting the different mechanisms such as biomechanical stretch, inflammation and myocyte injury that are involved in the pathophysiology and natural history of heart failure (HF). This article will focus specifically on the natriuretic peptides (NPs), and some of the newer biomarkers for acute kidney injury (AKI).

Natriuretic Peptides

NPs are released by cardiac myocytes in response to wall stress from intravascular volume expansion. Their design is to counteract the unfavorable hemodynamic profile resulting from cardiac dysfunction. Three distinct NPs (A-type, B-type, and C-type) are known to circulate in humans. Three NP receptors (NPR-A, NPR-B, NPR-C) are responsible for activating the cyclic-guanosine monophosphate (cGMP)-dependent signaling cascades. NPs convey a multitude of actions on the cardiovascular system including vascular smooth muscle cell relaxation, promotion of natriuresis, as well as direct myocardial effects. In concert, NPs achieve a more favorable neurohormonal and hemodynamic state [2].

A-type NP (ANP), first described by deBold, established the heart as an endocrine organ. ANP is stored in preformed granules in atrial tissue. The 151 amino acid (aa) preproANP is processed to the biologically active ANP (1–28) and inert proANP (1–98). B-type NP (BNP) is predominantly synthesized from increased wall stress. Its concentration is highest in atrial tissue, however more is released from the ventricles owed to its greater mass. The precursor preproBNP is cleaved to the 108 aa peptide proBNP. Corin cleaves this to the biologically active BNP [1–32 aa] and the inert amino-terminal pro-B-type NP (NT-proBNP) [1–76 aa] peptide in a 1:1 ratio (fig. 1). C-type NP (CNP) is primarily released by the central nervous tissue, vascular endothelium, and in minute amounts by cardiac tissues. Its role in HF has not been well defined.

Natriuretic Peptides in Heart Failure

NPs have established themselves as the premier laboratory test for the diagnosis and exclusion of clinically important HF. The utility of NPs was initially proven in the emergency department with patients presenting with acute dyspnea. The Breathing Not Properly trial first revealed that BNP could serve as an accurate

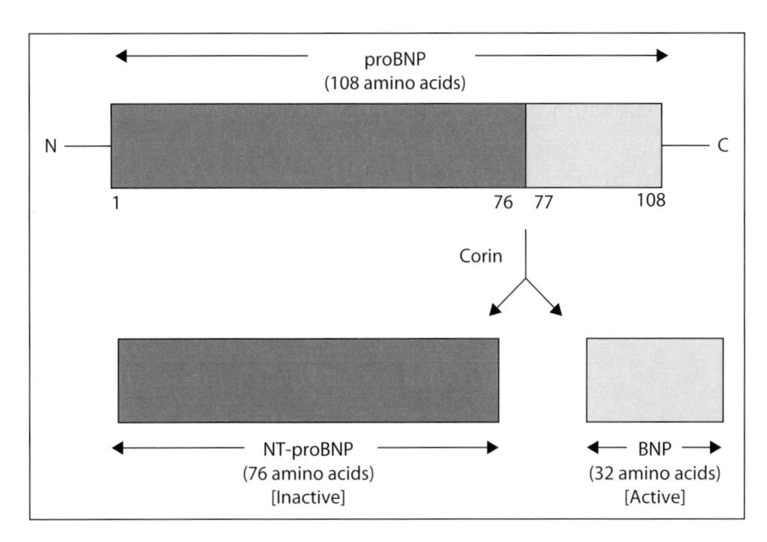

Fig. 1. Synthesis and processing of BNP.

marker for the diagnosis of HF in patients presenting to the emergency department with dyspnea. A BNP level ≤100 pg/ml yielded a 90% sensitivity and 76% specificity for separating cardiac from non-cardiac etiologies of dyspnea. BNP levels were a more accurate predictor of HF diagnosis than history, physical examination, and routine laboratory tests [3].

The ProBNP Investigation of Dyspnea in the Emergency Department (PRIDE) provided concrete evidence for the clinical usefulness of NT-proBNP in the assessment of 599 patients presenting to the emergency department with dyspnea. NT-proBNP proved to be more accurate in the diagnosis of HF than solely relying on clinical history alone (area under the receiver operator curve [AuROC] = 0.94). A cut-point of 300 ng/l provided a 99% negative predicative value in excluding a diagnosis of HF [4].

NPs are powerful predictors of mortality and non-fatal cardiac events [5, 6]. While those with New York Heart Association (NYHA) class IV HF are well known to have a poor prognosis, NP levels have been shown to be an independent predictor of mortality in more stable outpatients with lower NYHA classifications. Patients with a BNP level <130 pg/ml had a 1% risk of sudden cardiac death versus 19% risk at higher concentrations. The Acute Decompensated Heart Failure National Registry (ADHERE) of over 48,000 patients with ADHF showed a positive linear relationship of in-hospital mortality and levels of BNP, irrespective of whether the HF was secondary to systolic or diastolic dysfunction [5].

In managing hospitalized patients with HF, it is not always clear whether the selected therapy is achieving desired results, that is lowering cardiac filling pressure and relieving congestion and dyspnea. NP levels provide guidance in tailoring therapy to individual patients. Falling NP levels measured during treatment have been shown to be proportional to decreasing pulmonary capillary wedge pressure and inversely correlated with symptomatic improvement [7]. Thus, in order to prevent readmission to the hospital, congestion must be relieved by achieving a euvolemic status. NP levels are considered 'wet' on admission and will approach 'dry' or 'optivolemic' as congestion is decreased. This may prevent readmissions as well as improve symptoms at the time of discharge.

Elevated NPs in Other Clinical Settings
It is important to be aware of clinical scenarios other than acute HF that can cause NP levels to rise. Patients with a history of HF but without an acute exacerbation can have intermediate BNP levels, as shown in the Breathing Not Properly trial [3]. Acute coronary syndrome (ACS) is also associated with elevated levels of NPs: acute ischemia causes transient diastolic dysfunction resulting in increased left ventricular end-diastolic pressure, a rise in wall stress, and increased synthesis of BNP [7]. The role of BNP is perhaps most useful in differentiating between cardiac and pulmonary causes of shortness of breath. Although NP values can be elevated to intermediate levels in patients with underlying lung disease, they tend to remain significantly lower than in patients presenting with chronic heart failure (CHF) [7]. The Breathing Not Properly trial demonstrated that NP levels were useful in diagnosing HF, which was present in 87 out of the 417 patients with a history of COPD or asthma (BNP levels 587 vs. 109 pg/ml, $p < 0.0001$) [3]. In addition, right heart dysfunction from hemodynamically significant pulmonary embolism, severe lung disease, and pulmonary hypertension can lead to elevated levels of NPs [7]. Therefore, intermediate levels of NPs in an acutely dyspneic patient should trigger consideration of other life-threatening etiologies of dyspnea. Hyperdynamic states of sepsis, cirrhosis and hyperthyroidism can also be associated with elevated NPs. NP levels also are increased in the setting of atrial fibrillation.

NPs in Renal Disease
The association between NP levels and renal function is complex. In the setting of renal dysfunction, increased concentrations of NPs may stem from elevation in atrial pressure, systemic pressure or ventricular mass. Patients with renal disease often have hypertension that results in significant left ventricular hypertrophy, and cardiac comorbidities are also common. This interplay between the heart and kidney in patients with reduced renal function accounts for one component of the increase in NP levels, reflecting elevated 'true' physiologic NP levels [7]. The Breathing Not Properly study found a weak but significant correlation between glomerular filtration rate (GFR) and

BNP, and suggested higher cut-points for patients with GFR <60 ml/min/1.7 m^2.

Interpretation of NT-proBNP in the setting of renal dysfunction is more challenging given that clearance is not mediated by NPR-C or neutral endopeptidase and is more renally dependent [8]. GFR seems to be more strongly correlated with NT-proBNP ($r = -0.55$) than with BNP, though the discrepancy may be somewhat less prominent in patients with acute CHF ($r = -0.33$ for NT-proBNP, and $r = -0.18$ for BNP) [9]. An analysis from the PRIDE study demonstrated that NT-proBNP levels in patients with GFR <60 ml/min/1.7 m^2 were still the strongest predictors of outcome, and suggested a higher cut-point of >1,200 pg/ml for the diagnosis of HF [8]. Despite the relationship between NPs and GFR, both BNP and NT-proBNP still provide important diagnostic and prognostic information in patients with renal dysfunction.

Novel Renal Biomarkers for Cardiorenal Syndrome

A consensus definition and classification scheme for the cardiorenal syndrome (CRS) and its five specific subtypes has recently been proposed [9]. It is postulated that a number of cardiorenal biomarkers may potentially help to characterize these subtypes of CRS and suggest the timing of treatment initiation and its likely effectiveness [10]. Early diagnosis of renal involvement in patients with cardiac disease is of paramount importance. Serum creatinine, a commonly used biomarker, is insensitive and unreliable, especially during acute changes in kidney function [11]. Recently, newer biomarkers for AKI have been the focus of studies conducted in various clinical settings. In terms of CRS, such biomarker studies were often conducted in the setting of cardiac surgery. Cardiac surgery-associated AKI is considered a form of CRS type 1, in which an abrupt worsening of cardiac function leads to AKI. In this section, we briefly review some key biomarker studies in the context of CRS.

Cystatin C

All nucleated human cells produce cystatin C (Cys C), a cysteine protease inhibitor, and release it into plasma at a constant rate. This 13.3-kDa low molecule weight protein is freely filtered by the glomerulus, reabsorbed and catabolized completely by proximal convoluted tubules and not secreted in urine [12]. It can be measured both in serum and urine. Its levels are unaffected by age, gender, race, muscle mass, steroid therapy, infection, liver disease or inflammation [13]. For these reasons, in patients with CKD, it is proposed to perform better as a marker of GFR than serum creatinine. Its role as a biomarker in AKI has also been studied.

In a study of 72 adult patients undergoing cardiac surgery, urinary Cys C was able to detect AKI within 6 h after surgery [14]. Another study had showed serum Cys C as a better marker for detecting small temporary changes of GFR than serum creatinine or creatinine clearance in adults postcardiac surgery. This may allow better identification of patients who developed mild renal impairment [15]. In a study of 60 patients who underwent heart valve replacement surgery, serum Cys C was superior to creatinine in diagnosing renal dysfunction (AuROC for Cys C 0.876) [13]. Postoperative serum Cys C level also correlated well with duration and severity of AKI, duration of intensive care stay, need for renal replacement therapy and in-hospital mortality [16].

Apart from being a good marker for renal function, Cys C appears to be also a marker of cardiovascular risk in CRS types 2/4. Increased levels of Cys C were found to be an independent predictor of cardiac events at 1-year follow-up of 525 patients with non-ST elevation ACS (NSTE-ACS) [17]. In addition, Cys C also offered complementary prognostic information to other cardiac biomarkers like troponin T, high-sensitivity C-reactive protein and NT-proBNP, helping clinicians perform more accurate risk stratification of patients with acute HF or ACS [18, 19].

Neutrophil Gelatinase-Associated Lipocalin

Neutrophil gelatinase-associated lipocalin (NGAL), a 25-kDa protein of lipocalin superfamily, is emerging as one of the most promising biomarkers for AKI. It is expressed by neutrophils and epithelial cells in various human tissues like salivary glands, uterus, prostate, trachea, lungs, stomach, colon and kidneys [20]. After favorable reports from animal experiments [21], its role as an early AKI biomarker has been studied. It can be measured in blood as a point-of-care test, and in the urine using a standardized laboratory platform.

In a landmark study of 71 children undergoing cardiopulmonary bypass (CPB), both plasma and urinary NGAL at 2 h after CPB were found to be independent predictors of AKI with AuROC of 0.91 for plasma NGAL and 0.998 for urinary NGAL [22]. Using the creatinine criteria of >50% increase from baseline, a clinical diagnosis of AKI was possible only after 1–3 days. The predictive power of NGAL for AKI has also been confirmed in a small study 81 adult patients who developed AKI after cardiac surgery [23]. Mean urine NGAL peaked immediately and remained significantly higher 3 and 18 hours after surgery. Patients who did not experience AKI also had significant increases in urine NGAL in the early postoperative period, but to a lesser degree than those with AKI. NGAL has also been found to correlate with AKI course and clinical outcomes. Postoperative plasma [24] and urine NGAL levels [25] appeared to be reliable predictors of AKI severity and duration, length of hospital stay, dialysis requirement and and mortality in children undergoing CPB.

In terms of CRS types 2/4, Poniatowski et al. [26] have found serum and urine NGAL as sensitive early markers of renal dysfunction in patients with CHF. Urinary NGAL has also been found to correlate directly with urine albumin excretion and inversely with estimated GFR in patients with CHF [27].

Kidney Injury Molecule-1

Kidney injury molecule-1 (KIM-1), a transmembrane protein, is markedly overexpressed in proximal convoluted tubules in response to renal injury [28]. It can be measured in the urine using ELISA assays or microsphere-based Luminex xMAP Technology, and a point-of-care diagnostic dipstick is in development [29]. In a cohort of 40 children undergoing cardiac surgery, urinary KIM-1 levels were markedly increased at 12 h, with an AuROC of 0.83 for predicting AKI [30]. Liangos et al. [31] had also found that urinary KIM-1 performed best as an early biomarker for AKI in a group of 103 adult patients after cardiac surgery.

It has been hypothesized that a panel of markers will perform better than a single marker. Han et al. [32] evaluated three urinary biomarkers as early biomarkers of AKI after cardiac surgery singly and in combination: matrix metalloproteinase-9, *N*-acetyl-β-D-glucosaminidase (NAG) and KIM-1. In 90 adult patients, urinary KIM-1 is a sensitive marker for diagnosing AKI immediately after and 3 h postcardiac surgery (AuROC 0.68 and 0.65, respectively). However, combining the three biomarkers enhanced the sensitivity of early detection of postoperative AKI compared with individual biomarkers: the AuROCs for the combination of the three biomarkers were 0.75 (immediately postsurgery) and 0.78 (3 h postsurgery).

Other Biomarkers

Many novel biomarkers like interleukin (IL)-18, NAG, glutathione-*S*-transferase, liver fatty acid-binding protein, glutamyl transpeptidase, sodium hydrogen exchanger, matrix metalloproteinase-9, aprotinin, IL-6, IL-10, α_1-microglobulin have been studied as early biomarkers of AKI [10]. With continued research activity in this field, many more molecules are likely to emerge as potential early biomarkers for AKI.

Conclusion

In recent years there has been an exponential increase in studies and publications related to new biomarkers for AKI and HF. Although the NP-guided therapy has been shown to improve patient outcomes, this has yet to be demonstrated for the novel renal biomarkers. It will be critical to validate these new biomarkers in multicenter and prospective studies encompassing a broad spectrum of patients. Work with NPs has also shown that novel biomarkers are not to be used as 'stand-alone' tests; rather they are best used as adjuncts to everything else the healthcare provider brings to the table. Synthesis of clinically applicable information from relevant studies to form a panel of biomarkers for diagnosis and classification of CRS will be an important step forward.

References

1 Braunwald E: Biomarkers in heart failure. N Engl J Med 2008;358:2148–2159.
2 Nakao K, Ogawa Y, Suga S, Imura H: Molecular biology and biochemistry of the natriuretic peptide system. II. Natriuretic peptide receptors. J Hypertens 1992;10:1111–1114.
3 Maisel AS, Krishnaswamy P, Nowak RM, McCord J, Hollander JE, Duc P, Omland T, Storrow AB, Abraham WT, Wu AH, Clopton P, Steg PG, Westheim A, Knudsen CW, Perez A, Kazanegra R, Herrmann HC, McCullough PA: Rapid measurement of B-type natriuretic peptide in the emergency diagnosis of heart failure. N Engl J Med 2002;347:161–167.
4 Januzzi JL Jr, Camargo CA, Anwaruddin S, Baggish AL, Chen AA, Krauser DG, Tung R, Cameron R, Nagurney JT, Chae CU, Lloyd-Jones DM, Brown DF, Foran-Melanson S, Sluss PM, Lee-Lewandrowski E, Lewandrowski KB: The *N*-terminal Pro-BNP Investigation of Dyspnea in the Emergency Department (PRIDE) Study. Am J Cardiol 2005;95:948–954.
5 Fonarow GC, Peacock WF, Horwich TB, Phillips CO, Givertz MM, Lopatin M, Wynne J: Usefulness of B-type natriuretic peptide and cardiac troponin levels to predict in-hospital mortality from ADHERE. Am J Cardiol 2008;101:231–237.
6 Fonarow GC, Peacock WF, Phillips CO, Givertz MM, Lopatin M: Admission B-type natriuretic peptide levels and in-hospital mortality in acute decompensated heart failure. J Am Coll Cardiol 2007;49:1943–1950.
7 Daniels LB, Maisel AS: Natriuretic peptides. J Am Coll Cardiol 2007;50:2357–2368.
8 Anwaruddin S, Lloyd-Jones DM, Baggish A, Chen A, Krauser D, Tung R, Chae C, Januzzi JL Jr: Renal function, congestive heart failure, and amino-terminal pro-brain natriuretic peptide measurement: results from the ProBNP Investigation of Dyspnea in the Emergency Department (PRIDE) Study. J Am Coll Cardiol 2006;47:91–97.
9 Ronco C, McCullough P, Anker SD, Anand I, Aspromonte N, Bagshaw SM, Bellomo R, Berl T, Bobek I, Cruz DN, Daliento L, Davenport A, Haapio M, Hillege H, House AA, Katz N, Maisel A, Mankad S, Zanco P, Mebazaa A, Palazzuoli A, Ronco F, Shaw A, Sheinfeld G, Soni S, Vescovo G, Zamperetti N, Ponikowski P: Cardio-renal syndromes: report from the consensus conference of the Acute Dialysis Quality Initiative. Eur Heart J 2009 Dec 25 [Epub].
10 Soni SS, Fahuan Y, Ronco C, Cruz DN: Cardiorenal syndrome: biomarkers linking kidney damage with heart failure. Biomark Med 2009;3:549–560.
11 Bellomo R, Kellum JA, Ronco C: Defining acute renal failure: physiological principles. Intensive Care Med 2004;30:33–37.
12 Nguyen MT, Devarajan P: Biomarkers for the early detection of acute kidney injury. Pediatr Nephrol 2008;23:2151–2157.
13 Zhu J, Yin R, Wu H, Yi J, Luo L, Dong G, Jing H: Cystatin C as a reliable marker of renal function following heart valve replacement surgery with cardiopulmonary bypass. Clin Chim Acta 2006;374:116–121.
14 Koyner JL, Bennett MR, Worcester EM, Ma Q, Raman J, Jeevanandam V, Kasza KE, O'Connor MF, Konczal DJ, Trevino S, Devarajan P, Murray PT: Urinary cystatin C as an early biomarker of acute kidney injury following adult cardiothoracic surgery. Kidney Int 2008;74:1059–1069.

15 Wang QP Gu JW, Zhan XH, Li H, Luo XH: Assessment of glomerular filtration rate by serum cystatin C in patients undergoing coronary artery bypass grafting. Ann Clin Biochem 2009;46:495–500.
16 Haase M, Bellomo R, Devarajan P, Ma Q, Bennett MR, Mockel M, Matalanis G, Dragun D, Haase-Fielitz A: Novel biomarkers early predict the severity of acute kidney injury after cardiac surgery in adults. Ann Thorac Surg 2009;88:124–130.
17 Taglieri N, Fernandez-Berges DJ, Koenig W, Consuegra-Sanchez L, Fernandez JM, Robles NR, Sanchez PL, Beiras AC, Orbe PM, Kaski JC: Plasma cystatin C for prediction of 1-year cardiac events in Mediterranean patients with non-ST elevation acute coronary syndrome. Atherosclerosis 2010; 209:300–305.
18 Manzano-Fernandez S, Boronat-Garcia M, Albaladejo-Oton MD, Pastor P, Garrido IP, Pastor-Perez FJ, Martinez-Hernandez P, Valdes M, Pascual-Figal DA: Complementary prognostic value of cystatin C, N-terminal pro-B-type natriuretic peptide and cardiac troponin T in patients with acute heart failure. Am J Cardiol 2009;103:1753–1759.
19 Garcia Acuna JM, Gonzalez-Babarro E, Grigorian Shamagian L, Pena-Gil C, Vidal Perez R, Lopez-Lago AM, Gutierrez Feijoo M, Gonzalez-Juanatey JR: Cystatin C provides more information than other renal function parameters for stratifying risk in patients with acute coronary syndrome. Rev Esp Cardiol 2009;62:510–519.
20 Cowland JB, Borregaard N: Molecular characterization and pattern of tissue expression of the gene for neutrophil gelatinase-associated lipocalin from humans. Genomics 1997;45:17–23.
21 Mishra J, Ma Q, Prada A, Mitsnefes M, Zahedi K, Yang J, Barasch J, Devarajan P: Identification of neutrophil gelatinase-associated lipocalin as a novel early urinary biomarker for ischemic renal injury. J Am Soc Nephrol 2003;14:2534–2543.
22 Mishra J, Dent C, Tarabishi R, Mitsnefes MM, Ma Q, Kelly C, Ruff SM, Zahedi K, Shao M, Bean J, Mori K, Barasch J, Devarajan P: Neutrophil gelatinase-associated lipocalin as a biomarker for acute renal injury after cardiac surgery. Lancet 2005;365:1231–1238.
23 Wagener G, Jan M, Kim M, Mori K, Barasch JM, Sladen RN, Lee HT: Association between increases in urinary neutrophil gelatinase-associated lipocalin and acute renal dysfunction after adult cardiac surgery. Anesthesiology 2006;105:485–491.
24 Dent CL, Ma Q, Dastrala S, Bennett M, Mitsnefes MM, Barasch J, Devarajan P: Plasma neutrophil gelatinase-associated lipocalin predicts acute kidney injury, morbidity and mortality after pediatric cardiac surgery: a prospective uncontrolled cohort study. Crit Care 2007;11:R127.
25 Bennett M, Dent CL, Ma Q, Dastrala S, Grenier F, Workman R, Syed H, Ali S, Barasch J, Devarajan P: Urine NGAL predicts severity of acute kidney injury after cardiac surgery: a prospective study. Clin J Am Soc Nephrol 2008;3:665–673.
26 Poniatowski B, Malyszko J, Bachorzewska-Gajewska H, Malyszko JS, Dobrzycki S: Serum neutrophil gelatinase-associated lipocalin as a marker of renal function in patients with chronic heart failure and coronary artery disease. Kidney Blood Press Res 2009;32:77–80.
27 Damman K, van Veldhuisen DJ, Navis G, Voors AA, Hillege HL: Urinary neutrophil gelatinase-associated lipocalin, a marker of tubular damage, is increased in patients with chronic heart failure. Eur J Heart Fail 2008; 10:997–1000.
28 Ichimura T, Bonventre JV, Bailly V, Wei H, Hession CA, Cate RL, Sanicola M: Kidney injury molecule-1, a putative epithelial cell adhesion molecule containing a novel immunoglobulin domain, is upregulated in renal cells after injury. J Biol Chem 1998;273:4135–4142.
29 Vaidya VS FG, Waikar SS, Wang Y, Clement MB, Ramirez V, Glaab WE, Troth SP, Sistare FD, Prozialeck WC, Edwards JR, Bobadilla NA, Mefferd SC, Bonventre JV: A rapid urine test for early detection of kidney injury. Kidney Int 2009;76:108–114.
30 Vaidya VS, Ramirez V, Ichimura T, Bobadilla NA, Bonventre JV: Urinary kidney injury molecule-1: a sensitive quantitative biomarker for early detection of kidney tubular injury. Am J Physiol Renal Physiol 2006; 290:F517–F529.

31 Liangos O, Tighiouart H, Perianayagam MC, Kolyada A, Han WK, Wald R, Bonventre JV, Jaber BL: Comparative analysis of urinary biomarkers for early detection of acute kidney injury following cardiopulmonary bypass. Biomarkers 2009;14:423–431.

32 Han WK, Waikar SS, Johnson A, Betensky RA, Dent CL, Devarajan P, Bonventre JV: Urinary biomarkers in the early diagnosis of acute kidney injury. Kidney Int 2008;73:863–869.

Dinna N. Cruz, MD, MPH
Department of Nephrology, Dialysis & Transplantation, International Renal Research Institute
San Bortolo Hospital
Viale Rodolfi 37, IT–36100 Vicenza (Italy)
Tel. +39 0 444 753650, Fax +39 0 444 753973, E-Mail dinnacruzmd@yahoo.com

Ronco C, Bellomo R, McCullough PA (eds): Cardiorenal Syndromes in Critical Care.
Contrib Nephrol. Basel, Karger, 2010, vol 165, pp 93–100

How to Manage Cardiorenal Syndromes in the Emergency Room

Salvatore Di Somma · Chiara Serena Gori · Emiliano Salvatori

Emergency Department, Second Medical School, Sapienza University of Rome, Sant'Andrea Hospital, Rome, Italy

Abstract

In the emergency department (ED) a prompt diagnosis and appropriate treatment for all diseases improve a patient's outcome. Acute kidney injury (AKI) is defined as an abrupt deficiency of renal function over a period of hours to days resulting in a failure of the kidney to excrete nitrogenous waste products and to maintain fluid and electrolyte homeostasis. AKI diagnosis could be very challenging for ED physicians because it is often very difficult to obtain some anamnestic data such as daily urine output or a preexisting value of BUN and serum creatinine. The incidence of AKI is progressively increasing in EDs and the mortality rates of these patients range from 50 to 80% in multiorgan failure. For ED physicians it is also crucial to distinguish AKI from prerenal azotemia (volume depletion promptly resolved through administration of fluids) at the time of patient presentation. Moreover, a rapid diagnosis of AKI leads to stop the progressive kidney damage on the basis of an appropriate therapeutic approach. Recent studies have demonstrated that by using a new biomarker, neutrophil gelatinase-associated lipocalin (NGAL), it is possible to obtain an accurate and fast diagnosis of AKI. It is well known that in patients with cardiovascular diseases such as stroke, coronary artery diseases and congestive heart failure, high levels of creatinine are strictly related to a higher mortality. In the ED the occurrence of AKI in patients with acute worsening of cardiac function like acute decompensated heart failure is very common. Moreover, managing acute heart failure strictly depends on renal function. Therefore, a multimarker approach including NGAL+BNP (today easily obtained by a POCT system) could have a tremendous impact on an appropriate diagnosis, treatment and a supposed better patient outcome. Furthermore, an evaluation of total body fluid content is of great utility. We propose a new model of management for ED patients with cardiorenal syndromes using a multimarker approach and non-invasive evaluation of body fluid content by bioelectrical impedance vector analysis.

In the emergency department (ED) it is well known that a prompt diagnosis and appropriate treatment for all diseases is related to an improved patient's outcome. In particular, since heart disease and kidney are intricately interrelated, an immediate and accurate diagnosis of acute kidney injury (AKI) is critical for ED physicians. Owing to this relationship, people with renal dysfunction are more likely to have heart disease; besides heart disease is more common in people with renal dysfunction than in those of the same age and gender without renal disorders. Even patients with early renal dysfunction (as determined by mildly elevated serum creatinine (SCr) levels >1.5 mg/dl in men and >1.3 in women) are at increased risk for cardiovascular diseases, congestive heart failure (CHF) and mortality. In fact the higher SCr level correlates with a greater disease severity [1].

Acute Kidney Injury: Definition, Classification and Epidemiology

According to the Acute Kidney Injury Network (AKIN), AKI is defined as an abrupt reduction (within 48 h) in kidney function of either ≥0.3 mg/dl or an increase of ≥50% plus a reduction in urine output. Another definition of AKI has been proposed by the Acute Dialysis Quality Initiative (ADQI) Group, based on SCr, glomerular filtration rate (GFR) and urine output. The proposed classification is called RIFLE reflecting the terms Risk, Injury, Failure, Loss, and End-stage renal disease according to kidney function [2, 3]. The myriad causes of AKI are commonly categorized into prerenal, intrinsic (renal), and postrenal.

Prerenal causes of AKI are those that decrease effective blood flow to the kidney. These include low blood volume, low blood pressure, and heart failure (HF). Changes to the blood vessels supplying the kidney can also lead to prerenal AKI. These include renal artery stenosis, which is a narrowing of the renal artery that supplies the kidney, and renal vein thrombosis, which is the formation of a blood clot in the renal vein that drains blood from the kidney. Those causes that lead to damage the kidney itself are dubbed intrinsic. Intrinsic AKI can be due to damage to the glomeruli, renal tubules, or interstitium. Common causes of each are glomerulonephritis, acute tubular necrosis, and acute interstitial nephritis, respectively.

Postrenal AKI is a consequence of urinary tract obstruction. This may be related to benign prostatic hyperplasia, kidney stones, or an obstructed urinary catheter [4]. AKI is a common complication occurring in up to 5–20% of hospitalized patients, but more importantly 20–60% of them may require dialysis in short times and even the mortality is very high, reaching up to 80% in multiorgan failure. The incidence of AKI has been rising at an alarming rate in recent years, especially in patients with acute HF presenting in the ED [5].

Table 1. Cardiorenal syndrome: definition and classification [6]

CRS type 1 (acute CRS)	Abrupt worsening of cardiac function leading to AKI
CRS type 2 (chronic CRS)	Chronic abnormalities in cardiac function (e.g. chronic CHF) causing progressive and permanent CKD
CRS type 3 (acute renocardiac syndrome)	Abrupt worsening of renal function (e.g. acute kidney ischemia or glomerulonephritis) causing acute cardiac disorders (e.g. HF, arrhythmia and ischemia)
CRS type 4 (chronic renocardiac syndrome)	CKD (e.g. chronic glomerular disease) contributing to decreased cardiac function, cardiac hypertrophy, increased risk of adverse cardiovascular events or all
CRS type 5 (secondary CRS)	Systemic condition (e.g. diabetes mellitus and sepsis) causing both cardiac and renal dysfunction

Cardiorenal Syndromes

The cardiorenal syndrome (CRS) can generally be defined as a pathophysiological disorder of the heart and kidneys whereby acute or chronic dysfunction in one organ may induce acute or chronic dysfunction in the other organ. Ronco et al. [6] proposed a classification of CRS into five different subtypes, based on the pathophysiology, time frame and nature of the concomitant cardiac and renal dysfunction (table 1).

The strong association between heart and kidneys is clearly demonstrated in the setting of HF where excessive and inappropriate activation of the renin-angiotensin-aldosterone system leads to sodium retention, excess blood volume, increased preload volume and finally left ventricular failure. Renin-angiotensin-aldosterone system activation causes excess afterload which can lead to renal dysfunction [7]. So in patients with HF the presence of renal impairment will worsen their prognosis and this statement should be always considered by ED physicians.

Diagnosis of Cardiorenal Syndromes

A detailed and accurate history is crucial to aid in diagnosing the type of CRS and in determining its subsequent treatment. A detailed history and a physical examination in combination with routine laboratory tests, chest X-ray, bedside ultrasound and biomarkers (today easily obtained by a POCT system) are useful in making a correct diagnosis. In an acute setting, biomarkers can contribute to integrate the diagnosis of renal dysfunction with that of HF. For the ED physician it is also crucial to distinguish AKI from prerenal azotemia (volume depletion promptly resolved through administration of fluids) at the time of patient presentation. Moreover, a rapid diagnosis of AKI in the ED leads to stop the progressive kidney damage on the basis of an appropriate therapeutic approach.

Therefore, novel biomarkers such as neutrophil gelatinase-associated lipocalin (NGAL) can be used to predict AKI earlier than SCr which is not detectable until at least 2 days after the GFR has gone down. In a recent study by Nickolas et al. [8] of adult ED patients, a single POC measurement of NGAL at the time of arrival predicted AKI with excellent sensitivity and specificity. This single measurement of NGAL reliably distinguished prerenal azotemia from that of intrinsic AKI and of chronic kidney disease (CKD). NGAL was also predictive of poor patient outcomes, including dialysis requirement, admission to the intensive care unit and in hospital mortality. Cystatin C appears to be a better predictor of glomerular function than SCr in patients with CKD. In AKI, urinary excretion of cystatin C has been shown to predict the requirement for renal replacement therapy earlier than creatinine.

Kidney injury molecule-1 is a protein detectable in the urine after ischemic or nephrotoxic insults to proximal tubular cells. Urinary kidney injury molecule-1 seems to be highly specific for ischemic AKI and not for prerenal azotemia, CKD, or contrast-induced nephropathy.

Interleukin-18 is a pro-inflammatory cytokine detected in the urine after acute ischemic proximal tubular damage. It displays good sensitivity and specificity for ischemic AKI with an AUC 90% with increased levels 48 h prior to increases in SCr.

N-Acetyl-β-D-glucosaminidase is a lysosomal brush border enzyme found in proximal tubular cells. It has been shown to function as a marker of kidney injury, reflecting particularly the degree of tubular damage. It is not only found in elevated urinary concentrations in AKI and CKD, but also in diabetic patients, patients with essential hypertension, and HF [9].

Cardiac biomarkers have become an essential element in the diagnosis and prognosis of cardiovascular events. B-type natriuretic peptides are diagnostic tools in acute decompensated heart failure (ADHF) and are elevated in patients with type 1 CRS in which AKI occurs as a consequence of ADHF and potential applications in type 2 and type 4. Patients with CKD have higher levels of BNP and NT-proBNP than age- and gender-matched subjects without reduced renal function, even in the absence of clinical CHF [10]. Managing acute heart diseases is strictly dependent on renal function, therefore a multimarker approach including NGAL+BNP could have a tremendous impact on an appropriate diagnosis, treatment and a supposed better patient outcome.

New Perspectives: The Bioelectrical Impedance Vector Analysis

The bioelectrical impedance vector analysis (BIVA) is a non-invasive technique to estimate body mass and water composition by bioelectrical impedance measurements, resistance and reactance. BIVA evaluates quickly the body mass in situations of 'normality' state, extreme obesity, denutrition state and assesses the hydration state in situations of normal hydration (72.7–74.3%), hyperhydration and dehydration preventing worsening of both kidney and heart function. In the recent past this method was performed in limited medical areas only – actually it is performed in many critical areas such as emergency medicine, intensive care, dialysis, and cardiology departments. In EDs, achieving a normal hydration in critical patients should be an important clinical target. Although BIVA did not show relevant results in initial studies, the method showed in most recent studies the utility in fluid balance management. In patients with a hyperhydration state due to HF, some authors showed that the reactance is strongly related with the BNP values and NYHA functional classes. Other authors found in critically ill patients a correlation between impedance and central venous pressure [11].

Therapeutic Strategies of Cardiorenal Syndromes in the Emergency Department

Although there are clinical guidelines for managing both HF and CKD, there are no agreed guidelines for managing patients with cardiorenal and/or renocardiac syndromes.

Acute Cardiorenal Syndrome

Both abnormal renal function and also an early deterioration in the course of treatment of ADHF increase mortality. Thus, any treatment for HF should have a neutral effect or preferably improve renal function. Vasodilators and loop diuretics are widely recommended in cases of ADHF and in type 1 CRS, paying attention to side effects such as electrolyte imbalance, hypovolemia and reduced renal glomerular rate. Vasodilators are recommended as first-line therapy for patients with acute HF in the absence of hypotension in addition to diuretic therapy for relief of symptoms. Vasodilators will decrease preload, afterload, or both. If congestion coincides with low blood pressure, inotropic agents (dopamine, dobutamine, milrinone and levosimendan) are recommended. Extracorporeal ultrafiltration may be useful in ADHF associated with diuretic resistance. It should be mentioned that overtreatment with loop diuretics, angiotensin-converting enzyme inhibitor (ACEI), and/or spironolactone may induce AKI.

Chronic Cardiorenal Syndrome

Therapeutic approaches involve the elimination and treatment of the underlying causes and CHF progression. ACEIs, β-blockers, angiotensin receptor blockers (ARBs), and aldosterone antagonists significantly reduce mortality and morbidity in CHF. Guidelines suggest to combine ACEI and β-blocker, titrate dosages, to which either an ARB or aldosterone antagonist is subsequently added depending on clinical condition. Digoxin and diuretics improve symptoms in CHF but have no effect on mortality. Cardiac resynchronization therapy is now recommended for symptomatic CHF patients (NYHA III-IV) with poor left ventricular ejection fraction and QRS prolongation. Therapy of CHF with concomitant renal impairment is still not evidence-based, as these patients are generally excluded from CHF trials. Typically, these patients are hypervolemic, and more intensive diuretic treatment is needed. Thiazide diuretics may be less effective and loop diuretics are preferred. To improve natriuresis, loop diuretic infusions are more potent, and combinations with amiloride, aldosterone antagonists or metolazone may be considered, as increasing doses of loop diuretics are associated with worse outcomes. In refractory cases, renal replacement therapy may be required. ACEI and ARB initiation may cause deterioration in renal function, which is frequently transient and reversible [8].

Acute Renocardiac Syndrome

Since a typical clinical scenario would include AKI following contrast exposure, or following cardiovascular surgery (CSA-AKI), prevention likely affords a better chance to improve outcome than treating established disease. To prevent contrast nephropathy, many potential preventive strategies have been studied, and available evidence indicates that isotonic fluids have been the most successful intervention to date, with conflicting data surrounding *N*-acetylcysteine [9].

Chronic Renocardiac Syndrome

The etiology of HF in CKD is multifactorial. The medical care of patients with CKD should focus on delaying or halting the progression of CKD by treating the underlying conditions if possible. Systolic blood pressure control is considered more important and is also considered difficult to control in elderly patients with CKD. Use of ACEIs or ARBs is useful if tolerated, with close monitoring for renal deterioration and for hyperkalemia. Data support the use of ACEIs/ARBs in diabetic kidney disease with or without proteinuria. However, in non-diabetic kidney disease, ACEIs/ARBs are effective in retarding the progression of disease among patients with proteinuria <500 mg/day. An appropriate glycemic control in diabetic patients with a target HbA1C <7%, a protein restriction and the hyperlipidemia are recommended in nephropathic patients. Renal replacement therapy may be required in non-responder patients [12].

Case Report: An Example of a Multimarker Approach in Clinical Practice

C.S., a 78-year-old female, was referred to the ED for shortness of breath, fatigue, columnar leg edema and severe abdominal ascites. She suffered from HF, moderate mitral insufficiency, bicameral PMK, hypertension, mild chronic

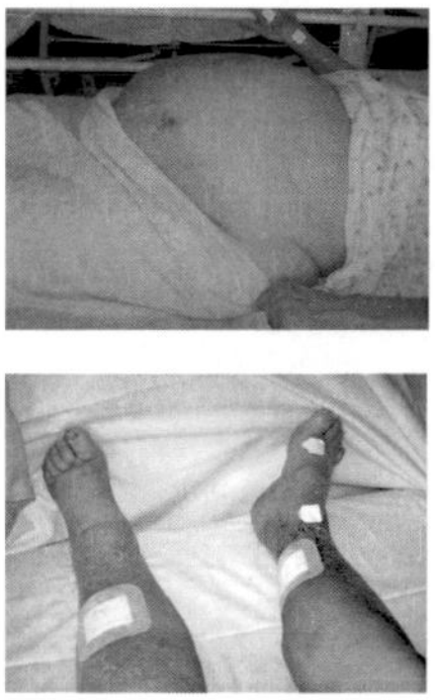

Tests	Before paracentesis	After paracentesis	Discharge (48 h after paracentesis)
NGAL ng/ml	233	148	141
NGAL pg/ml	776	654	356
Creatinine mg/dl	1.3	1.2	1.1
Hydration (BIVA) %	90.1%	92%	84.2%

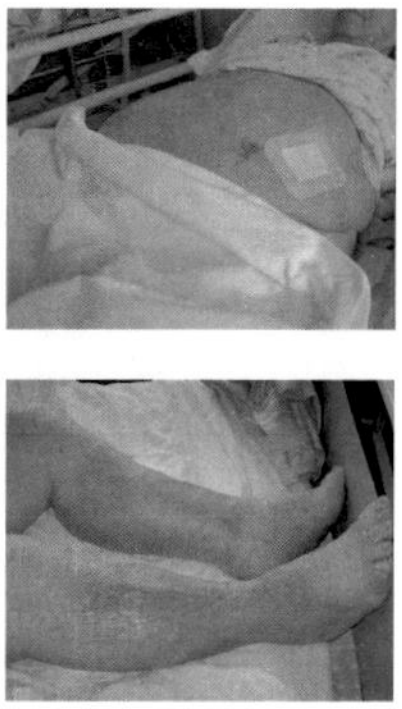

Fig. 1. A multimarker approach and BIVA before and after paracentesis and at discharge.

kidney failure, and hypothyroidism. At admission she presented blood pressure 135/65, respiratory rate 22/min, heart rate 60 bpm, and saturated O_2 100%. She underwent two blood exams for SCr 1.2 mg/dl (previous test creatinine 1.7 mg/dl) with GFR 48.5 ml/min, urea 34 mg/dl, BNP 807 pg/ml, Na^+ 135 mmol/l, and K^+ 4.4 mmol/l. At ECG she presented a PMK rhythm, while the echocardiogram revealed an increase in left ventricular dimension (dTD 57 mm), EF 50%, biatrial dilation, moderate mitral and tricuspidalic insufficiency. PAPs 30 mm Hg, no pericardial fluid. She also underwent evaluation of body fluid content by BIVA that showed 90.1% hydration and 169 ng/ml plasmatic NGAL. After 12 h of intensive diuretic therapy (high doses of furosemide) there was no urinary output improvement so that she underwent an evacuative and diagnostic paracentesis with a withdrawal of 6 liters of fluid. After this procedure we performed SCr, NGAL, BNP and BIVA measurements again at 12, 24 and 48 h and found a decrease of these biomarkers and a reduced congestion assessed by BIVA (fig. 1).

References

1 Fried LF, Shlipak MG, Crump C, Bleyer AJ, Gottdiener JS, Kronmal RA, Kuller LH, Newman AB: Renal insufficiency as a predictor of cardiovascular outcomes and mortality in elderly individuals. J Am Coll Cardiol 2003;41:1364–1372.

2 Bellomo R, Ronco C, Kellum JA, et al: Acute renal failure – definition, outcome measures, animal models, fluid therapy and information technology needs. The Second International Consensus Conference of the Acute Dialysis Quality Initiative (ADQI) Group. Crit Care 2004;8:R204–R212.

3 Acute Dialysis Quality Initiative Group: www.adqi.net (last accessed September 25, 2009).

4 Brenner & Rector Il Rene Oct 2002, Roma, Verduci Editore.
5 Xue JL, Daniels F, Star RA, Kimmel PL, Eggers PW, Molitoris BA, Himmelfarb J, Collins AJ: Incidence and mortality of acute renal failure in Medicare beneficiaries, 1992–2001. J Am Soc Nephrol 2006;17:1135–1142.
6 Ronco C, Haapio M, House AA, Anavekar N, Bellomo R: Cardiorenal syndrome. J Am Coll Cardiol 2008;52:1527–1539.
7 Breidthardt T, Mebazaa A, Mueller CE: Predicting progression in non-diabetic kidney disease: the importance of cardiorenal interactions. Kidney Int 2009;75:253–255.
8 Nickolas TL, O'Rourke MJ, Yang J, et al: Sensitivity and specificity of a single measurement of urinary neutrophil gelatinase associated-lipocalin for diagnosing acute kidney injury. Ann Intern Med 2008;148:810–819.
9 Ronco C, McCullough P, Anker SD, et al: Cardiorenal syndromes: report from the Consensus Conference of the Acute Dialysis Quality Initiative. Eur Heart J 2009 (in press).
10 Maisel A, Mueller C, Adams K Jr, Anker SD, Aspromonte N, Cleland JG, Cohen-Solal A, Dahlstrom U, DeMaria A, Di Somma S, Filippatos GS, Fonarow GC, Jourdain P, Komajda M, Liu PP, McDonagh T, McDonald K, Mebazaa A, Nieminen MS, Peacock WF, Tubaro M, Valle R, Vanderhyden M, Yancy CW, Zannad F, Braunwald E: State of the art: using natriuretic peptide levels in clinical practice. Eur J Heart Fail 2008;10:824–839.
11 Paterna S, Di Pasquale P, Parrinello G, et al: Changes in brain natriuretic peptide levels and bioelectrical impedance measurements in refractory congestive heart failure. J Am Coll Cardiol 2005;45:1997–2003..
12 Peterson JC, Adler S, Burkart JM, et al: Blood pressure control, proteinuria, and the progression of renal disease. The Modification of Diet in Renal Disease Study. Ann Intern Med 1995;123:754–762.

Prof. Salvatore Di Somma
Sapienza University, 2nd Faculty of Medicine, Ospedale S. Andrea
Via di Grottarossa 1035, IT–00189 Rome (Italy)
Tel. +39 06 337 75 581, Fax +39 06 337 75 890
E-Mail salvatore.disomma@ospedalesantandrea.it

Ronco C, Bellomo R, McCullough PA (eds): Cardiorenal Syndromes in Critical Care.
Contrib Nephrol. Basel, Karger, 2010, vol 165, pp 101–111

Prevention of Cardiorenal Syndromes

Peter A. McCullough

Department of Medicine, Divisions of Cardiology, Nutrition and Preventive Medicine, William Beaumont Hospital, Royal Oak, Mich., USA

Abstract

The cardiorenal syndromes (CRS) are composed of five recently defined syndromes which represent common clinical scenarios in which both the heart and the kidney are involved in a bidirectional injury process leading to dysfunction of both organs. Common to each subtype are multiple complex pathogenic factors, a precipitous decline in function and a progressive course. Most pathways that lead to CRS involve acute injury to organs which manifest evidence of chronic disease, suggesting reduced ability to sustain damage, maintain vital functions, and facilitate recovery. Prevention of CRS is an ideal clinical goal, because once initiated, CRS cannot be readily aborted, are not completely reversible, and are associated with serious consequences including hospitalization, complicated procedures, need for renal replacement therapy, and death. Principles of prevention include identification and amelioration of precipitating factors, optimal management of both chronic heart and kidney diseases, and future use of multimodality therapies for end-organ protection at the time of systemic injury. This paper will review the core concepts of prevention of CRS with practical applications to be considered in today's practice.

With the demographic changes in Western societies toward older, with proclivity towards type 2 diabetes and hypertension, there is increasing interest in the effects of combined chronic heart and kidney disease [1]. These changes are being rapidly driven by a progressive accumulation of excess adiposity experienced by most adults. It is estimated in the next few decades that over a third of adults will be diagnosed with type 2 diabetes mellitus, a principal common risk factor for both heart and kidney disease [2]. The number of individuals in the world who meet a definition of chronic kidney disease (CKD) based on a reduced estimated glomerular filtration rate or evidence of kidney damage by imaging studies or biomarkers, most commonly an elevated urine albumin:creatinine ratio, are expected to increase sharply over the next several decades. Since approximately half of all deaths in those with CKD are attributed to cardiovascular causes, and conversely, many cardiovascular hospitalizations are complicated by CKD and acute kidney injury (AKI), there is rationale to explore the intersection as a 'cardiovascular risk state' and understand the pathobiological evidence for changes that acutely affect both the heart and the kidney [3]. It should also be recognized that we are in the midst of a chronic heart failure (HF) epidemic [4]. There is increasing recognition that there is an overlap between CKD and HF, and in fact, a complex bidirectional pathophysiologic state [5] that worsens function of both organs as defined elsewhere in this article. The rationale for prevention of cardiorenal syndromes (CRS) is predicated on the concept that once the syndrome begins it is difficult to interrupt, is not completely reversible in all cases, and is associated with serious adverse outcomes including hospitalization, complicated procedures, need for renal replacement therapy, and death. Because the pathophysiology of CRS is believed to involve multiple mechanisms of organ injury and self-propagation, it is likely that multimodality preventive strategies working via several therapeutic targets will be needed (table 1). An approach to prevention is proposed to follow the classification system put forward by Ronco and colleagues [6, 7]. This discussion will utilize the RIFLE staging system [8] for the AKI components of CRS and the American College of Cardiology/American Heart Association (ACC/AHA) stages of HF [18].

Acute Cardiorenal Syndrome Complicating Decompensated Heart Failure

This syndrome is characterized by an initial abrupt change in cardiovascular function that leads to AKI, most commonly acutely decompensated, stage C, HF (ADHF). Since the cardiovascular event is primal, the prevention approach is to attenuate or potentially obviate the acute cardiac event in the first place. Of patients who present with ADHF, approximately one third are de novo cases in which the common precipitants are pneumonia, hypertension, atrial fibrillation, and acute cardiac ischemia [9, 10]. In the remaining two thirds with established HF, common reasons for decompensation include non-compliance with a low sodium diet and HF medications [11]. It is also recognized that some non-steroidal anti-inflammatory agents (NSAIDs) and thiazolidinediones can reduce renal blood flow, induce sodium and volume retention, and result in ADHF [12, 13]. An emerging complicating factor is obstructive sleep apnea syndrome in obese patients with HF, this syndrome works to superimpose right-sided pressure and volume overload and diastolic septal interactions on the left ventricle, further worsening ADHF. Of note, these patients have very high right-sided central and renal venous pressures which have been associated with the development of CRS [14]. The most important preventive approach in patients with de novo HF consists of the basic preventive strategies of the American Heart

Table 1. Potential normal regulatory and pathobiological targets for the prevention of CRS

Target	Rationale
Renal ischemic injury – Improve renal blood flow – Systemic and organ cooling	(1) Improve oxygen delivery and reduce tubular and interstitial damage
Neurohormonal activation – RAAS – Sympathetic nervous system – Endothelin – Tumor necrosis factor-α – Arginine vasopressin – Asymmetric dimethylarginine	(1) Excess neurohormonal activity leads to progression of LV and renal dysfunction by activating nuclear and cell pathophysiologic processes (2) A host of oral and parental agents are currently used in CKD and HF management
Chronic sodium and volume balance – 'Narrow optimal physiologic window'	(1) Volume depletion is a risk for AKI (2) Chronic volume overload contributes to the development of tissue and organ edema and may worsen both kidney and HF (3) Can be addressed with optimal outpatient diuretics (4) Inpatient strategies my include 'lower and slower' IV loop diuretic regimens or hemofiltration approaches
Inflammation – White blood cells – Interleukins – Antibodies – Complement	(1) Linked to plaque rupture and acute coronary syndromes (2) Associated with cardiac and renal fibrosis (3) Concurrent protein calorie malnutrition and cachexia strongly predicts mortality
Oxidative injury – Chelate trace metals (labile iron) that facilitate the transfer of hydroxyl groups and the generation of reactive oxygen species – Counteract oxygen free radicals with antioxidant compounds	(1) Secondary event in acute cardiac and organ injury that leads to greater zones of cell death (2) Labile iron chelators are in development (3) Weak oral/parental antioxidants are available
Adenosine endocrine/paracrine function – Primary regulator of renal blood flow	(1) Adenosine receptors modulate coronary and renal blood flow (2) Adenosine and its analogues can be given intravenously (3) Oral and IV adenosine receptor-modulating drugs are available
PG endocrine/paracrine function – Regulation of renal blood flow, sodium movement, and renal tubular function	(1) PG imbalance can lead to reduced renal blood flow, sodium retention, and cellular injury (2) Multiple PGs can be given orally or intravenously

Table 1. Continued

Target	Rationale
Natriuretic peptide endocrine/paracrine function – Augmentation of glomerular filtration, renal blood flow, natriuresis, diuresis, ventricular and vascular relaxation	(1) Endogenously protective of both heart and kidney function (2) Recombinant natriuretic peptides can be given intravenously and subcutaneously in supraphysiological doses
Chronic cardiac ischemia	(1) Possibly contributes to progressive LV diastolic and systolic dysfunction (2) Can be addressed with revascularization and antianginal therapy
Anemia – Combined iron re-utilization defect and impaired erythropoietin production – A partial contributor to renal ischemic injury and reduced O_2 delivery	(1) Associated with progression of kidney disease, LV hypertrophy, HF, and death (2) Can be reversed with supplemental iron and erythrocyte-stimulating agents
CKD bone and mineral disorder	(1) Hyperphosphatemia related to cardiovascular calcification, vascular stiffness, and mortality (2) Vitamin D deficiency and hyperparathyroidism associated with LV hypertrophy, cardiovascular events, and mortality (3) Can be partially controlled with phosphate binders, vitamin D analogues, and calcium-sensing receptor modulators
Uremia (elevated plasma urea, uric acid, phosphorus, etc.)	(1) Associated with multiple pathological alterations including decreased platelet aggregation, excess thrombin generation, myocyte dysfunction, damage to heart and cardiovascular tree and increased susceptibility to infection (2) Can be modulated with early dialytic techniques
Adiposity	(1) Associated with CKD and HF, related to increased myocyte accumulation of fatty acids, visceral deposits of epicardial fat, cardiac fibrosis, and diastolic dysfunction (2) Weight reduction has resulted in partial regression of LVH, improved diastolic function, and vascular compliance (3) Weight reduction and improved fitness confer a survival advantage in hospitalized patients after cardiac and general surgical procedures

Association/American College of Cardiology Stage A and B HF. These call for blood pressure control, use of drugs that block the renin-angiotensin-aldosterone system (RAAS), β-adrenergic blockers, and coronary artery disease risk factor modification. In addition, the treatment of chronic ischemic coronary disease and the reduction or elimination of coronary ischemia may be beneficial in HF patients, particularly those with large segments of hibernating myocardium [15]. In patients with established HF, a number clinical trials have demonstrated the HF care management is beneficial by emphasizing patient education, weight monitoring and responsive medical management, and compliance with diet and drugs [16–18]. In an extensive meta-analysis of 29 trials (5,039 patients), there was a 26% decrement in HF hospitalizations in the active intervention group when compared with controls [19]. Even remote telephone monitoring has been effective as shown in 14 trials in 4,264 patients resulting in a 21% decrease hospitalization and a 20% reduction on all-cause mortality [20]. It is believed a subset of patients in these trials had the CRS averted with this approach. In the setting incipient cardiogenic shock, the risk of acute cardiorenal syndrome (type 1) is high and can be only partially managed with prompt revascularization and hemodynamic support with inotropic agents, vasopressors, and intra-aortic balloon counterpulsation [21]. Thus the primary prevention of acute cardiorenal syndrome (type 1) occurs in the outpatient realm using evidence-based approaches to avoid decompensation and hospitalization of HF and prevention of superimposed coronary ischemia through revascularization and risk factor modification [22]. In those patients hospitalized with ADHF, prompt treatment is likely to prevent significant deterioration in renal function [23].

Intravenous Loop Diuretics as Precipitants of Cardiorenal Syndromes

There is considerable controversy concerning the role of intravenous (IV) loop diuretics in the precipitation of CRS. These drugs are nearly universally used in the setting of ADHF and are the one single departure from home medical therapy. Interestingly, CRS is rarely observed to occur in the home or outpatient setting. In the Finnish Acute Heart Failure Study (FINN-AKVA) study, of the 15.8% of ADHF patients who developed CRS after hospitalization, the baseline serum creatinine and cystatin C levels were 0.98 mg/dl and 1.44 mg/l, respectively [24]. IV loop diuretics induce a reduction in renal perfusion pressure, activation of the renin-angiotensin and sympathetic nervous systems, while facilitating natriuresis, diuresis, and relief of pulmonary congestion. These changes may be sufficiently abrupt to trigger the onset of CRS. High-dose versus low-dose, and rapid versus slower loop diuretic management strategies have been the subject of small, randomized trials. In one randomized trial involving patients presenting to the emergency department with ADHF and respiratory distress compared high-dose furosemide (80-mg bolus every 15 min) plus low-dose isosorbide dinitrate (1 mg/h doubled every 10 min) to high-dose isosorbide dinitrate (3-mg bolus every 5 min) alone. All patients initially received oxygen therapy and a 40-mg bolus of furosemide before being randomly assigned to the treatment groups; treatment was continued until oxygen saturation was 96% or the mean arterial pressure decreased by 30% or to <90 mm Hg. Among the 104 patients enrolled, mechanical ventilation was required by more patients in the high-dose furosemide group (40 vs. 13%, $p < 0.004$) [25]. The Dopamine in Acute Decompensated Heart Failure (DAD-HF) pilot trial enrolled 50 consecutive patients with ADHF and randomly assigned them to either a high dose of furosemide ($n = 25$) or a low dose of furosemide plus a low dose of dopamine ($n = 25$) [26]. Efficacy endpoints included changes in urine volume at 24 h, improvement of dyspnea at 24 h, worsening renal function at 24 h and 60-day mortality or rehospitalization from cardiovascular and other causes. There were no changes in hourly urine volume between the study groups for the duration of the study. Dyspnea improved at 24 h compared with baseline measurements in both the high-dose group (Borg scores 7.47 to 2.84; $p < 0.01$) and the low-dose group (7.17 to 2.50; $p < 0.01$), and there was an increase in blood urea nitrogen in the high-dose group (from 43.2 to 51.2 mg/dl; $p < 0.01$). Type 1 CRS occurred in 9 patients (36%) in the high-dose group versus 1 patient (4%) in the low-dose group ($p = 0.005$) and there were no differences in 60-day clinical outcomes. High-dose versus low-dose furosemide is one of several featured strategies listed in HF trials found in the ClinicalTrials.gov database. Most notably, the National Heart Lung and Blood Institute is sponsoring the Determining Optimal Dose and Duration of Diuretic Treatment in People with Acute Heart Failure (The DOSE-AHF Study) and will recruit 300 patients and randomize to (1) intensification ($2.5 \times$ oral dose) of IV furosemide by either q12 h bolus or continuous infusion or (2) low intensification ($1 \times$ oral dose) IV furosemide, again by either q12 h bolus or continuous infusion (ClinicalTrials.gov Identifier: NCT00577135).

In the mean time, the observational data, confounded by indication, clearly indicate that those patients who receive the highest incremental and cumulative doses of loop diuretics are at highest risk of CRS and subsequent mortality. If proven by randomized trials and allowed by payment systems, a future approach to the management of CRS, provided the patient is not in extremis, may include a more prolonged hospital stay, lower doses of loop diuretics, and a slower diuresis period allowing for the adjustment of disease-modifying drugs including β-blockers, agents that inhibit the RAAS. An integrated approach for practical management of ADHF is depicted in figure 1.

Chronic Cardiorenal Syndrome Influencing Outpatient Heart Failure Care

Chronic cardiorenal syndrome (type 2) relates to the common scenario in which longstanding HF leads to progressive CKD, possibly via episodes of intermittent

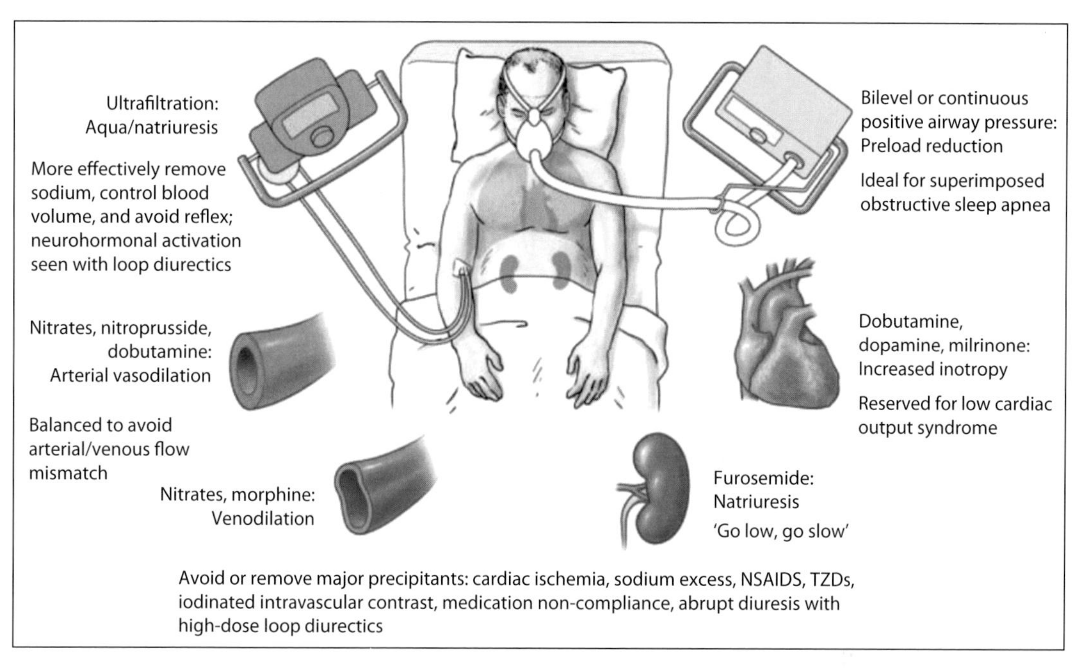

Fig. 1. Practical core prevention concepts in CRS. NSAIDs = Non-steroidal anti-inflammatory drugs; TZDs = thiazolidinediones.

AKI. In this setting, the therapies that improve the natural history of chronic HF include angiotensin-converting enzyme inhibitors, angiotensin receptor blockers, β-adrenergic blockers, and aldosterone receptor blockers, combination of nitrates and hydralazine, and cardiac resynchronization therapy [27, 28]. An important concept in prevention of chronic cardiorenal syndrome (type 2) is optimal management of sodium and extracellular fluid volume through low sodium diet and diuretics. Unlike ADHF, chronic diuretics in some form are likely beneficial in outpatient HF. In a meta-analysis of randomized controlled trials of diuretics in outpatients with HF, the three placebo-controlled trials reporting mortality data showed that the rate of death was lower among patients given diuretics than among those given placebo (n = 221, odds ratio 0.25, 95% CI 0.07–0.84) [29]. However, as should be recognized throughout this paper, therapies for stable congestive HF may not be as effective or even pose specific risks for CRS when applied to inpatient management of ADHF. Many studies indicate that the lowest chronic doses of loop diuretics necessary to maintain hemodynamics are optimal as discussed above [30]. Conversely, those patients requiring the highest doses of loop diuretics have the highest rates of mortality probably by further activation of neurohumoral pathways [31, 32]. It is important to consider the concept of diuretic resistance, being a confounder for high-dose diuretics and an indicator of the patient at high risk for mortality [33]. Randomized trials are planned to test both short and longer acting loop diuretics as well as the dosing strategy in the chronic management of HF [34]. Both hyper- and hypotension can precipitate CRS, but the optimal blood pressure window for patients with HF has not been defined [35, 36].

In chronic HF, there can be maladaptive changes including anemia-associated increases in cytokines and inflammatory factors which have been associated to both worsened cardiac and renal function [24]. By reducing oxygen delivery to the kidney it is conceivable that with low output HF, chronic renal ischemia could be a precipitant for AKI [37]. Small clinical trials using erythropoiesis-stimulating agents have demonstrated improved peak oxygen consumption, reduced left ventricular (LV) mass, and improved ejection fraction in patients with combined HF and CKD [38]. The large RED-HF Trial using darbepoetin to raise hemoglobin in chronic HF will hopefully give better guidance on both heart and renal outcomes in patients with stable systolic HF [39]. Many other prevention and therapeutic targets are listed in table 1. It is unlikely that manipulation of a single target will completely eliminate the risk of CRS, however it is likely that simultaneous management of multiple systems will prove the optimal guidance for the avoidance of CRS and reduced inpatient morbidity and mortality.

Lastly, avoidance of added insults (from the use of for example iodinated contrast media, NSAIDs, thiazolidinediones, and other nephrotoxic agents) to vulnerable patients is a key prevention measure [40]. Chronic HF is a risk factor for contrast-induced AKI, making the possibility of an intravascular contrast procedure tipping a patient into chronic cardiorenal syndrome (type 2) a possibility [41]. In chronic renal hypoperfusion states, renal blood flow is dependent on intact ability to synthesize prostaglandins (PGs) and maintain afferent arteriolar flow. Thus, NSAIDs, which impair the constitutive and inducible production of PGs, have been associated with worsened HF and likely CRS [42].

Conclusions

Type 1 and 2 CRS account for major morbidly in ADHF and chronic HF patients. The avoidance of CRS renders the HF patient much more amenable to evidence-based management with pharmaceutical and device treatments. There are no proven treatments for CRS once it develops in the setting of HF. Thus, prevention of these syndromes through avoidance of precipitants possibly including rapid diuresis with high-dose loop diuretics and leveraging of disease-modifying therapy is the best approach at this time. Future strategies testing the conventional wisdom of IV diuretics, novel therapies regulating renal hemodynamics, and early forms of volume control with renal replacement therapy hold the hope for preventing the initiation of CRS and facilitating management of this common cardiovascular condition.

References

1 Whaley-Connell AT, Sowers JR, McFarlane SI, Norris KC, Chen SC, Li S, Qiu Y, Wang C, Stevens LA, Vassalotti JA, Collins AJ, Kidney Early Evaluation Program Investigators: Diabetes mellitus in CKD: Kidney Early Evaluation Program (KEEP) and National Health and Nutrition and Examination Survey (NHANES) 1999–2004. Am J Kidney Dis 2008;51(suppl 2):S21–S29.

2 Narayan KM, Boyle JP, Thompson TJ, Sorensen SW, Williamson DF: Lifetime risk for diabetes mellitus in the United States. JAMA 2003;290:1884–1890.

3 McCullough PA, Li S, Jurkovitz CT, Stevens LA, Wang C, Collins AJ, Chen SC, Norris KC, McFarlane SI, Johnson B, Shlipak MG, Obialo CI, Brown WW, Vassalotti JA, Whaley-Connell AT, Kidney Early Evaluation Program Investigators: CKD and cardiovascular disease in screened high-risk volunteer and general populations: the Kidney Early Evaluation Program (KEEP) and National Health and Nutrition Examination Survey (NHANES) 1999–2004. Am J Kidney Dis 2008;51(suppl 2):S38–S45.

4 McCullough PA, Philbin EF, Spertus JA, Kaatz S, Sandberg KR, Weaver WD: Confirmation of a heart failure epidemic: findings from the Resource Utilization Among Congestive Heart Failure (REACH) Study. J Am Coll Cardiol 2002;39:60–69.

5 McCullough PA: Cardiorenal risk: an important clinical intersection. Rev Cardiovasc Med 2002;3:71–76.

6 Ronco C, House AA, Haapio M: Cardiorenal syndrome: refining the definition of a complex symbiosis gone wrong. Intensive Care Med 2008;34:957–962.

7 Ronco C, Haapio M, House AA, Anavekar N, Bellomo R: Cardiorenal syndrome. J Am Coll Cardiol 2008;52:1527–1539.

8 Endre ZH: Acute kidney injury: definitions and new paradigms. Adv Chronic Kidney Dis 2008;15:213–221.

9 Opasich C, Rapezzi C, Lucci D, Gorini M, Pozzar F, Zanelli E, Tavazzi L, Maggioni AP, Italian Network on Congestive Heart Failure (IN-CHF) Investigators: Precipitating factors and decision-making processes of short-term worsening heart failure despite 'optimal' treatment (from the IN-CHF Registry). Am J Cardiol 2001;88:382–387.

10 Fonarow GC, Abraham WT, Albert NM, Stough WG, Gheorghiade M, Greenberg BH, O'Connor CM, Pieper K, Sun JL, Yancy CW, Young JB, OPTIMIZE-HF Investigators and Hospitals: Factors identified as precipitating hospital admissions for heart failure and clinical outcomes: findings from OPTIMIZE-HF. Arch Intern Med 2008;168:847–854.

11 Ghali JK, Kadakia S, Cooper R, Ferlinz J: Precipitating factors leading to decompensation of heart failure. Traits among urban blacks. Arch Intern Med 1988;148:2013–2016.

12 Krum H, Curtis SP, Kaur A, Wang H, Smugar SS, Weir MR, Laine L, Brater DC, Cannon CP: Baseline factors associated with congestive heart failure in patients receiving etoricoxib or diclofenac: multivariate analysis of the MEDAL program. Eur J Heart Fail 2009;11:542–550.

13 Home PD, Pocock SJ, Beck-Nielsen H, Curtis PS, Gomis R, Hanefeld M, Jones NP, Komajda M, McMurray JJ, RECORD Study Team: Rosiglitazone evaluated for cardiovascular outcomes in oral agent combination therapy for type 2 diabetes (RECORD): a multicentre, randomised, open-label trial. Lancet 2009;373:2125–2135.

14 Mullens W, Abrahams Z, Francis GS, Sokos G, Taylor DO, Starling RC, Young JB, Tang WH: Importance of venous congestion for worsening of renal function in advanced decompensated heart failure. J Am Coll Cardiol 2009;53:589–596.

15 Shanmugam G, Légaré JF: Revascularization for ischaemic cardiomyopathy. Curr Opin Cardiol 2008;23:148–152.

16 Giordano A, Scalvini S, Zanelli E, Corrà U, Longobardi GL, Ricci VA, Baiardi P, Glisenti F: Multicenter randomised trial on home-based telemanagement to prevent hospital readmission of patients with chronic heart failure. Int J Cardiol 2009;131:192–199.

17 López Cabezas C, Falces Salvador C, Cubí Quadrada D, Arnau Bartés A, Ylla Boré M, Muro Perea N, Homs Peipoch E: Randomized clinical trial of a postdischarge pharmaceutical care program vs. regular follow-up in patients with heart failure. Farm Hosp 2006;30:328–342.
18 Murray MD, Young J, Hoke S, Tu W, Weiner M, Morrow D, Stroupe KT, Wu J, Clark D, Smith F, Gradus-Pizlo I, Weinberger M, Brater DC: Pharmacist intervention to improve medication adherence in heart failure: a randomized trial. Ann Intern Med 2007;146:714–725.
19 McAlister FA, Stewart S, Ferrua S, McMurray JJ: Multidisciplinary strategies for the management of heart failure patients at high risk for admission: a systematic review of randomized trials. J Am Coll Cardiol 2004;44:810–819.
20 Clark RA, Inglis SC, McAlister FA, Cleland JG, Stewart S: Telemonitoring or structured telephone support programmes for patients with chronic heart failure: systematic review and meta-analysis. BMJ 2007;334:942.
21 Gowda RM, Fox JT, Khan IA: Cardiogenic shock: basics and clinical considerations. Int J Cardiol 2008;123:221–228.
22 Becker RC, Meade TW, Berger PB, Ezekowitz M, O'Connor CM, Vorchheimer DA, Guyatt GH, Mark DB, Harrington RA, American College of Chest Physicians: The primary and secondary prevention of coronary artery disease: American College of Chest Physicians Evidence-Based Clinical Practice Guidelines, ed 8. Chest 2008;133(suppl):S776–S814.
23 Fonarow GC: Acute decompensated heart failure: challenges and opportunities. Rev Cardiovasc Med 2007;8(suppl 5):S3–S12.
24 Lassus J, Harjola VP, Sund R, Siirilä-Waris K, Melin J, Peuhkurinen K, Pulkki K, Nieminen MS, for the FINN-AKVA Study Group: Prognostic value of cystatin C in acute heart failure in relation to other markers of renal function and NT-proBNP. Eur Heart J 2007;28:1841–1847.
25 Cotter G, Metzkor E, Kaluski E, et al: Randomised trial of high-dose isosorbide dinitrate plus low-dose furosemide versus high-dose furosemide plus low-dose isosorbide dinitrate in severe pulmonary oedema. Lancet 1998;351:389–393.
26 Triposkiadis F: The Dopamine in Acute Decompensated Heart Failure (DAD-HF) pilot trial LBCT. II. 13th Annual Scientific Meeting of the Heart Failure Society of America, Boston, Mass. 2009.
27 Hunt SA, American College of Cardiology; American Heart Association Task Force on Practice Guidelines (Writing Committee to Update the 2001 Guidelines for the Evaluation and Management of Heart Failure). ACC/AHA 2005 guideline update for the diagnosis and management of chronic heart failure in the adult: a report of the American College of Cardiology/American Heart Association Task Force on Practice Guidelines (Writing Committee to Update the 2001 Guidelines for the Evaluation and Management of Heart Failure). J Am Coll Cardiol 2005;46:e1–e82.
28 Boerrigter G, Costello-Boerrigter LC, Abraham WT, St John Sutton, MG, Heublein D, Kruger KM, Hill MR, McCullough PA, Burnett JC: Cardiac resynchronization therapy with biventricular pacing improves renal function in heart failure patients with reduced glomerular filtration rate. Circulation 2007;116:ii–405.
29 Faris R, Flather M, Purcell H, et al: Current evidence supporting the role of diuretics in heart failure: a meta-analysis of randomised controlled trials. Int J Cardiol 2002;82:149–158.
30 Hasselblad V, Gattis Stough W, Shah MR, Lokhnygina Y, O'Connor CM, Califf RM, Adams KF Jr: Relation between dose of loop diuretics and outcomes in a heart failure population: results of the ESCAPE trial. Eur J Heart Fail 2007;9:1064–1069.
31 Metra M, Nodari S, Parrinello G, Bordonali T, Bugatti S, Danesi R, Fontanella B, Lombardi C, Milani P, Verzura G, Cotter G, Dittrich H, Massie BM, Dei Cas L: Worsening renal function in patients hospitalised for acute heart failure: clinical implications and prognostic significance. Eur J Heart Fail 2008;10:188–195.
32 Ahmed A, Husain A, Love TE, Gambassi G, Dell'Italia LJ, Francis GS, Gheorghiade M, Allman RM, Meleth S, Bourge RC: Heart failure, chronic diuretic use, and increase in mortality and hospitalization: an observational study using propensity score methods. Eur Heart J 2006;27:1431–1439.

33 Neuberg GW, Miller AB, O'Connor CM, et al, PRAISE Investigators: Diuretic resistance predicts mortality in patients with advanced heart failure. Am Heart J 2002;144:31–8.
34 J-MELODIC Program Committee: Rationale and design of a randomized trial to assess the effects of diuretics in heart failure: Japanese Multicenter Evaluation of Long- vs. Short-Acting Diuretics in Congestive Heart Failure (J-MELODIC). Circ J 2007;71:1137–1140.
35 Goebel JA, Van Bakel AB: Rational use of diuretics in acute decompensated heart failure. Curr Heart Fail Rep 2008;5:153–162.
36 Silvers SM, Howell JM, Kosowsky JM, Rokos IC, Jagoda AS, American College of Emergency Physicians: Clinical policy: critical issues in the evaluation and management of adult patients presenting to the emergency department with acute heart failure syndromes. Ann Emerg Med 2007;49:627–669.
37 McCullough PA, Lepor NE: The deadly triangle of anemia, renal insufficiency, and cardiovascular disease: implications for prognosis and treatment. Rev Cardiovasc Med 2005;6:1–10.
38 Jones M, Schenkel B, Just J: Epoetin-alfa's effect on left ventricular hypertrophy and subsequent mortality. Int J Cardiol 2005;100:253–265.
39 Anand IS: Anemia and chronic heart failure implications and treatment options. J Am Coll Cardiol 2008;52:501–511.
40 Mukherjee D: Non-steroidal anti-inflammatory drugs and the heart: what is the danger? Congest Heart Fail 2008;14:75–82.
41 McCullough PA, Adam A, Becker CR, Davidson C, Lameire N, Stacul F, Tumlin J, CIN Consensus Working Panel: Risk prediction of contrast-induced nephropathy. Am J Cardiol 2006;98(suppl 1):27–36.
42 DiBona GF: Prostaglandins and non-steroidal anti-inflammatory drugs. Effects on renal hemodynamics. Am J Med 1986;80:12–21.

Peter A. McCullough, MD, MPH
Division of Nutrition and Preventive Medicine, William Beaumont Hospital
4949 Coolidge, Royal Oak, MI 48073 (USA)
Tel. +1 248 655 5948, Fax +1 248 655 5901
E-Mail peteramccullough@gmail.com

Ronco C, Bellomo R, McCullough PA (eds): Cardiorenal Syndromes in Critical Care.
Contrib Nephrol. Basel, Karger, 2010, vol 165, pp 112–128

Acute Heart Failure Treatment: Traditional and New Drugs

Mihai Gheorghiade[a] · Alberto Palazzuoli[b] · Claudio Ronco[c]

[a]Division of Cardiology, Northwestern University Feinberg School of Medicine, Chicago, Ill., USA; [b]Department of Internal Medicine and Metabolic Diseases, Cardiology Section University of Siena, Siena, and [c]Department of Nephrology, Dialysis & Transplantation, International Renal Research Institute, San Bortolo Hospital, Vicenza, Italy

Abstract

Goals of treatment in acute heart failure (AHF) are different with regard to the clinical presentation, etiology and hemodynamic profiles. Therefore AHF is a colorful picture that encompasses several syndromes with particular characteristics. Unfortunately the protocol treatments proposed until now regarding old and new drugs, showed neutral or negative results in most of trials. This could depend on patient population selection, syndrome heterogeneity, drugs used, administration and posology, study design and study target. Most of the trials showed positive results in the early period but any improvement in mortality and re-hospitalization at a later stage. In this review we summarize the drugs more commonly used in AHF syndromes on the basis of pressure values, organ perfusion, pulmonary and venous congestion. However, a significant improvement in outcome needs to consider not only the acute phase but the effect of each drug on cardiac damage, organ perfusion, neurohormonal activation and renal injury after a short period. The most important goal of treatment should be an outcome improvement at the post-discharge phase. This could be obtained in the future assuming an 'holistic' point of view looking for all the potential consequences and benefits at both cardiac and systemic levels.

Hospitalization for acute heart failure syndromes (AHFS) is one of the most important predictors of post-discharge mortality and readmission in patients with chronic heart failure (HF). Management of AHFS is challenging given the heterogeneity of the patient population, absence of a universally accepted definition, incomplete understanding of its pathophysiology, and lack of robust evidence-based guidelines. The majority of patients appear to respond well to initial therapies consisting of loop diuretics and vasoactive agents [1]. However, post-discharge mortality and rehospitalization rates reach 10–20 and 20–30%, respectively, within 3–6 months. Although this may reflect the severity of HF, myocardial injury and/or renal impairment occurring in AHFS may contribute to this grim prognosis. Improving post-discharge mortality and prevention of readmissions are the most important goals in AHFS. This syndrome can be defined as new onset or gradual or rapidly worsening HF signs and symptoms requiring urgent therapy. Regardless of the underlying cause (e.g., ischemic event) or precipitant (e.g., severe hypertension), pulmonary and systemic congestion due to elevated ventricular filling pressures with or without a decrease in cardiac output is a nearly universal finding in AHFS. Coronary artery disease, hypertension, valvular heart disease, and/or atrial fibrillation, as well as noncardiac conditions such as renal dysfunction, diabetes, anemia, and medications (i.e., nonsteroidal anti-inflammatory drugs, glitazones), may also contribute to these abnormalities [1–3].

Evaluation Phases and Aims in AHFS Patients

Early Phase

This phase of AHFS management typically takes place in the emergency department, where 80% of all hospitalized patients initially present. Evaluation and management often proceeds concomitantly. After stabilization/treatment of life-threatening conditions, improving hemodynamics and symptoms are key goals. Abnormal hemodynamics often results from conditions such as hypertension, ischemia, and/or arrhythmias. These conditions, as well as any other precipitants of HF, should be treated for optimal results. The downstream impact of early therapy on outcomes for AHFS has not been well studied. Intravenous loop diuretics with or without vasoactive agents (inotropes and/or vasodilators) improve symptoms in most patients. The potential deleterious effects of these therapies, if any, on the myocardium and kidney have not been well studied [1]. Intravenous inotropes and vasodilators that initially improve signs and symptoms may adversely affect post-discharge outcomes (fig. 1).

In-Hospital Phase

Further improvement in signs and symptoms, achieving euvolemia, and targeted initiation and/or up-titration of evidence-based therapies for chronic HF based on a comprehensive assessment are the goals of this phase [2, 4].

Early phase	In-hospital phase	Pre-discharge phase
• Improve symptoms • Restore oxygenation • Improve organ perfusion and hemodynamics • Limit cardiac/renal damage • Minimize ICU length of stay	• Stabilize patient and optimize treatment strategy • Initiate appropriate (life-saving) pharmacological therapy • Consider device therapy in appropriate patients • Minimize hospital length of stay	• Plan follow-up strategy • Educate and initiate appropriate lifestyle adjustments • Provide adequate secondary prophylaxis • Prevent early readmission • Improve quality of life and survival

Fig. 1. Phases in acute heart failure management.

Monitoring for potential cardiac injury and renal function is important. The role of serial B-type natriuretic peptide (BNP)/N-terminal pro-BNP measurements in this setting remains to be determined. Because a dissociation between clinical (dyspnea, edema) and hemodynamic congestion (high left ventricular filling pressure) may be present after initial therapy, assessment of filling pressures is important. The level of BNP/N-terminal pro-BNP has also been proposed as a 'measure' of congestion. A tailored approach with evidence-based therapy in response to BNP levels in chronic HF was associated with better outcomes in the outpatient setting [5]. This approach remains to be investigated in AHFS. Currently, evidence and/or guidelines to assess congestion during hospitalization or pre-discharge are not well established. Refractory or advanced HF should be managed according to published guidelines [6]. Thromboembolic events and myocardial ischemia should be considered in patients not responding to standard therapy. A thorough assessment to ensure implementation of evidence-based guidelines (pharmacological, surgical, interventional, and implantable cardiac defibrillator/cardiac resynchronization therapy) should occur during this phase or soon after discharge. The Acute Decompensated National Heart Failure Registry (ADHERE) and Organized Program to Initiate Life-Saving Treatment in Hospitalized Patients with Heart Failure (OPTIMIZE-HF) registries demonstrated the relative paucity of comprehensive assessment [7, 8]. Hospitalization presents opportunities to optimize management, given the resources available in hospital versus in an outpatient setting. The traditional focus during hospitalization has been on alleviating congestion (e.g., improving symptoms and decreased body weight), rather than optimization of therapies known to improve outcomes in patients. Appropriate management of comorbidities (e.g., coronary artery disease, atrial fibrillation, hypertension, diabetes mellitus) based on evidence-based guidelines may also improve post-discharge outcomes. Of current HF quality measures (angiotensin-converting enzyme, ACE, inhibitor/angiotensin receptor blocker, ARB, anticoagulant at discharge for HF patients with atrial fibrillation, assessment of the ejection fraction (EF), smoking cessation, and adequate discharge instructions), only ACE inhibitor/ARB has been shown to improve outcomes in AHFS [7].

Pre-Discharge Phase

The goals at discharge are to improve signs and symptoms, to preserve euvolemia with successful transition to oral diuretics, to implement current HF guidelines, and to plan the post-discharge phase with clear instructions regarding weight monitoring, medications, and telephone and clinic follow-up.

Early Post-Discharge Phase ('Vulnerable' Phase)

Recent data have demonstrated deterioration in signs and symptoms, neurohormonal profile, and renal function during the first few weeks after discharge in patients who die or are rehospitalized within 60–90 days [9]. This deterioration occurs despite standard therapy, including β-blockers, ACE inhibitors, or ARB, and often aldosterone-blocking agents [9]. Assessment of these variables in the early post-discharge period may provide unique opportunities to further optimize standard therapy (up-titration) and/or introduce additional therapy known to improve outcomes (e.g., hydralazine/nitrates, aldosterone-blocking agents, cardiac resynchronization therapy). In addition, the use of novel intravenous therapies that are known to improve hemodynamics or to preserve myocardial and renal function should be studied in this vulnerable period [10].

Treatment Options According to Blood Pressure Values

Patients with AHFS can be classified into 3 groups based on their systolic blood pressure levels at the time of presentation. (1) The hypertensive group: these patients are more likely to be female and have preserved systolic function. Their inhospital mortality rate is approximately 2%, with 5% mortality and 30% readmission rates within 60–90 days of discharge. (2) The normotensive group: these patients tend to have a lower left ventricular (LV) EF (LVEF) and signs and symptoms of pulmonary/systemic congestion (edema) before and at the time of admission. Their inhospital mortality rate is approximately 3%, with 7% mortality and 30% readmission rates within 60–90 days of discharge. (3) The hypotensive group: this group consists of patients with systolic blood pressure levels of <120 mm Hg at the time of admission. In general, they have a very low LVEF, and >90% have a history of HF. These 3 groups not only differ prognostically but also require pharmacologic treatments tailored accordingly.

Available Agents for Patients with AHFS Presenting with Congestion and High or Normal Blood Pressure

Most patients admitted to hospital for AHFS present with systemic and/or pulmonary congestion associated with normal-to-high systolic blood pressure. Treatments for this group of patients primarily target congestion (increased pulmonary capillary wedge pressure, PCWP). Pharmacologic agents that are commonly used in this setting include diuretics, vasodilators (e.g., nitroglycerin, nitroprusside, nesiritide), and intravenous ACE inhibitors.

Diuretics
Loop diuretics (furosemide, bumetanide, torsemide) represent the first-line approach to rapidly reduce fluid overload and relieve symptoms. These drugs block the sodium-potassium-chloride transporter and increase the urinary excretion of sodium, potassium, chloride, and hydrogen ions, and, consequently, free water. They may be associated with adverse effects including hypotension, electrolyte abnormalities (potentially leading to arrhythmia), renal dysfunction, and neurohormonal abnormalities, especially when used at high doses. Despite their clinical use for decades, large-scale randomized controlled trials, to define the best strategy for their use and their effects on clinical outcomes, have not been conducted. Clear dosing guidelines are lacking, and dosing choices and administration methods tend to be empiric. Continuous infusion of furosemide preceded by a loading dose has been shown to be superior to a single or intermittent intravenous bolus injection of an equal dosage in small studies. The Heart Failure Society of America (HFSA) and the European Society of Cardiology (ESC) guidelines on the use of diuretics are based on recommendation classes and levels of evidence. The ESC 2005 guideline on acute HF24 states, 'Administration of diuretics is indicated in patients with acute and acutely decompensated HF in the presence of symptoms secondary to fluid retention *[classI/level B]*. Intravenous administration of loop diuretics with a strong and brisk diuretic effect is the preferred choice in patients with acute HF *[class IIb/level C]*. Thiazides and spironolactone can be used in association with loop diuretics, the combination in low doses being more effective and having fewer secondary effects than the use of higher doses of a single drug (class IIb/level C). Combination of loop diuretics with dobutamine, dopamine, or nitrates is also a therapeutic approach that is more effective and produces fewer secondary effects than increasing the dose of the diuretic *[class Iib/level C]*'.

Vasodilators
Nitrates. Nitroglycerin affects the vascular smooth muscle by stimulating guanylate cyclase, leading to venodilation at low doses (30–40 μg/min) and arteriolar dilation at higher doses (250 μg/min). It reduces left venticular filling

pressures and pulmonary congestion without compromising stroke volume or increasing myocardial oxygen demand. The beneficial hemodynamic effects of intravenous nitroglycerin are associated with neurohumoral activation and are often limited by the development of pharmacologic tolerance, especially with continuous long-term use. The adverse effects of nitroglycerin therapy include headache, hypotension, and methemoglobinemia (rare) [11].

Nitroprusside. Via its active metabolite nitric oxide, nitroprusside has equally potent vasodilatory effects on arterial and venous vessels [11]. The reduction in preload and afterload leads to decreased LV filling pressures and usually increased LV stroke volume and cardiac output. The adverse effects include hypotension and cyanide toxicity. Early studies involving small cohorts of patients presenting with HF or acute myocardial infarction complicated by an increase in PCWP documented favorable hemodynamic effects.

Nesiritide. The natriuretic peptide (NP) family consists of atrial (A-type), brain (B-type), and C-type NPs with many cardiovascular, renal, and neurohormonal actions that contribute to cardiovascular homeostasis [12]. Among these effects, they prevent excess salt and water retention, promote vascular relaxation, and suppress sympathetic outflow. Nesiritide is a recombinant form of human BNP. Nesiritide exerts vasodilatory effects on arterial, venous, and coronary vessels, leading to increased cardiac output. The adverse effects include hypotension, headache, and renal dysfunction. A meta-analysis showed that nesiritide significantly increased the risk of worsening renal function (defined as an increase in serum creatinine of >0.5 mg/dl, compared with inotrope- and noninotrope-based therapies [12]. The ESC 2005 guideline on acute HF comments on nitrates and nitroprusside (but not on nesiritide): 'Two randomized trials in acute HF have established the efficacy of intravenous nitrates in combination with furosemide and have demonstrated that titration to the highest hemodynamically tolerable dose of nitrates with low dose furosemide is superior to high dose diuretic treatment alone *[class 1/ level B]*. Sodium nitroprusside (0.3 μg/kg/min up-titrating carefully to 1–5 μg/kg/min) is recommended in patients with severe HF, and in patients with predominantly increased afterload such as hypertensive HF or mitral regurgitation *[class I/level C]*'.

Uralitide. Uralitide is a synthetic form of urodilatin. Urodilatin is a renal NP. Ularitide has similar effects to BNP by inducing diuresis, natriuresis and vasodilation. In healthy volunteers, ularitide infusion reduces cardiac preload, pulmonary arterial pressure, PCWP and systemic vascular resistance and improves the cardiac index and renal filtration fraction by increasing glomerular filtration rate, suggesting a potential benefit to patients with AHFS. The clinical efficacy of short-term infusion of ularitide in patients with ADHF was first investigated in a pilot study (SIRIUS I) [13], followed by SIRIUS II, a double-blind, parallel-dose phase II trial in which all three ularitide doses compared with placebo are associated with significantly greater reductions in mean PCWP and right atrial pressure, suggesting improvement in left and right ventricular and diastolic pressure.

Relaxin. This is a naturally occurring peptide hormone that plays a central role in the hemodynamic and renovascular adaptive changes that occur during pregnancy. Triggering similar changes could potentially be beneficial in the treatment of patients with HF. The effects of relaxin include the production of nitric oxide, inhibition of endothelin (ET), inhibition of angiotensin II, production of vascular endothelial growth factor and matrix metalloproteinases. These effects lead to systemic and renal vasodilation, increased arterial compliance, and other vascular changes. The recognition of this has led to the study of relaxin for the treatment of HF. An initial pilot study has shown favorable hemodynamic effects in patients with HF, including a reduction in ventricular filling pressures and increased cardiac output. The demonstrated effects of relaxin suggest that it may be therapeutic in clinical settings where reductions in systemic vascular resistance and improved renal blood flow may be beneficial. On a large scale the ADHERE and OPTIMIZE registries have confirmed earlier observations that a substantially higher proportion of acute HF events than previously recognized is associated with elevated blood pressure, most probably induced by vasoconstriction and vasomotor nephropathy, or renal vasoconstriction. The ongoing RELAX-AHF clinical program is designed to evaluate the effects of relaxin on the symptoms and outcomes in a large group of patients admitted to hospital for acute HF [14].

Available Agents for Patients with AHFS Presenting with Low Blood Pressure With or Without Congestion

In theory, inotropic agents improve hemodynamic parameters, increasing cardiac output and reducing left and right ventricular filling pressure, through direct enhancement of myocardial contractility. Accordingly, they are indicated for the treatment of patients with both peripheral hypoperfusion and fluid retention caused by impaired cardiac contractility. In the American College of Cardiology Federation/American Heart Association guidelines, inotropic agents are indicated to improve symptoms and end-organ function in patients with low output syndrome, LV systolic dysfunction and systolic blood pressure of <90 mm Hg despite adequate filling pressure [15]. In the ESC guidelines, patients with acute HF are stratified on the basis of their systolic blood pressure at presentation, and inotropic agents are indicated in patients with blood pressure values of ≤100 mm Hg [2]. Based on clinical assessment and systolic blood pressure values, these indications clearly limit the use of the currently available intravenous inotropic agents only to those patients most likely to benefit from their administration. Systolic blood pressure values of <90–100 mm Hg have been reported in less than 10% of the patients with AHFS [16].

Dobutamine

Dobutamine stimulates β_1- and β_2-adrenergic receptors, leading to upregulation of adenylyl cyclase and, ultimately, an increase in intracellular calcium concentration. In the heart, it exerts relatively more prominent inotropic than chronotropic effects compared with isoproterenol. Low-dose dobutamine induces mild arterial vasodilation resulting in afterload reduction and stroke volume augmentation, whereas higher doses induce vasoconstriction. Adverse effects include increases or decreases in blood pressure, increased heart rate, increased ventricular response rate (in patients with atrial fibrillation), and increased myocardial oxygen demand. The PRECEDENT study used 24-hour Holter recording to compare the pro-arrhythmic and tachycardic effects of dobutamine or nesiritide in patients with AHFS [17]. Dobutamine significantly increased the mean number of ventricular tachycardia events, repetitive ventricular beats, premature ventricular beats, and heart rate, whereas nesiritide reduced ventricular ectopy or had a neutral effect.

Dopamine

The cardiovascular effects of dopamine are mediated by several types of receptors in a dose-dependent manner. At low doses (≤2 μg/kg per min), dopamine primarily acts on the vascular D1 receptor in the coronary, renal, and splanchnic beds, leading to vasodilation and natriuresis. At intermediate doses (2–5 μg/kg per min), dopamine interacts with the β_1-receptor, producing positive inotropic effects; it usually increases systolic blood pressure and pulse pressure. At high doses (5–15 μg/kg per min), dopamine causes generalized vasoconstriction. In a small study of dopamine in 9 patients with chronic severe HF, a dose of 2.1 μg/kg per min caused renal blood flow to nearly double and cardiac index to increase only by 21%, whereas at a dose of 4.0 μg/kg per min, the cardiac index was increased maximally with no further augmentation of renal blood flow. Adverse effects due to overdosage are generally related to excessive sympathomimetic activity. These include tachycardia, arrhythmia, hypertension, anginal pain, nausea, vomiting, and headache. The ESC guideline recommends, 'Dopamine may be used as an inotrope (>2 μg/kg per min, intravenous) in acute HF with hypotension. Infusion of low doses of dopamine (≤2–3 μg/kg per min) may be used to improve renal blood flow and diuresis in decompensated HF with hypotension and low urine output'.

Milrinone

Milrinone (a derivative of amrinone), often termed an inodilator, is a phosphodiesterase inhibitor that increases cardiac contractility and produces balanced arterial and venous dilation, by increasing intracellular concentrations of cyclic adenosine monophosphate and, consequently, calcium. Milrinone causes an increase in cardiac output and a decrease in pulmonary and systemic vascular resistance, without increasing myocardial oxygen consumption [18]. The

Prospective Randomized Milrinone Survival Evaluation (PROMISE) study showed that long-term therapy with oral milrinone increases the morbidity and mortality of patients with chronic HF compared with placebo when given in addition to conventional therapy (digoxin, diuretics, and ACE inhibitors). In the median 6.1 months of follow-up, compared with placebo, milrinone therapy was associated with a 28% increase in all-cause mortality and a 34% increase in cardiovascular mortality, with the greatest adverse effects observed in patients with the most severe symptoms (NYHA class IV), who had a 53% increase in mortality. Compared with patients in the placebo group, the milrinone group had more hospitalizations (44 vs. 39%) and more frequent serious adverse cardiovascular reactions, including hypotension and syncope. The HFSA 2006 HF Practice Guideline states, 'Intravenous inotropes (dobutamine or milrinone) may be considered to relieve symptoms and improve end-organ function in patients with advanced HF characterized by LV dilation, reduced LVEF, and diminished peripheral perfusion or end-organ dysfunction (low output syndrome), particularly if these patients have marginal systolic blood pressure (<90 mm Hg), have symptomatic hypotension despite adequate filling pressure, or are unresponsive to, or intolerant of, intravenous vasodilators *[strength of evidence: C/ strength of recommendation: May be considered]*. These agents may be considered in similar patients with evidence of fluid overload if they respond poorly to intravenous diuretics or manifest diminished or worsening renal function *[strength of evidence: C/ strength of recommendation: May be considered]*. When adjunctive therapy is needed in other patients with acute decompensated HF, administration of vasodilators should be considered instead of intravenous inotropes *[strength of evidence: B/ strength of recommendation: Should be considered]*. Intravenous inotropes are not recommended unless left heart filling pressures are known to be elevated based on direct measurement or clear clinical signs *[strength of evidence: B/ strength of recommendation: Is not recommended]*. Administration of intravenous inotropes (dobutamine or milrinone) in the setting of acute decompensated HF should be accompanied by continuous or frequent blood pressure monitoring and continuous monitoring of cardiac rhythm *[strength of evidence: C/ strength of recommendation: Should be considered]*. If symptomatic hypotension or worsening tachyarrhythmias develop during administration of these agents, discontinuation or dose reduction should be considered *[strength of evidence: C/ strength of recommendation: Should be considered]*.

Milrinone may be preferred to dobutamine in patients on concomitant β-blocker therapy, and/or with an inadequate response to dobutamine *[class of recommendation: IIa/ level of evidence: C]*'.

Digoxin

The evidence on the beneficial properties of digoxin in the setting of HF has been gathered from over 200 years of clinical and experimental research [19].

Digoxin acts through inhibition of the sarcolemmal Na-K ATPase pump thus leading to increased intracellular sodium that is then exchanged with calcium. The increase in intracellular calcium causes the inotropic effect of the drug. The beneficial hemodynamic effects of digoxin are attained in the absence of any hypotension or tachycardia and with associated favorable effects on neurohormones including a decrease in the sympathetic drive and renin-angiotensin-aldosterone activation and an increase in vagal stimulation. Despite these favorable results, few studies have assessed the effects of digoxin in patients with acute HF [19]. Conversely, digoxin use is rather common in patients with chronic HF (approximately 50–65% of patients overall, range 55–91%). The landmark Digitalis Investigators Group (DIG) trial assessed the effects of digoxin on mortality in a total of 7,788 patients with chronic stable HF (85% NYHA class II–III) randomized to placebo or digoxin with treatment targeting to serum digoxin concentrations of 0.8–2.5 ng/ml. The study included 6,800 HF patients with EF B 45% (DIG-Main) and 988 patients with EF 45% (DIG-Ancillary). The median daily dose of digoxin was 0.25 mg. Although digoxin did not reduce the primary end-point of all-cause mortality during the 37 months of follow-up, it did significantly reduce HF-related deaths or rehospitalization. Further data analyses showed the pivotal role of the serum digoxin concentrations achieved during treatment. This issue was first addressed in a post hoc analysis by Gheorghiade et al. [27]. This study showed that while the effects of digoxin on LV function are dose-dependent, increasing serum digoxin concentrations have no relation with changes in exercise tolerance and neurohormonal parameters. Thus, digoxin administration targeted to low serum levels may be associated with hemodynamic and clinical improvements, reduced morbidity and possibly increased survival. Implementation of this agent is, however, limited by its narrow risk to benefit ratio. Lastly, digoxin, the oldest drug currently used for the treatment of HF, may be the agent to open new tracks for the research on inotropic agents.

New Inotropic Agents

Levosimendan (TEERL)

Levosimendan is a calcium-sensitizing agent with two complimentary mechanisms of action, enhanced cardiac troponin C sensitivity to intracellular calcium and peripheral vasodilation through opening of smooth muscle ATP-dependent potassium channels, as well as some potential PDE3-I activity. These mechanisms mediate its inotropic and vasodilating effects, respectively. The recent Randomized Multicenter Evaluation of Intravenous Levosimendan Efficacy Versus Placebo in the Short-Term Treatment of Decompensated Heart Failure (REVIVE) trial, consisting of 2 consecutive phases REVIVE-1 (n = 100) and 2 (n = 600), investigated the effect on levosimendan versus placebo on primary combined end-points. Particularly, the primary end-point of REVIVE-2 was a composite based on patients' self-rating as moderately or markedly improved at 6 and 24 h and 5 days in the absence of any criteria for worsening [20]. In the REVIVE-2, despite a significant improvement in the primary end point in patients randomized to levosimendan (overall $p = 0.015$), active drug treatment was associated with higher rates of adverse effects including hypotension (49 vs. 36% on placebo), ventricular tachycardia (24 vs. 17% on placebo) and atrial fibrillation (4 vs. 1%). Mortality at 90 days was 15% in the levosimendan group versus 12% with placebo ($p = 0.210$) [20]. Despite interesting new mechanisms of action and favorable results in preliminary trials, more recent studies have thus failed to yield conclusive results regarding the risk to benefit ratio of levosimendan in patients with AHFS.

Istaroxime

Istaroxime is a novel agent that exerts its effects through calcium cycle modulation and has both inotropic and lusitropic properties. Istaroxime inhibits Na^+/K^+ adenosine triphosphatase activity while stimulating sarcoplasmic reticulum Ca^{2+} adenosine triphosphatase isoform 2a. The combined mechanism of istaroxime allows cytosolic calcium accumulation during systole creating an inotropic response, as well as rapid sequestration of calcium during diastole creating a lusitropic response. The effects of istaroxime infusion on hemodynamic parameters and the safety profile of istaroxime were studied in animal models and human subjects. These positive results were demonstrated by HORIZON-HF, a randomized, placebo-controlled trial evaluating the short-term effects of 3 different doses (0.5, 1 and 1.5 μg/kg/min) of istaroxime in 120 patients hospitalized for AHFS and EF ≤35%. All patients underwent pulmonary artery catheterization and comprehensive 2-dimensional/Doppler and tissue Doppler echocardiography at baseline and at the end of the 6-hour infusion. The administration of intravenous istaroxime, when added to standard therapy resulted in a rapid improvement in PCWP (primary end-point).

Cardiac Myosin Activators

These agents directly target myocardial myosin ATPase, increasing the rate of effective myosin cross-bridge formation, and hence the duration and amount of myocyte contraction with increased myocyte energy utilization, and no effect on intracellular calcium or cAMP. Active research in the recent past has led to the development of the selective cardiac myosin activator CK-1827452, now known as omecamtiv mecarbil, the first agent to be tested in humans [21]. In a pivotal phase I trial by Teerlink [21] on 34 healthy volunteers, omecamtiv mecarbil, at the dose of 0.5 mg/kg/min given as a 6-hour continuous infusion, induced a 6.8 and 9.2% absolute increase in EF and in fractional shortening, respectively ($p = 0.0001$ for both) [21]. Systolic ejection time was prolonged

Table 1. Therapies for AHFS

	Congestion with normal to high blood pressure	Low blood pressure with or without congestion

Current agents	Diuretics
	Vasodilators
	  Nitroglycerin
	  Nitroprusside
	  Nesiritide
	ACE inhibitors (hypertension)
		Dobutamine
		Dopamine
		Milrinone
		Digoxin IV

Investigational agents	Vasopressin antagonists Adenosine antagonists (Rolofillyne) Endothelin antagonists Ularitide	Cardiac myosin activators Metabolic modulators Levosimendan Istaroxime

by a mean 84 ms (p = 0.0001). These findings, as well as those observed in experimental models, are consistent with a unique positive inotropic effect elicited through a direct increase in systolic ejection time rather than through an increase in contraction velocity. No difference in these effects has been found between patients with ischemic and non-ischemic cardiomyopathy. To date, this agent has been safe and well tolerated. Cardiac myosin activators may be expected to play an active role in the quest for the ideal, safe and effective inotropic agent, and the availability of a highly bioavailable oral formulation suggests that these benefits may be extended to therapy of chronic HF.

Investigational Agents for AHFS Patients Presenting with Congestion and Normal to High Blood Pressure

Given the limitations of high-dose diuretics and vasodilators, a variety of new agents are under investigation for the treatment of pulmonary and/or systemic congestion in the setting of AHFS (table 1).

Vasopressin Receptor Antagonist: Tolvaptan
Vasopressin is a neurohormone produced by the central nervous system in response to changes in serum osmolarity, severe hypovolemia or hypotension. Vasopressin levels are inappropriately high in both acute and chronic HF, and non-osmotic release of vasopressin represents a negative adaptation that is central to the pathophysiology of HF. There are two main types of

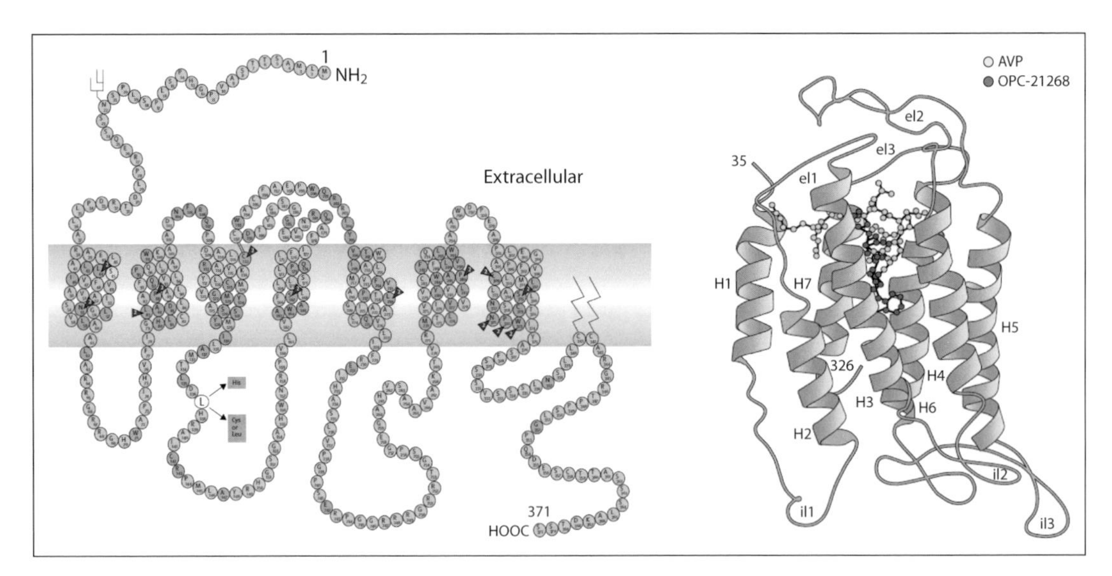

Fig. 2. V2 receptor structure and structure of its antagonists. Reprinted with permission from The Lancet [Decaux G, Soupart A, Vassart G: Non-peptide arginine-vasopressin antagonists: the vaptans. Lancet 2008;371:1624–1632] with permission from Elsevier.

vasopressin receptors: V1 located in vasculature, and V2 located in the distal nephron segments and mediating water reabsorption (fig. 2). Tolvaptan is a selective V2-receptor antagonist. In a randomized double-blind controlled study, tolvaptan was well tolerated without significant changes in heart rate, blood pressure, serum potassium or renal function. The Acute and Chronic Therapeutic Impact of a Vasopressin Antagonist in Congestive Heart Failure (ACTIV in CHF) trial was conducted to evaluate the clinical effects of tolvaptan in patients hospitalized for HF. This study had the unique aspect of including both short- and long-term end points: change in body weight at 24 h, and worsening HF (defined as death, hospitalization, or unscheduled visits for HF) at 60 days.

All doses of tolvaptan resulted in significant weight reduction over the placebo, without significant adverse effects. Although there was no difference in the 60-day outcome measures, in retrospect there was a trend toward lower mortality in the tolvaptan group reaching statistical significance in patients with a high BUN level or severe systemic congestion. A post hoc analysis confirmed that BUN was a statistically significant predictor of both mortality and the composite end-point of death or HF hospitalization at 60 days. In terms of clinical outcomes, the data from the ACTIV in CHF study should be interpreted with caution: this was a phase II study, with a relatively small database (limiting the available number of events) and perhaps all the factors that could affect outcomes may not have been analyzed and this could have confused the association between blood urea nitrogen levels and outcomes. The EVERST trial was designed to investigate the short- and long-term effects of tolvaptan in patients hospitalized with AHFS and documented evidence of impaired lLVEF. The two short-term trials showed a significant clinical improvement (dyspnea, edema, body weight and serum sodium); the longer-term safety outcome trial showed no effect, either favorable or unfavorable, on its primary outcome [9].

Adenosine Antagonist: Rolofylline

Adenosine plays its physiologic effects through interaction with several receptor subtypes. The two major receptor subtypes, A1 and A2, differ in their affinity to adenosine receptor antagonists. Rolofylline, an adenosine A1 receptor antagonist, induces diuresis via inhibition of sodium absorption in the proximal tube, blocks tubuloglomerular feedback and therefore preserves or increases glomerular filtration rate in HF. Phase I and II studies of oral rolofillyne demonstrate diuretic and natriuretic effects of adenosine A1 receptor blockade in healthy subjects and in patients with renal impairment. It has been demonstrated that rolofylline can enhance diuresis and prevent worsening renal function, while improving symptoms and short-term outcomes in patients with AHFS. The results of this phase III pilot study firstly showed that preserving renal function may translate into benefit on outcome. This hypothesis is now being tested in a large outcome trial (the PROTECT) [22].

Endothelin Receptor Antagonist: Tezosentan

In HF patients significant elevation of plasma ET is observed. The overexpression of the ET system is likely to potentiate the progression of HF through interactions with other neurohormones, hemodynamic effects, vascular and cardiac remodeling, as well as through its renal actions [23]. Moreover in patients with AHFS, ET concentrations were shown to be highly predictive of arrhythmias and are among the strongest predictors of death [24]. Tezosentan is a intravenously administrated ET receptor antagonist with high affinity to both receptors (ETa/ETb), and was specifically developed for the treatment of AHFS. By antagonizing the effect of ET, this drug may improve the outcome of HF by reducing preload and afterload, delaying myocyte hypertrophy, increasing the contractility of the failing myocardium and reducing the incidence of arrhythmias. Two phase II clinical trials showed that in patients with AHFS tezosentan significantly increased cardiac output, decreased PCWP and had a clinical profile similar to those of normal volunteers. In a recent placebo-controlled dose-ranging study [25], it was demonstrated that a low dose of tezosentan reduced PCWP and systemic vascular resistance as well as the plasma concentration of BNP. VERITAS 1 and 2 were two randomized, double-blind, placebo-controlled studies designed to evaluate morbidity and mortality and dyspnea improvement in patients with AHFS. Although hemodynamic effects were observed in the treatment arm, low-dose tezosentan had no effect on the co-primary end points [26]. Earlier trials in chronic HF showed that long-term treatment with oral ET receptor antagonist had no effect on left ventricular remodeling or clinical outcome.

Conclusions

Many studies have consistently shown that current therapies are associated with increased mortality in patients with both acute and chronic HF. All these agents continue to be used owing to the lack of effective and safe alternatives. There is an unmet need for developing new agents to safely improve cardiac performance in these patients. A significant improvement in outcome could be obtained in the future assuming an 'holistic' point of view looking for the effect of each drugs on cardiac damage, organ perfusion, neurohormonal activation and renal injury after short period. New agents with more favorable effects on myocardial perfusion and cardiac efficiency are likely going to open new pathways to improve quality of life and outcomes of patients with acute and worsening decompensated HF.

References

1 Gheorghiade M, Zannad F, Sopko G, et al: Acute heart failure syndromes: current state and framework for future research. Circulation 2005;112:3958–3968.

2 Dickstein K, Cohen-Solal A, Filippatos G, et al: ESC guidelines for the diagnosis and treatment of acute and chronic heart failure 2008: the Task Force for the Diagnosis and Treatment of Acute and Chronic Heart Failure 2008 of the European Society of Cardiology. Developed in collaboration with the Heart Failure Association of the ESC (HFA) and endorsed by the European Society of Intensive Care Medicine (ESICM). Eur Heart J 2008;29:2388–2442.

3 Levy P, Compton S, Welch R, et al: Treatment of severe decompensated heart failure with high-dose intravenous nitroglycerin: a feasibility and outcome analysis. Ann Emerg Med 2007;50:144–152.

4 Hunt SA: ACC/AHA 2005 guideline update for the diagnosis and management of chronic heart failure in the adult: a report of the American College of Cardiology/American Heart Association Task Force on Practice Guidelines (Writing Committee to Update the 2001 Guidelines for the Evaluation and Management of Heart Failure). J Am Coll Cardiol 2005;46:e1–e82.

5 Jourdain P, Jondeau G, Funck F, et al: Plasma brain natriuretic peptide-guided therapy to improve outcome in heart failure: the STARS-BNP multicenter study. J Am Coll Cardiol 2007;49:1733–1739.

6 Metra M, Ponikowski P, Dickstein K, et al, on behalf of Heart Failure Association of the European Society of Cardiology: Advanced chronic heart failure: a position statement from the Study Group on Advanced Heart Failure of the Heart Failure Association of the European Society of Cardiology. Eur J Heart Fail 2007;9:684–694.

7 Gheorghiade M, Filippatos G: Reassessing treatment of acute heart failure syndromes: the ADHERE registry. Eur Heart J 2005; 7(suppl):B13–B19.
8 Fonarow GC, Stough WG, Abraham WT, et al: Characteristics, treatments, and outcomes of patients with preserved systolic function hospitalized for heart failure: a report from the OPTIMIZE-HF Registry. J Am Coll Cardiol 2007;50:768–777.
9 Gheorghiade M, Filippatos G, Pang PS, et al: Changes in clinical, neurohormonal, electrolyte, renal, and hepatic profiles during and after hospitalization for acute decompensated heart failure: analysis from the EVEREST trial (late breaking clinical trial). Paper presented at European Society of Cardiology, September 2, 2008.
10 Gheorghiade M, Pang PS: Acute heart failure syndromes. J Am Coll Cardiol 2008;53:557–573.
11 Moazemi K, Chana JS, Willard AM, Kocheril AG: Intravenous vasodilator therapy in congestive heart failure. Drugs Aging 2003;20: 485–508.
12 Sackner-Bernstein JD, Skopicki HA, Aaronson KD: Risk of worsening renal function with nesiritide in patients with acutely decompensated heart failure. Circulation 2005;111:1487–1491.
13 Mitrovic V, Lüss H, Nitsche K, Forssmann K, Maronde E, Fricke K, Forssmann WG, Meyer M: Effects of the renal natriuretic peptide urodilatin (ularitide) in patients with decompensated chronic heart failure: a double-blind, placebo-controlled, ascending-dose trial. Am Heart J 2005;150:1239.
14 Teichman SL, Unemori E,Dschietzig T, Conrad K, Voors A, Teerlink JR, Felker GM, Metra M, Cotter G: Relaxin, a pleiotropic vasodilator for the treatment of heart failure. Heart Fail Rev 2009;14:321–329.
15 Jessup M, Abraham WT, Casey DE, Feldman AM, Francis GS, Ganiats TG, Konstam MA, Mancini DM, Rahko PS, Silver MA, Stevenson LW, Yancy CW: 2009 focused update: ACCF/AHA Guidelines for the Diagnosis and Management of Heart Failure in Adults: a report of the American College of Cardiology Foundation/American Heart Association Task Force on Practice Guidelines: developed in collaboration with the International Society for Heart and Lung Transplantation. Circulation 2009;119:1977–2016.
16 Adams KF Jr, Fonarow GC, Emerman CL, LeJemtel TH, Costanzo MR, Abraham WT, Berkowitz RL, Galvao M, Horton DP, ADHERE Scientific Advisory Committee and Investigators: Characteristics and outcomes of patients hospitalized for heart failure in the United States: rationale, design, and preliminary observations from the first 100,000 cases in the Acute Decompensated Heart Failure National Registry (ADHERE). Am Heart J 2005;149:209–216.
17 Burger AJ, Horton DP, LeJemtel T, Ghali JK, Torre G, Dennish G, Koren M, Dinerman J, Silver M, Cheng ML, Elkayam U, for the Prospective Randomized Evaluation of Cardiac Ectopy with Dobutamine or Natrecor Therapy: Effect of nesiritide (B-type natriuretic peptide) and dobutamine on ventricular arrhythmias in the treatment of patients with acutely decompensated congestive heart failure: the PRECEDENT study. Am Heart J 2002;144:1102–1108.
18 Anderson JL: Hemodynamic and clinical benefits with intravenous milrinone in severe chronic heart failure: results of a multicenter study in the United States. Am Heart J 1991; 121:1956–1964.
19 Gheorghiade M, van Veldhuisen DJ, Colucci WS: Contemporary use of digoxin in the management of cardiovascular disorders. Circulation 2006;113:2556–2564.
20 Cleland JG, Freemantle N, Coletta AP, Clark AL: Clinical trials update from the American Heart Association: REPAIRAMI, ASTAMI, JELIS, MEGA, REVIVE-II, SURVIVE, and PROACTIVE. Eur J Heart Fail 2006;8:105–110.
21 Teerlink JR: A novel approach to improve cardiac performance: cardiac myosin activators. Heart Fail Rev 2009;14:289–298.
22 Cotter G, Dittrich HC, Weatherly BD, et al: The PROTECT Pilot Study: a randomized, placebo-controlled, dose-finding study of the adenosine A1 receptor antagonist rolofylline in patients with acute heart failure and renal impairment. J Card Fail 2008;14:631–640.
23 Teerlink JR: The role of endothelin in the pathogenesis of heart failure. Curr Cardiol Rep 2002;4:206–212.

24 Aronson D, Burger AJ: Neurohormonal prediction of mortality following admission for decompensated heart failure. Am J Cardiol 2003;91:245–248.
25 Cotter G, Kaluski E, Stangl K, Pacher R, Richter C, Milo-Cotter O, et al: The hemodynamic and neurohormonal effects of low doses of tezosentan (an endothelin A/B receptor antagonist) in patients with acute heart failure. Eur J Heart Fail 2004;6:601–609.
26 McMurray JJ, Teerlink JR, Cotter G, Bourge RC, Cleland JG, Jondeau G, Krum H, Metra M, O'Connor CM, Parker JD, Torre-Amione G, van Veldhuisen DJ, Lewsey J, Frey A, Rainisio M, Kobrin I; VERITAS Investigators: Effects of tezosentan on symptoms and clinical outcomes in patients with acute heart failure: the VERITAS randomized controlled trials. JAMA 2007;298:2009–2019.
27 Gheorghiade M, Braunwald E: Reconsidering the role for digoxin in the management of acute heart failure syndromes. JAMA 2009;302:2146–2147.

Dr. Alberto Palazzuoli
Department of Internal Medicine and Metabolic Diseases, University of Siena
Viale Bracci
IT–53100 Siena
Tel. +39 0 577585363, Fax +39 0 577233446, E-Mail palazzuoli2@unisi.it

Ronco C, Bellomo R, McCullough PA (eds): Cardiorenal Syndromes in Critical Care.
Contrib Nephrol. Basel, Karger, 2010, vol 165, pp 129–139

Management of Chronic Cardiorenal Syndrome

Philipp Attanasio[a] · Claudio Ronco[b] · Markus S. Anker[a] · Piotr Ponikowski[d] · Stefan D. Anker[a,c]

[a]Division of Applied Cachexia Research, Department of Cardiology, Campus Virchow Clinic, Charité – Universitätsmedizin Berlin, Berlin, Germany; [b]Department of Nephrology, Dialysis & Transplantation, International Renal Research Institute, San Bortolo Hospital, Vicenza, and [c]Centre for Clinical and Basic Research, IRCCS San Raffaele, Rome, Italy, and [d]Cardiac Department, Faculty of Public Health, Medical University, Military Hospital, Wroclaw, Poland

Abstract

Patients with heart failure (HF) often have renal dysfunction and patients with kidney disease develop congestive HF, therefore the concept of cardiorenal syndromes evolved which can be a chronic or acute cardiorenal syndrome. Both chronic renal and heart disease share comorbidities like anemia which cause symptoms, disease progression and increase the risk of hospital admission and mortality. Even though there are clinical guidelines for managing both dysfunctions separately, patients with severe HF or severe kidney disease have often been excluded from clinical trials of the respective other disorder. We outline here a summary of the current state of the art of the clinical practice to manage patients with chronic cardiorenal syndrome using drug therapy. We will furthermore focus on that management of anemia and iron deficiency in particular as there have been recent advances.

While there are standard treatment protocols for both patients with heart failure (HF) and kidney dysfunction, there are no evidence-based clinical targets for managing patients with both cardiac and renal disease. The question is as to whether management of patients with kidney failure requires modification for those patients with HF and vice versa. We reviewed the currently available literature to bring out an expert opinion treatment strategy for these patient groups focusing on the chronic forms of cardiorenal and renocardiac syndrome and management of anemia as an important comorbidity in both entities.

Definitions

Comprising both cardiac and renal dysfunction, the cardiorenal or renocardiac syndromes are a heterogeneous group of conditions [1]. Cardiorenal syndrome is defined as HF causing renal dysfunction. Acute cardiorenal syndrome is caused by acute worsening of cardiac function like in acute coronary syndrome, valvular heart disease, hypertension, arrhythmias, infection and non-compliance with HF treatment [2]. Coronary artery disease and hypertension, followed by valvular heart disease and cardiomyopathies [3], are the most common causes of chronic heart disease and cause chronic cardiorenal syndrome. The term renocardiac syndrome is reserved for a decline in kidney function causing cardiac dysfunction. Acute and chronic renocardiac syndrome is likewise defined by acute and chronic deterioration of renal function.

Staging of chronic kidney disease (CKD) is done according to the estimated glomerular filtration rate (eGFR) (table 1). Acute renal dysfunction and hence CKD staging is possible by estimating the rate of glomerular filtration and by analyzing the change in creatinine and urine output (table 2) [4].

Table 1. Staging of CKD by degree of impaired eGFR

CKD stage	Description	eGFR ml/min · 1.73 m^2
1	kidney damage with normal or increased eGFR	≥90
2	kidney damage with mild reduced eGFR	60–89
3a	moderately reduced eGFR	45–59
3b	moderately reduced eGFR	30–45
4	severely reduced eGFR	15–29
5	kidney failure	<15 (or on dialysis)

Management of Chronic Cardiorenal Syndrome

Therapeutic approaches to patients with chronic heart failure (CHF) are complex and include pharmacological and non-pharmacological management. Increasing survival remains the key endpoint and major goal in clinical trials coupled with therapies directed towards improvement in quality of life. Non-

Table 2. Staging of acute kidney injury according to GFR [adapted from 4]

RIFLE stage	GFR or serum creatinine criterion	Urine output
Risk	↑ serum creatinine × 1.5 or ↓ GFR >25%	<5 ml/kg · h for 6 h
Injury	↑ serum creatinine × 2.0 or ↓ GFR >50%	<5 ml/kg · h for 12 h
Failure	↑ serum creatinine × 3.0 or ↓ GFR >75% or serum creatinine ≥4 mg/dl or absolute ↑ serum creatinine ≥0.5 mg/dl	<3 ml/kg · h for 24 h or anuria × 12 h
Loss	dialysis dependence >4 weeks	–
End-stage kidney disease	dialysis dependence >3 months	–

Serum creatinine (in mg/dl): to convert to μmol/l multiply by 0.88.

pharmacological management of CHF includes symptom recognition, adherence to treatment, lifestyle changes regarding nutrition and diet, exercise training, education and smoking [5, 6].

Digoxin and diuretics attenuate CHF symptoms, but have no effect on mortality [7]. The following pharmacological therapies have been proven to significantly improve morbidity and/or mortality: angiotensin-converting enzyme (ACE) inhibitors, β-blockers, angiotensin II receptor blockers (ARBs), and aldosterone antagonists. The optimal approach is to start treatment with a β-blocker and add an ACE inhibitor and subsequently an aldosterone antagonist or ARB, depending on the patient's response [8]. ARBs have been shown to be non-inferior to ACE inhibitors, and the addition of an ARB to an ACE inhibitor has been shown to improve morbidity and mortality endpoints. Hydralazine and nitrates may be an alternative treatment in symptomatic patients unable to tolerate ACE inhibitors or ARBs. For selected CHF patients, implantation devices are an additional effective treatment, but there is no evidence that this improves kidney function and hence is not the focus of this review.

There are very few studies involving patients with right-sided HF due to pulmonary hypertension [9]. Treatment depends upon whether the right ventricle is over- or underfilled, but should also be directed towards the underlying cause [10]. For those with ascites, paracentesis should be considered.

Management of Renal Dysfunction in Patients with Chronic Heart Failure (Summary Based on the Current Guidelines of the European Society of Cardiology – ESC)

Renal dysfunction is a strong independent predictor of increased morbidity and mortality of CHF [11, 12]. Prevalence of renal dysfunction increases with CHF age, severity, and other comorbidities (such as diabetes mellitus and hypertension). CHF patients with renal dysfunction require more intensive diuretic treatment [13]. In patients with creatinine clearance <30 ml/min/1.73 m^2, infusions of diuretics are more potent than intermittent boluses, and loop diuretics are preferred over thiazide diuretics [14, 15]. Combinations with epithelial sodium channel blockers, metolazone or aldosterone antagonists should be considered in the management of these patients [16, 17], as it is now established that increasing doses of loop diuretics are associated with worse outcome [18].

Therapy with ARBs and ACE inhibitors is often associated with transient mild deterioration in renal function. Patients with renal artery stenosis and CKD are at a higher risk of developing renal dysfunction in response to introducing ARBs and ACE inhibitors in patient management. ARBs and ACE inhibitors also cause potassium retention, which can be exacerbated by aldosterone antagonists – thus CKD patients are advised to restrict dietary sodium and potassium [19]. Using potassium binders may be required in some patients.

There is no absolute serum creatinine level as contraindication to use of ARBs and ACE inhibitors, but the ESC recommends a specialist supervision [3] when serum creatinine exceeds 2.75 mg/dl (250 μmol/l). A recent report indicated that in a high-risk patient group an ARB alone was not different to placebo alone [20], and that the combination of an ARB with an ACE inhibitor failed to reduce the rate of worsening of renal function compared to ACE inhibitor alone [21]. For CHF patients becoming refractory to diuretics, renal replacement therapy may be needed [22] and initial results are encouraging [23]. Of note, since renal dysfunction is associated with impaired clearance of many drugs used in CHF, dose adjustments and careful monitoring of plasma levels may be required in order to avoid toxicity (like for digoxin).

Management of Heart Failure in Patients with Chronic Kidney Disease

Correction of anemia has been shown to reduce left ventricular hypertrophy in CKD patients [24]. In CKD, hypertension is common [25]. Recent studies have shown a survival benefit for patients with CKD who have lower home blood pressure [26]. To prevent volume overload and HF, interdialytic weight gains should be minimized [27]. Lower sodium dialysates [28] and dietary sodium restriction [19] are effective in reducing interdialytic weight gains. This approach lowers ultrafiltration requirements, intradialytic hypotension and repetitive ischemic stunning to the heart [29, 30]. Only a relatively small proportion of CKD patients are prescribed cardioprotective ACE inhibitors, β-blockers, even though these drugs have been shown to reduce cardiovascular mortality and morbidity [31–33]. In CKD patients, prescribing of ACE inhibitors or β-blockers did not increase the incidence of intradialytic hypotension [27].

Additional management strategies include correcting anemia, providing adequate dialysis [34], controlling phosphate and calcium product and parathyroid hormone to minimize vascular calcification [35, 36].

Anemia in Chronic Heart and Chronic Kidney Disease

Anemia is an important comorbidity in both chronic heart and renal failure [37]. Studies have shown the prognostic relevance of anemia because of its association with increased risk of hospitalization and mortality. Patients with anemia also reported worse quality of life and showed reduced exercise tolerance [38]. The etiology of anemia in both syndromes certainly is multifactorial. It includes iron deficiency [39] and lack of erythropoesis stimulating agents [40]. Anemia is related to impaired exercise capacity [41] and poor prognosis [42] in patients with CHF with and without CKD.

In the last few years, some smaller studies have reported that correction of low hemoglobin by erythropoiesis-stimulating proteins such as recombinant human erythropoietin or darbepoetin alfa may significantly improve cardiac and renal function and reduce the number of hospitalizations [41, 42]. An alternative approach is the repletion of iron stores with intravenous iron. Three smaller studies showed that intravenous iron was able to improve exercise capacity and quality of life in both chronic heart and kidney failure patients [43–45]. Recently a larger study with more than 450 participants enrolled investigated the effects of ferric carboxymaltose in CHF in anemic and non-anemic patients with iron deficiency. These patients also had CKD as evidenced by their median eGFR of 39 ml/min. Results showed a significant improvement in quality of life, symptoms as well as functional capacity. Interestingly, results were similar in patients with and without anemia [46].

Conclusion

For patients with both CKD and CHF, drugs commonly recommended for the treatment of CHF alone are not completely accepted. Careful monitoring is required for potential adverse effects, including reduction in GFR and renal blood flow. In these HF patients with CKD, drugs need to be titrated slowly to avoid worsening of renal function. Further studies aimed to assess appropriate dosages and titration of drugs, according to both the severity of underlying renal dysfunction and also the etiology of HF, are needed. To allow better understanding of the impact of the current treatments available in patients with cardiorenal syndromes, future research trials in HF should include more patients with renal insufficiency of a more severe degree.

References

1 Ronco C, McCullough P, Anker SD, et al, for the Acute Dialysis Quality Initiative (ADQI) Consensus Group (2009): Cardiorenal syndromes: report from the consensus conference of the Acute Dialysis Quality Initiative. Eur Heart J 2010;31:703–711.

2 Zannad F, Mebazaa A, Juillière Y, et al, EFICA Investigators: Clinical profile, contemporary management and one-year mortality in patients with severe acute heart failure syndromes: the EFICA study. Eur J Heart Fail 2006;8:697–705.

3 Dickstein K, Cohen-Solal A, Filippatos G, et al, European Society of Cardiology; Heart Failure Association of the ESC (HFA); European Society of Intensive Care Medicine (ESICM): ESC guidelines for the diagnosis and treatment of acute and chronic heart failure: 2008. Eur J Heart Fail 2008;10:933–989.

4 Heywood JT, Khan TA: The use of vasoactive therapy for acute decompensated heart failure: hemodynamic and renal considerations. Rev Cardiovasc Med 2007;8(suppl 5): S22–S29.

5 Anker SD, John M, Pedersen PU, et al: European Society for Parenteral and Enteral Nutrition. ESPEN Guidelines on Enteral Nutrition: Cardiology and Pulmonology. Clin Nutr 2006;25:311–318.

6 Jaarsma T, van der Wal MH, Lesman-Leegte I, et al, Coordinating Study Evaluating Outcomes of Advising and Counseling in Heart Failure (COACH) Investigators: Effect of moderate or intensive disease management program on outcome in patients with heart failure. Arch Intern Med 2008;168: 316–324.

7 Digitalis Investigation Group: The effect of digoxin on mortality and morbidity in patients with heart failure. N Engl J Med 1997;336:525–533.

8 Weir RA, McMurray JJ, Puu M, et al, CHARM Investigators: Efficacy and tolerability of adding an angiotensin receptor blocker in patients with heart failure already receiving an angiotensin-converting inhibitor plus aldosterone antagonist, with or without a β-blocker. Findings from the Candesartan in Heart failure: Assessment of Reduction in Mortality and morbidity (CHARM)-Added trial. Eur J Heart Fail 2008;10:157–163.

9 Matthews JC, Dardas TF, Dorsch MP, Aaronson KD: Right-sided heart failure: diagnosis and treatment strategies. Curr Treat Options Cardiovasc Med 2008;10: 329–341.

10 Ryan MJ, Tuttle KR: Elevations in serum creatinine with RAAS blockade: why isn't it a sign of kidney injury? Curr Opin Nephrol Hypertens 2008;17:443–449.

11 Metra M, Nodari S, Parrinello G, et al: Worsening renal function in patients hospitalised for acute heart failure: clinical implications and prognostic significance. Eur J Heart Fail 2008;10:188–195.

12 Perkovic V, Ninomiya T, Arima H, et al: Chronic kidney disease, cardiovascular events, and the effects of perindopril-based blood pressure lowering: data from the PROGRESS study. J Am Soc Nephrol 2007;18:2766–2772.

13 Komukai K, Ogawa T, Yagi H, et al: Decreased renal function as an independent predictor of re-hospitalization for congestive heart failure. Circ J 2008;72:1152–1157.

14 Mahesh B, Yim B, Robson D, et al: The comparison of the diuretic and natriuretic efficacy of continuous and bolus intravenous furosemide in patients with chronic kidney disease. Nephrology (Carlton) 2008;13:247–250.

15 Ostermann M, Alvarez G, Sharpe MD, Martin CM: Frusemide administration in critically ill patients by continuous compared to bolus therapy. Nephron Clin Pract 2007;107:c70–c76.

16 Cotter OM, Sasimangalam AN, Arumugham PS, et al: Diuretics – a panacea for acute heart failure? Different formulations, doses, and combinations. Heart Fail Monit 2008;6:9–19.

17 Teiwes J, Toto RD: Epithelial sodium channel inhibition in cardiovascular disease. A potential role for amiloride. Am J Hypertens 2007;20:109–117.

18 Hasselblad V, Gattis Stough W, Shah, MR, et al: Relation between dose of loop diuretics and outcomes in a heart failure population: results of the ESCAPE trial. Eur J Heart Fail 2007;9:1064–1069.

19 Cano N, Fiaccadori E, Tesinsky P, et al: ESPEN guidelines on enteral nutrition: adult renal failure. Clin Nutr 2006;25:295–310.

20 Mann JFE, Schmieder RE, Dyal, et al: Effect of telmisartan on renal outcomes: a randomised trial. Ann Intern Med 2009;151: 1–10.

21 Mann JFE, Schmieder RE, McQueen M, et al, on behalf of the ONTARGET Investigators: Renal outcomes with telmisartan, ramipril or both, in people at high vascular risk (the ONTARGET study): a multicentre, randomised, double-blind, controlled trial. Lancet 2008;372:547–553.

22 Kazory A, Ross EA: Contemporary trends in the pharmacological and extracorporeal management of heart failure: a nephrologic perspective. Circulation 2008;117:975–983.

23 Rogers HL, Marshall J, Bock J, et al: A randomized, controlled trial of the renal effects of ultrafiltration as compared to furosemide in patients with acute decompensated heart failure. J Card Fail 2008;14:1–5.

24 Palazzuoli A, Silverberg D, Iovine F, et al: Effects of β-erythropoietin treatment on left ventricular remodelling, systolic function and B-type natriuretic peptide levels in patients with the cardiorenal anemia syndrome. Am Heart J 2007;154:645.e9–e15.

25 Davenport A: Audit of the effect of dialysate sodium concentration on interdialytic weight gains and blood pressure control in chronic haemodialysis patients. Nephron Clin Pract 2006;104:c120–c125.

26 Alborzi P, Patel N, Agarwal R: Home blood pressures are of greater prognostic value then a hemodialysis unit recordings. Clin J Am Soc Nephrol 2007;2:1228–1234.

27 Davenport A, Cox C, Thuraisingham R: Achieving blood pressure targets during dialysis improves control but increases intradialytic hypotension. Kidney Int 2008;73:759–764.

28 Davenport A, Cox C, Thuraisingham R: The importance of dialysate sodium concentration in determining interdialytic weight gains in chronic hemodialysis patients: the PanThames Renal Audit. Int J Artif Organs 2008;31:411–417.

29 Davenport A, Cox C, Thuraisingham R: Blood pressure control and symptomatic intradialytic hypotension in diabetic haemodialysis patients: a cross-sectional survey. Nephron Clin Pract 2008;109:c65–c71.

30 Selby NM, McIntyre CW: The acute cardiac effects of dialysis. Semin Dial 2007;20:220–228.

31 De Silva R, Nikitin NP, Witte KK, et al: Effects of applying a standardised management algorithm for moderate to severe renal dysfunction in patients with chronic stable heart failure. Eur J Heart Fail 2007;9:415–423.

32 Gowdak LH, Arantes RL, de Paula FJ, et al: Underuse of American College of Cardiology/American Heart Association Guidelines in hemodialysis patients. Ren Fail 2007;29:559–565.

33 Gowdak LH, Arantes RL, de Paula FJ, et al: β-Blocker use in long-term dialysis patients: association with hospitalized heart failure and mortality. Arch Intern Med 2004;164:2465–2471.

34 Spalding EM, Chandna SM, Davenport A, Farrington K: Kt/V underestimates the haemodialysis dose in women and small men. Kidney Int 2008;74:348–355.

35 Sigrist MK, Taal MW, Bungay P, McIntyre CW: Progressive vascular calcification over 2 years is associated with arterial stiffening and increased mortality in patients with stages 4 and 5 chronic kidney disease. Clin J Am Soc Nephrol 2007;2:1241–1248.

36 Takei T, Otsubo S, Uchida K, et al: Effects of sevelamer on the progression of vascular calcification in patients on chronic hemodialysis. Nephron Clin Pract 2008;108:c278–c288.
37 Sharma R, Francis DP, Pitt B, et al: Haemoglobin predicts survival in patients with chronic heart failure: a substudy of the ELITE II trial. Eur Heart J 2004;25:1021–1028.
38 Kalra PR, Bolger AP, Francis DP, et al: Effect of anemia on exercise tolerance in chronic heart failure in men. Am J Cardiol 2003;91:888–891.
39 Ezekowitz JA, McAlister FA, Armstrong PW: Anemia is common in heart failure and is associated with poor outcomes: insights from a cohort of 12,065 patients with new-onset heart failure. Circulation 2003;107:223–225.
40 Nangaku M, Fliser D: Erythropoiesis-stimulating agents: past and future. Kidney Int Suppl 2007;107:S1–S3.
41 Ponikowski P, Anker SD, Szachniewicz J, et al: Effect of darbepoetin alfa on exercise tolerance in anemic patients with symptomatic chronic heart failure: a randomized, double-blind, placebo-controlled trial. J Am Coll Cardiol 2007;49:753–762.
42 Silverberg DS, Wexler D, Blum M, et al: The use of subcutaneous erythropoietin and intravenous iron for the treatment of the anemia of severe, resistant congestive heart failure improves cardiac and renal function and functional cardiac class, and markedly reduces hospitalizations. J Am Coll Cardiol 2000;35:1737–1744.
43 Usmanov RI, Zueva EB, Silverberg DS, Shaked M: Intravenous iron without erythropoietin for the treatment of iron deficiency anemia in patients with moderate to severe congestive heart failure and chronic kidney insufficiency. J Nephrol 2008;21:236–242.
44 Toblli JE, Lombrana A, Duarte P, Di Gennaro F: Intravenous iron reduces NT-pro-brain natriuretic peptide in anemic patients with chronic heart failure and renal insufficiency. J Am Coll Cardiol 2007;50:1657–1665.
45 Bolger AP, Bartlett FR, Penston HS, et al: Intravenous iron alone for the treatment of anemia in patients with chronic heart failure. J Am Coll Cardiol 2006;48:1225–1227.
46 Anker SD, Comin Colet J, Filippatos G, et al: Ferric carboxymaltose in patients with heart failure and iron deficiency. N Engl J Med 2009;361:2436–2448.

Recommended Reading

Banerjee A, Davenport A: Changing patterns of pericardial disease in patients with end-stage renal disease. Hemodial Int 2006;10:249–255.
Bangalore S, Messerli FH, Kostis JB, Pepine CJ: Cardiovascular protection using β-blockers: a critical review of the evidence. J Am Coll Cardiol 2007;50:563–572.
Barnett D, Phillips S, Longson C: Cardiac resynchronisation therapy for the treatment of heart failure: NICE technology appraisal guidance. Heart 2007;93:1134–1135.
Basile C, Lomonte C, Vernaglione L, et al: The relationship between the flow of arteriovenous fistula and cardiac output in haemodialysis patients. Nephrol Dial Transplant 2008;23:282–287.
Bishara B, Shiekh H, Karram T, et al: Effects of novel vasopressin receptor antagonists on renal function and cardiac hypertrophy in rats with experimental congestive heart failure. J Pharmacol Exp Ther 2008;326:414–422.
Blair JE, Khan S, Konstam MA, et al: Weight changes after hospitalisation for worsening heart failure and subsequent rehospitalisation and mortality in the EVEREST trial. Eur Heart J 2009;30:1166–1173.
Cardiac Insufficiency Bisoprolol Study II (CIBIS-II): A randomised trial. Lancet 1999;353:9–13.
Cice G, Ferrara L, D'Andrea A, et al: Carvedilol increases two-year survival in dialysis patients with dilated cardiomyopathy, a prospective placebo controlled trial. J Am Coll Cardiol 2003;41:1438–1444.
Cingolani HE, Ennis IL: Sodium-hydrogen exchanger, cardiac overload, and myocardial hypertrophy. Circulation 2007;115:1090–1000.
Cohn JN, Tognoni G, Valsartan Heart Failure Trial Investigators: A randomized trial of the angiotensin-receptor blocker valsartan in chronic heart failure. N Engl J Med 2001;345:1667–1675.

CONSENSUS Trial Study Group: Effects of enalapril on mortality in severe congestive heart failure. Results of the Cooperative North Scandinavian Enalapril Survival Study (CONSENSUS). N Engl J Med 1987;316:1429–1435.

Costanzo MR: The role of ultrafiltration in the management of heart failure. Congest Heart Fail 2008;14:19–24.

Costanzo MR, Guglin ME, Saltzberg MT, et al, UNLOAD Trial Investigators: Ultrafiltration versus intravenous diuretics for patients hospitalized for acute decompensated heart failure. J Am Coll Cardiol 2007;49:675–683.

Costanzo MR, Johannes RS, Pine M, et al: The safety of intravenous diuretics alone versus diuretics plus parenteral vasoactive therapies in hospitalized patients with acutely decompensated heart failure: a propensity score and instrumental variable analysis using the Acutely Decompensated Heart Failure National Registry (ADHERE) database. Am Heart J 2007;154:267–277.

Cowan BR, Young AA, Anderson C, et al: Left ventricular mass and volume with telmisartan, ramipril, or combination in patients with previous atherosclerotic events or with diabetes mellitus (from the Ongoing Telmisartan Alone and in combination with Ramipril Global Endpoint Trial (ONTARGET). Am J Cardiol 2009;104:1484–1489.

Cridlig J, Selton-Suty C, Alla F, et al: Cardiac impact of the arteriovenous fistula after kidney transplantation: a case-controlled, match-paired study. Transpl Int 2008;21: 948–954.

Davenport A: Ultrafiltration in diuretic-resistant volume overload in nephritic syndrome and patients with ascites due to chronic liver disease Cardiology 2001;96:190–195.

Davenport A: Blood pressure measurements in haemodialysis patients. Kidney Int 2008;73:1092–1093.

Elliott MJ, Zimmerman D, Holden RM: Warfarin anticoagulation in hemodialysis patients: a systematic review of bleeding rates. Am J Kidney Dis 2007;50:433–440.

Ershow AG, Costello RB: Dietary guidance in heart failure: a perspective on needs for prevention and management. Heart Fail Rev 2006;11:7–12.

Fellstrom BC, Jardine AG, Schmieder RE, et al, AURORA Study Group: Rovustatin and cardiovascular events in patients undergoing haemodialysis. N Engl J Med 2009;360: 1395–1407.

Gibney N, Cerda J, Davenport A, et al: Volume management by renal replacement therapy in acute kidney injury. Int J Artif Organs 2008;31:145–155.

GISSI-HF Investigators: Effect of n–3 polyunsaturated fatty acids in patients with chronic heart failure (the GISSI-HF trial): a randomised double-blind placebo-controlled trial. Lancet 2008;372:1223–1230.

GISSI-HF Investigators: Effect of rosuvastatin in patients with chronic heart failure (the GISSI-HF trial): a randomised double-blind placebo-controlled trial. Lancet 2008;372:1231–1239.

Gura V, Ronco C, Nalesso F, et al: A wearable hemofilter for continuous ambulatory ultrafiltration. Kidney Int 2008;73:497–502.

Hampl H, Hennig L, Rosenberger C, et al: Optimized heart failure therapy and complete anemia correction on left-ventricular hypertrophy in nondiabetic and diabetic patients undergoing hemodialysis. Kidney Blood Press Res 2005;28:353–362.

Heart Failure Society of America: Nonpharmacologic management and healthcare maintenance in patients with chronic heart failure. J Card Fail 2006;12:e29–e37.

Ho JE, Teerlink JR: Role of tolvaptan in acute decompensated heart failure. Expert Rev Cardiovasc Ther 2008;6:601–608.

Holden RM, Harman GJ, Wang M, et al: Major bleeding in hemodialysis patients. Clin J Am Soc Nephrol 2008;3:105–110.

Kjekshus J, Apetrei E, Barrios V, et al, CORONA Group: Rosuvastatin in older patients with systolic heart failure. N Engl J Med 2007;357:2248–2261.

Lakhdar R, Al-Mallah MH, Lanfear DE: Safety and tolerability of angiotensin-converting enzyme inhibitor versus the combination of angiotensin-converting enzyme inhibitor and angiotensin receptor blocker in patients with left ventricular dysfunction: a systematic review and meta-analysis of randomized controlled trials. J Card Fail 2008;14: 181–188.

Levin NW, Kotanko P, Eckardt KU, et al: Blood pressure in chronic kidney disease stage 5D. Report from a Kidney Disease: Improving Global Outcomes controversies conference. Kidney Int 2010;77:273–284.

Libetta C, Sepe V, Zucchi M, et al: Intermittent haemodiafiltration in refractory congestive heart failure: BNP and balance of inflammatory cytokines. Nephrol Dial Transplant 2007;22:2013–2019.

Lüss H, Mitrovic V, Seferovic PM, et al: Renal effects of ularitide in patients with decompensated heart failure. Am Heart J 2008;155:1012.e1–8.

McAlister FA, Stewart S, Fettua S, McMurray JJ: Multidisciplinary strategies for the management of heart failure patients at high risk for admission: a systematic review of randomised trials. J Am Coll Cardiol 2004;44:810–819.

McMurray J, Solomon S, Pieper K, et al: The effect of valsartan, captopril, or both on atherosclerotic events after acute myocardial infarction: an analysis of the Valsartan in Acute Myocardial Infarction Trial (VALIANT). J Am Coll Cardiol 2006;47:726–733.

Metoprolol CR/XL Randomised Intervention Trial in Congestive Heart Failure (MERIT-HF): Effect of metoprolol CR/XL in chronic heart failure. Lancet 1999;353: 2001–2007.

Mielniczuk LM, Tsang SW, Desai AS, et al: The association between high-dose diuretics and clinical stability in ambulatory chronic heart failure patients. J Card Fail 2008;14:388–393.

Okonko DO, Anker SD: Anemia in chronic heart failure: pathogenetic mechanisms. J Card Fail 2004;10(suppl):S5–S9.

Owan TE, Chen HH, Frantz RP, et al: The effects of nesiritide on renal function and diuretic responsiveness in acutely decompensated heart failure patients with renal dysfunction. J Card Fail 2008;14:267–275.

Parker M, Coats AJ, Fowler MB, et al: Effect of carvediliol on survival in severe chronic heart failure. N Engl J Med 2001;344: 1651–1658.

Pfeffer MA, McMurray JJ, Velazquez EJ, et al, Valsartan in Acute Myocardial Infarction Trial Investigators: Valsartan, captopril, or both in myocardial infarction complicated by heart failure, left ventricular dysfunction, or both. N Engl J Med 2003;349:1893–1906.

Pfeffer MA, Swedberg K, Granger CB, et al, CHARM Investigators and Committees: Effects of candesartan on mortality and morbidity in patients with chronic heart failure: the CHARM-Overall programme. Lancet 2003;362:759–766.

Piccini JP, Hernandez AF, Dai D, et al, for the Get With the Guidelines Steering Committee and Hospitals: Use of cardiac resynchronization therapy in patients hospitalized with heart failure. Circulation 2008;118:926–933.

Pitt B, Zannad F, Remme WJ, et al, Randomized Aldactone Evaluation Study Investigators: The effect of spironolactone on morbidity and mortality in patients with severe heart failure. N Engl J Med 1999;341:709–717.

Port FK, Hulbert-Shearon TE, Wolfe RA, et al: Predialysis blood pressure and mortality risk in a national sample of maintenance haemodialysis patients. Am J Kidney Dis 1999;33:507–517.

Rajaram V, Joseph J: Role of adenosine antagonism in the cardiorenal syndrome: pathophysiology and therapeutic potential. Curr Heart Fail Rep 2007;4:153–157.

Sackner-Bernstein J, Skopicki HA, Aaronson KD: Risk of worsening renal function in patients with acutely decompensated heart failure. Circulation 2005;111:1487–1491.

SOLVD Investigators: Effect of enalapril on survival in patients with reduced left ventricular ejection fraction and congestive heart failure. N Engl J Med 1991;325:293–302.

Struthers A, Krum H, Williams GH: A comparison of the aldosterone-blocking agents eplerenone and spironolactone. Clin Cardiol 2008;31:153–158.

Taylor AL, Ziesche S, Yancy C, et al, African-American Heart Failure Trial Investigators: Combination of isosorbide dinitrate and hydralazine in blacks with heart failure. N Engl J Med 2004;351:2049–2057.

To AC, Yehia M, Collins JF: Atrial fibrillation in hemodialysis patients: do the guidelines for anticoagulation apply? Nephrology (Carlton) 2007;12:441–447.

Vardas PE, Auricchio A, Blanc JJ, et al: European Society of Cardiology; European Heart Rhythm Association: Guidelines for cardiac pacing and cardiac resynchronization therapy: the task force for cardiac pacing and cardiac resynchronization therapy of the European Society of Cardiology. Developed in collaboration with the European Heart Rhythm Association. Eur Heart J 2007;28:2256–2295.

Veeraveedu PT, Watanabe K, Ma M, et al: Effects of V2-receptor antagonist tolvaptan and the loop diuretic furosemide in rats with heart failure. Biochem Pharmacol 2008;75: 1322–1330.

Wanner C, Krane V, März W, et al, German Diabetes and Dialysis Study Investigators: Atorvastatin in patients with type 2 diabetes mellitus undergoing hemodialysis. N Engl J Med 2005;353:238–248.

Wasse H, Gillen DL, Ball AM, et al: Risk factors for upper gastrointestinal bleeding among end-stage renal disease patients. Kidney Int 2003;64:1455–1461.

Whellan DJ, O'Connor CM, Lee KL, et al: HF-ACTION Trial Investigators: Heart failure and a controlled trial investigating outcomes of exercise training (HF-ACTION): design and rationale. Am Heart J 2007;153:201–211.

Yang JH, Cheng LT, Gu Y, et al: Volume overload in patients treated with continuous ambulatory peritoneal dialysis associated with reduced circadian blood pressure variation. Blood Purif 2008;26:399–403.

Zager PG, Nikolic J, Brown RH, et al: 'U' curve association of blood pressure and mortality in hemodialysis patients. Medical Directors of Dialysis Clinic, Inc. Kidney Int 1998;54:561–569.

Prof. Stefan D. Anker, MD, PhD
Applied Cachexia Research, Department of Cardiology
Charité, Campus Virchow-Klinikum
Augustenburger Platz 1, DE–13353 Berlin (Germany)
Tel. +49 30 450 553463, Fax +49 30 450 553951, E-Mail s.anker@cachexia.de

Ronco C, Bellomo R, McCullough PA (eds): Cardiorenal Syndromes in Critical Care.
Contrib Nephrol. Basel, Karger, 2010, vol 165, pp 140–144

Congestion and Cardiorenal Syndromes

Alexandre Mebazaa

Department of Anesthesiology and Critical Care, Hôpital Lariboisière, University Paris 7, Inserm U942, Paris, France

Abstract

Acute right ventricular failure is increasingly seen in the emergency room and intensive care unit. Acute right ventricular failure is associated with high central venous pressure (CVP) and altered liver and kidney enzymes. Several studies suggested that high CVP might markedly reduce renal blood flow by increasing renal backpressure. A strong relationship was observed between CVP and renal blood flow both in the acute and chronic heart failure setting. Likewise, reducing CVP markedly improved renal function. Accordingly, in patients with heart failure, congestion is the major determinant of renal dysfunction. Thus, vasodilators might be best indicated in heart failure patients with predominant signs of congestion. Their effect to reduce CVP might be very beneficial as it may improve renal blood flow.

Relationships between heart and kidney functions are complex. Several scenarios have been described under the umbrella of cardiorenal syndromes. In case of acute heart failure leading to kidney dysfunction, the predominant role of cardiorenal decreased kidney perfusion and/or increased venous congestion remains uncertain.

Case Report

A patient was admitted for hemodynamic instability related to decompensated heart failure [1]. The patient had a history of systolic heart failure with a left ventricular ejection fraction (LVEF) of 25%. At admission to the intensive care unit (ICU), blood pressure was 70/40/52 mm Hg. A central line insertion allowed measurement of right atrial pressure (RAP) at 19 mm Hg, leading to the introduction of 10 μg/kg/min dobutamine, rapidly followed by the introduction of 1 mg/h epinephrine in order to restore an acceptable coronary perfusion pressure. Simultaneously a pulmonary artery catheter was inserted which showed a physiologic pulmonary arterial pressure of 20/8/13 mm Hg, low pulmonary capillary wedge pressure of 2 mm Hg, diminished cardiac output (CO) of 2.3 liters/min, and mixed venous oxygen saturation of 45%. Biology showed elevated lactates at 8.5 IU/l, acute hepatocellular failure with hepatic cytolysis (elevated transaminases ASAT from 48 to 354 IU/l, ALAT 47–73 IU/l), TP 40%, factor V 35% and an acute oligo-anuric renal failure (plasma creatinine 74–94 μmol/l). This multi-organ failure can be explained by organ hypoperfusion, but also by hepatic and renal venous congestion. The analysis of this picture of severe hemodynamic instability with a reduced CO, and greatly elevated RAP, and organ perfusion pressures restored and maintained by vasoconstrictor and inotropic catecholamines point to an acute right ventricular dysfunction. It was decided to introduce inhaled nitric oxide (NO) in an attempt to reduce to the greatest possible degree the afterload of the failing right ventricle. At a rate of 8 ppm, hemodynamics were restored: blood pressure 90/40/55 mm Hg; RAP 8 mm Hg; mean pulmonary artery pressure 13 mm Hg; CO 5 liters/min, while the pulmonary capillary wedge pressure remained at 4 mm Hg.

Physiology of Right Ventricular Function

The purpose of the right ventricle is to maintain the lowest possible pressures in the right heart cavities to optimize venous return. To do so, the right ventricle generates flow. It ejects blood into a low-resistance, high-compliance circuit, as the pressures and resistance in the pulmonary circulation are six times less than those of the systemic circulation. Also the distension of the pulmonary vascular bed is greater, and pulmonary vessel rigidity is not altered within the lungs. A pressure gradient of 5 mm Hg is sufficient for the blood to perfuse across the pulmonary vascular bed [1]. In contrast, the left ventricle's main function is to generate pressure. Additionally, the left ventricle has a weak parietal compliance, a result of its wall thickness, and also ejects into a systemic vascular network with low compliance and a high degree of reflection. Indeed, the diameter of the arterial blood vessels decreases as we progress along the arterial branch, which has important effects on blood flow characteristics.

One of the major differences between acute right and left ventricular failure is that acute right ventricular failure further degenerates into a vicious cycle of auto-aggravation. In addition to its own effect on CO, right ventricle failure induces tricuspid insufficiency and a decrease in left ventricle preload that both aggravate the deterioration in CO.

In acute right ventricular failure, right atrial and ventricular end-diastolic pressures increase causing right ventricular dilatation and subsequent tricuspid insufficiency. The appearance of tricuspid insufficiency is not only important because of its additive deleterious effect on CO. More importantly it signifies the moment where due to a sudden rise in hepatic and renal congestion; the auto-aggravation turns into an irreversible vicious cycle.

The diagnosis of acute right ventricular failure is not easy to make, particularly if it presents in a patient already in the ICU with other apparent organ dysfunctions.

Congestion and Renal Dysfunction

Several experts have suggested that the pathophysiology of impaired renal function in cardiovascular disease is multifactorial. However, few papers have addressed the issue of venous congestion as an important mechanism of heart failure-induced renal failure. Damman et al. [2] sought to investigate the relationship between increased central venous pressure (CVP), renal function, and mortality in a broad spectrum of cardiovascular patients. A total of 2,557 patients who underwent right heart catheterization at the University Medical Center Groningen, the Netherlands, between January 1, 1989, and December 31, 2006, were identified, and their data were extracted from electronic databases. Estimated glomerular filtration rate (eGFR) was assessed with the simplified modification of diet in renal disease formula. The mean age was 59 ± 15 years, and 57% were men. Mean eGFR was 65 ± 24 ml/min/1.73 m^2, with a cardiac index of 2.9 ± 0.8 liters/min/m^2 and CVP of 5.9 ± 4.3 mm Hg. We found that CVP was associated with cardiac index ($r = -0.259$, $p < 0.0001$) and eGFR ($r = -0.147$, $p < 0.0001$). The cardiac index was also associated with eGFR ($r = 0.123$, $p < 0.0001$). In multivariate analysis CVP remained associated with eGFR ($r = -0.108$, $p < 0.0001$). In a median follow-up time of 10.7 years, 741 (29%) patients died. Interestingly, the authors found that CVP was an independent predictor of reduced survival (HR 1.03/mm Hg increase, 95% CI 1.01–1.05, $p = 0.0032$).

At the same time, Mullens et al. [3] explored the relationship between CVP and renal function in patients admitted to the ICU for acutely decompensated heart failure. Their objective was to determine whether venous congestion, rather than impairment of CO, is primarily associated with the development of worsening renal function (WRF) in patients with advanced decompensated heart failure. A total of 145 consecutive ICU patients admitted with advanced decompensated heart failure treated with intensive medical therapy guided by pulmonary artery catheter were studied. Here we define WRF as an increase in serum creatinine of ≥0.3 mg/dl during hospitalization. In the study cohort (age 57 ± 14 years, cardiac index 1.9 ± 0.6 liters/min/m^2, left ventricular ejection fraction 20 ± 8%, serum creatinine 1.7 ± 0.9 mg/dl), 58 patients (40%) developed WRF. Patients who developed WRF had a greater CVP on admission (18 ± 7 vs. 12 ± 6 mm Hg, $p < 0.001$) and after intensive medical therapy (11 ± 8 vs. 8 ± 5 mm Hg, $p = 0.04$). The development of WRF occurred less frequently in patients who achieved a CVP of <8 mm Hg ($p = 0.01$). Furthermore, the ability of CVP to stratify risk for development of WRF was apparent across the spectrum of systemic blood pressure, pulmonary capillary wedge pressure, cardiac index, and eGFRs.

Those two studies strongly suggested that venous congestion is the most important hemodynamic factor driving WRF in decompensated patients with advanced heart failure. The mechanism of congestion-induced renal dysfunction is unclear. It is however likely that high CVP leads to increased renal backpressure reducing renal blood flow.

In collaboration with Uthoff et al. [4] in Basel, Switzerland, we recently investigate the relationship between CVP at presentation to the emergency room and the risk of cardiac rehospitalization and mortality in patients with decompensated heart failure. Interestingly, CVP was determined noninvasively using high-resolution compression sonography at presentation in 100 patients with decompensated heart failure. Cardiac hospitalizations and cardiac and all-cause mortality were assessed as a function of continuous CVP levels and predefined CVP categories (low <6 cm H_2O, intermediate 6–23 cm H_2O and high >23 cm H_2O). Endpoints were adjudicated blinded to CVP. At presentation the mean age was 78 ± 11 years, 60% of patients were male, mean B-type natriuretic peptide level was 1,904 ± 1,592 pg/ml and mean CVP was 13.7 ± 7.0 (range 0–33) cm H_2O. During follow-up (median 12 months), 25 cardiac rehospitalizations, 26 cardiac deaths and 7 non-cardiac deaths occurred. Univariate and stepwise multivariate Cox regression analysis revealed an independent relationship between CVP and cardiac rehospitalization (HR 1.09, 95%CI 1.01–1.18, $p = 0.034$). Kaplan-Meier analyses confirmed a stepwise increase in cardiac rehospitalization for low to high CVP (log rank test $p = 0.015$). No association between CVP and (cardiac) mortality was detectable. Thus, congestion and high CVP at emergency room presentation in patients with acute heart failure is an independent predictor of cardiac rehospitalization but not of cardiac and all-cause mortality.

Conclusion

Acute right ventricular failure is a pathology that is more frequently seen in the emergency room and in the ICU than in the past. It leads to congestion with high CVP. The latter might reduce renal blood flow by increasing renal backpressure. A strong relationship has repeatedly been seen between CVP and renal blood flow in both acute and chronic heart failure setting. Likewise, reducing CVP markedly improved renal function.

References

1 Mebazaa A, Karpati P, Renaud E, Algotsson L: Acute right ventricular failure – from pathophysiology to new treatments Intensive Care Med 2004;30:185–196.
2 Damman K, van Deursen VM, Navis G, Voors AA, van Veldhuisen DJ, Hillege HL: Increased central venous pressure is associated with impaired renal function and mortality in a broad spectrum of patients with cardiovascular disease. J Am Coll Cardiol 2009;53:582–588.
3 Mullens W, Abrahams Z, Francis GS, Sokos G, Taylor DO, Starling RC, Young JB, Tang WH: Importance of venous congestion for worsening of renal function in advanced decompensated heart failure. J Am Coll Cardiol 2009;53:589–596.
4 Uthoff H, Thalhammer C, Potocki M, Reichlin T, Noveanu M, Aschwanden M, Staub D, Arenja N, Socrates T, Twerenbold R, Mutschmann S, Heinisch C, Jaeger K, Mebazaa A, Mueller C: Central venous pressure at emergency room presentation predicts cardiac rehospitalization in patients with decompensated heart failure. Eur J Heart Fail, in press.

Alexandre Mebazaa, MD, PhD
Département d'Anesthésie-Réanimation, Hôpital Lariboisière
2, rue A. Paré
FR–75475 Paris Cedex 10 (France)
Tel. +33 149958085, Fax +33 149958073, E-Mail alexandre.mebazaa@lrb.ap-hop-paris.fr

Ronco C, Bellomo R, McCullough PA (eds): Cardiorenal Syndromes in Critical Care.
Contrib Nephrol. Basel, Karger, 2010, vol 165, pp 145–152

Cardiorenal Syndromes – Recommendations from Clinical Practice Guidelines: The Cardiologist's View

Piotr Ponikowski[a] · Claudio Ronco[b] · Stefan D. Anker[c,d]

[a]Cardiac Department, Faculty of Public Health, Medical University, Military Hospital, Wroclaw, Poland; [b]Department of Nephrology, Dialysis & Transplantation, International Renal Research Institute, San Bortolo Hospital, Vicenza, [c]Centre for Clinical and Basic Research, IRCCS San Raffaele, Rome, Italy, and [d]Division of Applied Cachexia Research, Department of Cardiology, Charité-Universitätsmedizin, Berlin, Germany

Abstract

In the past two decades European cardiologists have always been told to follow the evidence-based guidelines in their everyday clinical practice. In the case of patients with heart failure and concomitant renal disease, this universal rule is not easily applicable simply because there is a lack of specific trials in this field. Patients with cardiorenal syndromes are at risk of complications and have high morbidity and mortality. However, the management of these patients is often empirical. Drugs commonly recommended for the treatment of chronic and acutely decompensated heart failure are not always accepted for patients with concomitant renal disease. Future research trials in heart failure should include patients with renal insufficiency to allow better understanding of the impact of the current treatments available in patients with cardiorenal syndromes.

Renal dysfunction is a common comorbidity complicating the natural course of heart failure (HF), and similarly patients with kidney disease often present accompanying heart disease. This has led to the concept of cardiorenal syndrome (CRS), which is defined as 'pathophysiologic disorder of the heart and kidneys whereby acute or chronic dysfunction in one organ induces acute or chronic dysfunction in the other' [1]. Although, guidelines for the management of patients with HF and kidney disease are well developed and regularly updated by the cardiology and nephrology societies, there are no agreed guidelines/recommendations for the management of patients with cardiorenal and/or renocardiac syndromes, as these patients have typically been excluded from clinical trials. In practice, however, such patients are commonly seen by either cardiologists or nephrologists and often become a real clinical challenge. The question which often arises is whether management of patients with HF requires modification for those with impaired renal function and vice versa.

Below, 'a cardio-centric approach' for the management of CRS based on the recommendations from the most recent European Society of Cardiology guidelines [1] and the CRS classification proposed by Ronco et al. [2] will be discussed.

Patients with Acute Cardiorenal Syndrome (CRS Type 1 – Acute Worsening in Cardiac Function Leading to Acute Impairment in Renal Function)

According to the ESC guidelines [1] acute HF is defined as a rapid onset or gradual change in the symptoms and signs of HF, resulting in the need for urgent therapy or/and hospitalization for relief of symptoms. Dynamic changes in the signs and symptoms of HF also constitute a crucial element of the clinical presentation in acute HF. Regardless of the precipitating factor, pulmonary and/or systemic congestion due to elevated ventricular filling pressures, with or without a decrease in cardiac output, is a nearly universal finding. Most frequently, acute HF develops as decompensation in the presence of chronic HF; in 15–30% of all cases it may present as new onset HF. Acute HF comprises the following clinical scenarios (some of which may overlap) [2]: (a) worsening or decompensated chronic HF (with peripheral edema/congestion); (b) pulmonary edema; (c) hypertensive HF (with elevated systemic blood pressure, congestion and often with preserved left ventricular ejection fraction); (d) cardiogenic shock (with evidence of tissue hypoperfusion induced by HF after adequate correction of preload); (e) right HF, and (f) HF in the course of acute coronary syndrome. It must be remembered that some guidelines [3] do not introduce the term 'acute HF' instead referring to a 'hospitalized patient' as clinical scenario when acute or progressive symptoms of HF develop and require urgent hospitalization.

The most common causes underlying acute HF include arrhythmias (most often atrial fibrillation), infection, acute coronary syndrome, uncontrolled hypertension, and noncompliance with HF management [4].

The majority of patients admitted to hospital due to acute HF suffer from renal dysfunction either developing 'de novo' as a complication of the heart disease or from an exacerbation of previously diagnosed chronic renal disease. Both impaired baseline renal function and deterioration in renal function early in the course of treatment are strong adverse prognostic factors for mortality and morbidity. Thus, any treatment initiated for acute HF should at least have a neutral effect or preferably improve renal function.

The management of acute HF has remained essentially unchanged over the last two decades and is still empiric in most cases. It is simply a consequence of a lack of positive results from large clinical trials. It is noteworthy that current guidelines regarding the management of patients with CRS type 1 are even less evidence-based, as it is simply assumed that general recommendations can be applied. The following issues are worth discussing here [2].

The initial goals of therapy should comprise: improvement in symptoms; restoration of peripheral oxygenation; enhancement in the organ perfusion; stabilization in hemodynamics, and limitation in cardiac and renal damage. Clinical assessment early on admission enables identification of whether a patient presents clinical signs and symptoms of congestion (peripheral or central) and whether peripheral hypoperfusion is present. Baseline systolic blood pressure is an important marker either for further decision making and to assess prognosis. Those with normal elevated blood pressure tend to have a better outcome [2, 5]. Persistent low blood pressure (<90 mm Hg) is uniformly related to a risk of subsequent complications, including deterioration in renal function and usually requires inotropic support [2, 5].

Patients should be oxygenated to achieve a peripheral oxygen saturation of >90%, and some of them may require noninvasive ventilation. Morphine should be considered in the early stage of treatment for those who demonstrate restlessness, dyspnea, anxiety, or chest pain. For the initial therapy a combination of loop diuretic and vasodilator (nitroglycerine, isosorbide dinitrate and nitroprusside) is recommended and widely used in clinical practice [2].

Diuretics are recommended in the treatment of acute HF in the presence of symptoms secondary to congestion and volume overload. Typically, patients admitted with acutely decompensated chronic HF have some underlying renal impairment and previously prescribed diuretics, so larger doses of loop diuretics are often required, and infusions are more potent than simple boluses [2]. High doses of diuretics may potentially predispose to electrolyte disturbances (hypokalemia, hyponatremia), increase the risk of hypovolemia and hypotension which leads to worsening renal function [2].

Vasodilators (nitroglycerine, isosorbide dinitrate and nitroprusside) are recommended at an early stage of acute HF for patients without symptomatic hypotension or serious obstructive valvular disease. Vasodilators are efficient to relieve pulmonary congestion usually without compromising hemodynamics [6]. However, if used in excessive uncontrolled dosages they can precipitate hypotension and exacerbate renal dysfunction. Nesiritide, a recombinant of B-type natriuretic peptide, a potent vasodilator with modest natriuretic effects has been recently introduced into the management of acute HF [7]. However, recent analyses have raised concern that it may cause deterioration in renal function when administered with diuretics [7].

Some patients fail to adequately respond to therapy with vasodilators and diuretics and may require ultrafiltration to reduce volume overload. Ultrafiltration may also be considered in those resistant to standard doses of diuretics.

In patients with low output states and signs of hypoperfusion or congestion despite the use of vasodilators and/or diuretics, inotropic agents should be considered. Among them dobutamine with inotropic and chronotropic actions (via stimulation of β1-adrenegic receptors) and dopamine (a dopaminergic, α1- and β-adrenergic receptor agonist) are most frequently used in Europe [4]. Phosphodiesterase inhibitors, milrinone and enoximone, may also be considered, but they should be used cautiously in patients with ischemic heart disease. Levosimendam, a lusitropic agent which increases cardiac contractility without affecting intracellular calcium, is another option in patients requiring support. In this case though, the inotropic effect is independent of β-adrenergic stimulation, and it represents an alternative for patients on β-blocker therapy [2]. When needed, inotropic agents should be administered as early as possible and withdrawn as soon as adequate organ perfusion is restored and/or congestion reduced [2].

Low cardiac output syndromes (e.g. cardiogenic shock) are rare presentations of acute HF, but related to very poor outcome and often leading to acute kidney injury. Treatments are designed to increase cardiac output, stabilize patient hemodynamics and restore renal blood flow. A fluid challenge, with saline, may be indicated clinically, if blood pressure and organ perfusion do not respond to inotropes. In such cases inotropic support as discussed above is recommended. If the systemic blood pressure remains low, cautious introduction of vasopressors (norepinephrine) may be considered. If clinical response is poor, an elective ventilation, insertion of an intra-aortic balloon pump may be required. Depending upon preexisting comorbidity and underlying etiology, a left ventricular assist device as a bridge to transplantation, cardiac surgery or less often clinical recovery may be appropriate [2].

Adenosine via stimulation of its renal A1 receptors may cause constriction of the afferent arteriole and deterioration in renal function (a phenomenon called 'tubuloglomerular feedback'). Additionally, adenosine increases sodium uptake at the level of the proximal and distal tubule and collecting duct. Thus, adenosine A1 receptor blockers may become an interesting tool in the treatment of acute HF with impaired renal function, and trials with these agents are currently ongoing [8]. Hyponatremia is common in patients with acute HF, being associated with more severe clinical status, renal dysfunction, and high morbidity and mortality. Vasopressin receptor 2 antagonists have been studied as an option to treat hyponatremia in acute HF. They tend to increase renal free water clearance and recent studies reported improvement in hyponatremia and weight loss, however without survival benefit [9].

In all patients who adequately responded to initial therapy, the next step in management should comprise further clinical stabilization, initiation of lifesaving pharmacological therapies, and consideration of device therapy in selected patients.

As all patients with CRS type 1 constitute a very high risk population, before hospital discharge a well-designed follow-up strategy should be prepared and discussed with the patient and his/her relatives. Only, a comprehensive HF management program initiated in the hospital and continued after discharge can prevent early readmission, improve quality of life, and survival.

Patients with Chronic Cardiorenal Syndrome (CRS Type 2 – Chronic Abnormalities in Cardiac Function Causing Progressive and Permanent Chronic Kidney Disease)

Coronary artery disease and hypertension are the commonest causes of chronic HF, followed by valvular heart disease and cardiomyopathies [2]. Renal dysfunction is a common finding in patients with HF, strongly associated with increased morbidity and mortality [10]. It is estimated that more than 50% of HF patients may have at least moderately impaired renal function and its prevalence increases with HF severity, age, a history of hypertension or diabetes mellitus [10]. Conversely, even mild renal dysfunction constitutes a risk factor for HF development and this association becomes stronger with deterioration in renal function.

The underlying cause of renal dysfunction should always be sought in order to detect potentially reversible causes (such as hypotension and/or dehydration due to drug overdosing, deterioration in renal function due to ACE inhibitors, angiotensin receptor blockers [ARBs] or other concomitant medications, e.g. non-steroidal anti-inflammatory drugs) [2]. When coincidental renal disease is suspected, further renal investigations are indicated. Renal dysfunction is not only a consequence of HF, but itself may play a key role in the pathophysiology of the HF syndrome. Thus, in all HF patients renal function should be regularly monitored by measurements of blood urea nitrogen, serum creatinine levels and/or estimation of glomerular filtration rate. Therapy in HF patients with concomitant renal dysfunction is not evidence-based, as these patients are not satisfactorily represented in randomized clinical trials of HF.

HF patients with renal dysfunction often have excessive salt and water retention, which requires more intensive diuretic treatment. In patients with a creatinine clearance of <30 ml/min, thiazide diuretics are less effective and loop diuretics are preferred. Patients may be at a higher risk of further deterioration in renal function when high doses of diuretics are used [2]. More potent options may be diuretic infusions instead of intravenous boluses and a combination of loop diuretic and thiazide/metolazone.

Therapy with ACE inhibitors and ARBs is usually associated with a mild deterioration in renal function as evidenced by a minor increase in blood urea nitrogen and creatinine levels and a decrease in estimated glomerular filtration rate. These changes are frequently transient and reversible. Both agents will preserve glomerular filtration and in the majority of cases therapy can be safely continued. However, patients with preexisting renal insufficiency or renal artery stenosis are at a higher risk of developing clinically relevant deterioration in renal function. Careful monitoring is recommended. If renal deterioration continues, other secondary causes such as excessive diuresis, persistent hypotension, other nephrotoxic therapies or concurrent renovascular disease should be excluded. There is no absolute level of creatinine which precludes the use of ACE inhibitors/ARBs. However, if the serum creatinine level exceeds 250 μmol/l (~2.5 mg/dl), specialist supervision is recommended. In patients with a serum creatinine of >500 μmol/l (~5 mg/dl), hemofiltration or dialysis may be needed to control fluid retention and minimize the risk of uremia [2].

Retrospective analyses from randomized trials suggest that β-blockers are equally effective in those with preserved and impaired renal function [11].

Spironolactone should be used with caution in patients with renal dysfunction as it may cause significant hyperkalemia.

In the HF population with impaired systolic function, a combination of ACE inhibitor with β-blocker constitutes a background therapy. If symptoms persist either ARBs or aldosterone antagonist should be subsequently added. As patients with renal dysfunction treated with a combination of 3 neurohormonal blockers are at risk of hyperkalemia and a further drop in renal function, they need to be carefully monitored [2]. At the same time a therapy consisting of all classes of neurohormonal blockers, i.e. ACE inhibitors, β-blockers, ARBs and aldosterone antagonists, is not recommended.

Patients with chronic HF of ischemic origin are often prescribed anti-platelet therapies (aspirin combined with clopidogrel) which may increase bleeding at fistula needling sites and the risk of gastrointestinal hemorrhage in patients with chronic kidney disease. Anticoagulation with warfarin is recommended for HF patients with atrial fibrillation to reduce the risk of stroke, but at the same time the risk of spontaneous hemorrhage for patients with moderate-severe kidney disease. Statins do not improve survival in HF or those with advanced chronic kidney disease.

Renal dysfunction is associated with impaired clearance of digoxin; to avoid toxicity, the maintenance dose of the drug should be reduced and plasma levels monitored.

For selected patients an implantation of devices may become an additional option to improve morbidity and mortality. Regardless of renal function, cardiac resynchronization therapy is recommended for symptomatic HF patients (NYHA III–IV) with poor left ventricular systolic function and QRS prolongation, and implantable cardiac defibrillators for survivors of cardiac arrest or sustained ventricular arrhythmias, but also for all symptomatic HF patients with poor left ventricular systolic function [2].

For all HF patients, regardless of renal function, non-pharmacological management or 'self-care management' is an integral part of successful chronic HF treatment [2]. Adherence to treatment, proper symptom recognition, lifestyle changes regarding diet and nutrition, smoking, exercise training and education are all elements that need to be clearly explained to patients and relatives [2].

Anemia is a common comorbidity affecting HF patients with chronic kidney disease. A low hemoglobin level is associated with decreased aerobic capacity, a subjective experience of fatigue and reduced functional status, poor quality of life and anemia has consistently been shown to be an independent risk factor for hospital admission and mortality in HF. Although correction of anemia has not been established as routine therapy in chronic HF the use of erythropoietin-stimulating agents and/or intravenous iron currently represent an unproven option [2, 12].

References

1 European Society of Cardiology; Heart Failure Association of the ESC (HFA); European Society of Intensive Care Medicine (ESICM), Dickstein K, Cohen-Solal A, Filippatos G, et al: ESC guidelines for the diagnosis and treatment of acute and chronic heart failure 2008 of the European Society of Cardiology. Developed in collaboration with the Heart Failure Association of the ESC (HFA) and endorsed by the European Society of Intensive Care Medicine (ESICM). Eur J Heart Fail 2008;10:933–989.

2 Ronco C, Anker S, McCullough P, et al: Cardio-renal syndromes: report from the consensus conference of the Acute Dialysis Quality Initiative. Eur Heart J 2009 [Epub ahead of print].

3 Jessup M, Abraham WT, Casey DE, et al: 2009 focused update: ACCF/AHA Guidelines for the Diagnosis and Management of Heart Failure in Adults: a report of the American College of Cardiology Foundation/American Heart Association Task Force on Practice Guidelines: developed in collaboration with the International Society for Heart and Lung Transplantation. Circulation 2009;119:1977–2016.

4 Nieminen MS, Brutsaert D, Dickstein K, et al; EuroHeart Survey Investigators; Heart Failure Association, European Society of Cardiology: EuroHeart Failure Survey II (EHFS II): a survey on hospitalized acute heart failure patients: description of population. Eur Heart J 2006;27:2725–2736.

5 Alla F, Zannad F, Filippatos G: Epidemiology of acute heart failure syndromes. Heart Fail Rev 2007;12:91–95.

6 Hollenberg SM: Vasodilators in acute heart failure. Heart Fail Rev 2007;12:143–147.

7 Sackner-Bernstein J, Skopicki HA, Aaronson KD: Risk of worsening renal function with nesiritide in patients with acutely decompensated heart failure. Circulation 2005; 111:1487–1491.

8 Rajaram V, Joseph J: Role of adenosine antagonism in the cardio-renal syndrome: pathophysiology and therapeutic potential. Curr Heart Fail Rep 2007;4:153–157.

9 Konstam MA, Gheorghiade M, Burnett JC Jr, et al: Effects of oral tolvaptan in patients hospitalized for worsening heart failure: the EVEREST Outcome Trial. JAMA 2007;297: 1319–1331.

10 Smith GL, Lichtman JH, Bracken MB, et al: Renal impairment and outcomes in heart failure: systematic review and meta-analysis. J Am Coll Cardiol 2006;47:1987–1996.

11 Erdmann E, Lechat P, Verkenne P, Wiemann H: Results from post-hoc analyses of the CIBIS II trial: effect of bisoprolol in high-risk patient groups with chronic heart failure. Eur J Heart Fail 2001;3:469–479.

12 Anker SD, Comin Colet J, Filippatos G, et al: Ferric carboxymaltose in patients with heart failure and iron deficiency. N Engl J Med 2009;361:2436–2448.

Dr. Piotr Ponikowski
Department of Heart Diseases, Wroclaw Medical University
ul. Weigla 5
PL–50981 Wroclaw (Poland)
E-Mail piotrponikowski@4wsk.pl

Ronco C, Bellomo R, McCullough PA (eds): Cardiorenal Syndromes in Critical Care.
Contrib Nephrol. Basel, Karger, 2010, vol 165, pp 153–158

The Changing Pattern of Acute Kidney Injury: From One to Multiple Organ Failure

Mark D. Okusa

Division of Nephrology and the Center for Immunity, Inflammation and Regenerative Medicine,
University of Virginia, Charlottesville, Va., USA

Abstract

Acute kidney injury (AKI) is frequently encountered in the intensive care unit, and from its inception, morbidity and mortality increase in these patients compared to those without AKI. Despite numerous clinical trials and newer pharmacological agents, very little progress has been made to reduce the deaths that occur in this population. An important emerging concept is that AKI does not occur in isolation and it frequently involves other organs. Clinical conditions such as shock, trauma, and sepsis lead to an increase in fluid volume, cytokines/chemokines, uremic toxins and other soluble mediators that are known to affect distant organs. This critical loss of balance of these mediators appears to be due both to a reduction in clearance and increase in production as demonstrated by experimental studies of bilateral nephrectomy and ischemia-reperfusion, respectively. The evidence and mechanisms for distant organ injury following AKI will be discussed.

Mortality in the critically ill population with acute kidney injury (AKI) remains high and has been reported to be 40–60% [1]. AKI is more prevalent as a result of advances in medical and surgical techniques, aging population and increase in drugs for the treatment of various medical conditions. Even after controlling for comorbid conditions, AKI has an independent negative impact on mortality. After adjusting for comorbidities, patients undergoing radiocontrast procedures have a 5- to 6-fold risk of dying. Small rises in serum creatinine in the setting of cardiopulmonary bypass surgery were independently associated with an increase in 30-day mortality. Thus if small increases in serum creatinine are, in fact, independent predictors of increased mortality, why do these patients with AKI die? It is evident that in humans, AKI is seldom an isolated event. Recently, both clinical and experimental have data suggested that AKI may have remote deleterious effects on other organs. Thus it is not surprising that in the intensive care unit, the mortality associated with patients with AKI is 3-fold greater than patients with a similar degree of illness but without AKI. In these patients, AKI is likely to lead to failure of multiple organs including lung, heart, intestines, brain, liver and bone marrow [2]. In support of this concept, many in a cohort of patients with AKI following radiocontrast developed complications after the onset of AKI including sepsis, hemorrhage, central nervous system manifestation and respiratory failure. Thus, AKI can be considered as a complex and multisystemic condition that is thought to lead to a remote organ dysfunction syndrome contributing to a high rate of mortality. Experimental studies provide some insight as to the mechanism by which isolated events leading to the loss of GFR can injure distant organs. Many potential factors may lead to distant organ dysfunction, including circulating factors such as cytokines and chemokines, activated leukocytes, and adhesion molecules leading to immune cell infiltration. Oxidative injury, apoptosis and cellular necrosis contribute to the final pathway of organ dysfunction. Thus, a better understanding of the systemic consequences of AKI will likely lead to appropriate and optimal strategies in AKI management.

AKI and Pulmonary Dysfunction

In critically ill patients, coexistence of AKI and acute lung injury is associated with high mortality of up to 80%. AKI leads to fluid overload and increased mortality, events thought to be due to organ dysfunction. Experimental studies demonstrated increased pulmonary vascular permeability, lung edema, alveolar hemorrhage, and leukocyte trafficking following ischemic AKI [2]. In addition, clinical studies demonstrate a bidirectional interaction between kidney and lung. Whereas AKI is thought to lead to pulmonary dysfunction, pulmonary dysfunction leading to mechanical ventilation is thought to lead to aggravation of AKI.

AKI Leads to Pulmonary Dysfunction. Pulmonary dysfunction from AKI may be due to impaired fluid excretion leading to increased pulmonary hydrostatic pressure and pulmonary edema or due to the production (and decreased clearance) of mediators of lung injury. Hassoun and colleagues [2] addressed the functional and genomic response of the lung following acute renal ischemia-reperfusion injury (IRI) or bilateral nephrectomies. These authors found bilateral renal IRI and bilateral nephrectomies produced disparate effects on lung structure and function. Their data provide clear distinctions of ischemia-related changes compared with bilateral nephrectomy-related changes in the lung genomic profile.

In contrast, acute cessation of kidney function achieved by either bilateral nephrectomies or by bilateral renal pedicle clamps was associated with an increase in multiple serum cytokines leading to pulmonary vascular congestion and neutrophil infiltration leading to acute lung injury. It is likely that both loss of clearance of systemic mediators (bilateral nephrectomy) and increase in production of systemic mediators through kidney injury lead to lung permeability and injury, and the increase in morbidity associated with AKI.

Experimental studies indicate that AKI induced lung leukocyte infiltration and cytokines lead to cellular apoptosis and cell death. Neutrophil infiltration in the kidney is a key finding in kidney IRI and similarly following AKI, lung neutrophil content is increased [3]. Further studies to assess compartmentation of neutrophils in lung by flow cytometry and by microscopy demonstrated that neutrophils are marginated (attached to the endothelium) but have not trans-migrated [3], suggesting that intraluminal and not intersititial neutrophils are contributing to lung injury following kidney IRI. Neutrophils are likely to contribute to injury through phagocytosis and through the production and release of numerous mediators upon degranulation or death including reactive oxygen species and proteases. Furthermore, danger-associated molecular patterns, endogenous intracellular molecules, also referred to as alarmins that lead to activation of dendritic cells, augment the immune response.

Proinflammatory cytokines, including tumor necrosis factor (TNF-α), IL-6, IL1β and others, have been demonstrated in both kidney IRI and bilateral nephrectomies [4]. Klein et al. [5] showed that IL-6 is a direct mediator of AKI-induced increase in vascular permeability, leukocyte trafficking, and increased edema following bilateral IRI or nephrectomies [5]. In contrast, administration of the anti-inflammatory cytokine IL-10 before bilateral nephrectomy reduced lung injury and inflammatory markers [4].

Cellular mechanisms for altered water transport in the lungs in response to AKI have been identified by recent studies demonstrating reduced expression of epithelial sodium channel, sodium-potassium ATPase and aquaporin 5 [2]. The reduced expression of these ion and water transport proteins may lead to reduced fluid removal by injured lungs.

Pulmonary Dysfunction Exacerbates AKI. While it is evident that AKI induces distant organ injury, it is also thought that acute lung injury can aggravate renal function [6]. In patients with AKI that require mechanical ventilation, the mortality rate for AKI was higher than for those not requiring mechanical ventilation. Mechanical ventilation may induce AKI through effects on arterial blood gases, systemic and renal blood flow and pulmonary inflammatory reactions releasing cytokines [6]. Maintenance of normal gas exchange is often not achieved in severely ill patients with acute lung injury leading to hypoxemia and hypercapnia. Both human and animals studies have demonstrated that both hypoxemia and hypercapnia together or separately can induce reductions in renal blood flow. Mechanical ventilation, especially with increasing positive end-expiratory pressure, also affects cardiac performance through decreasing cardiac output, preload, pulmonary vascular volume, resistance and right ventricular afterload, effects that alter renal blood flow [6] as well as the distribution of cortical and medullary blood flow. These studies suggest that the kidney lung interaction is bidirectional and an important contributor to the mortality associated with AKI.

AKI and Central Nervous System Dysfunction

Central nervous system (CNS) dysfunction frequently accompanies acute loss of kidney function. Systemic effects of altered electrolyte, acid-base balance and brain inflammation in response to AKI contribute to the CNS manifestations of AKI. Mice with AKI exhibited increased brain vascular permeability and an increase in the level of glial fibrillary acidic protein, a marker for activated glial cells during brain inflammation and activated microglial cells (brain macrophages) that were associated with an increased number of pyknotic neurons [2]. Thus, CNS manifestations of AKI may be the result of distant effects of AKI-induced inflammation.

AKI and Cardiac Dysfunction

AKI is frequently associated with cardiac dysfunction, which likely contributes to the overall high morbidity and mortality. Recently the interaction between the heart and kidney was described as 'cardiorenal syndrome' of which there are five types (types 1–5) based upon a new nomenclature [7]. Type 3 cardiorenal syndrome is characterized by AKI leading to acute cardiac dysfunction. AKI can lead to cardiac decompensation by fluid overload, hyperkalemia, uremia, activation of the sympathetic nervous system, pericarditis and acidemia [7]. AKI promotes cardiac injury characterized by hypertrophy and fibrosis. Following ischemic AKI, macrophage number and gene expression for the macrophage chemokine osteopontin were increased in the heart, resulting in pathological proliferation and inflammation. Kelly [8] used an animal model of bilateral renal ischemia and found increased levels of TNF-α, IL-1 and intercellular adhesion molecule-1 mRNA in the heart after renal ischemia-reperfusion. These changes likely contributed to the observed apoptosis of myocardial cells by renal ischemia insufficiency but not by bilateral nephrectomy. These results suggest that renal ischemia-reperfusion but not uremia is necessary for the apoptosis observed. Furthermore, following 48 h reperfusion there was an increase in left ventricular end-diastolic diameter and decreased fractional shortening by M-echocardiography. So, not surprisingly, circulating factors and volume changes are likely to considerably affect cardiac function following AKI.

AKI and Liver Dysfunction

The liver plays a critical role in routine homeostatic functions – synthesis of various molecules including coagulation factors, albumin, and glucose via gluconeogenesis and clearing blood of bilirubin and drugs. Hepatic dysfunction is frequently associated with AKI, which complicates effective treatments in the intensive care unit. Following acute renal IRI, the enzymatic and non-enzymatic antioxidant defense systems in hepatic tissue have been shown to be significantly impaired. Liver superoxide dismutase, glutathione peroxidase, and catalase activities decreased and serum ALT and AST activities increased after kidney reperfusion injury [9]. Both IRI and bilateral nephrectomies induce an increase in serum ALT and AST within 6 h [9]. Furthermore, hepatic levels of TNF-α and malondialdehyde, an index of lipid peroxidation, increased significantly after 6 and 24 h of renal ischemia or nephrectomy suggesting that activation of oxidative stress induces tissue damage in hepatocytes. Severe liver dysfunction may in turn lead to further exacerbation of kidney injury through affects that produce renal vasoconstriction and hepatorenal syndrome.

AKI and the Immune System

Immune cells contribute to injury and repair of the kidney following IRI as recently reviewed [10]. In response to chemokines, bone marrow cells are mobilized and infiltrate into the kidney. Monocyte/macrophages through CCR2 and CX3CR1 receptors are mobilized from bone marrow in response to chemokines, leading to inflammation and injury following acute kidney IRI [10]. Additionally, ischemic kidneys release SDF1 and uric acid leading to the release of bone marrow-derived progenitor cells that might contribute to tissue protection and repair. AKI may also lead to abnormal leukocyte function and immune suppression increasing the susceptibility of patients to infections and death.

Distant Organ Injury following AKI: Implications for Therapy

High mortality rates from AKI are likely due to multiorgan system dysfunction. Recent studies from rodent models have identified multiple mechanisms leading to organ dysfunction although human data are still lacking. The growing literature demonstrating pronounced distant organ effects following AKI necessitates human studies to confirm what has been identified in animal models. Therapeutic interventions should focus on attenuating kidney injury but should also be aimed at the effects of AKI on distant organs. Such therapies might include drugs that broadly target multiple organs, which may lead to disease modification and reduction in morbidity and mortality in patients with AKI.

Other therapies may be targeted to mediators such as cytokines and chemokines that are thought to induce distant organ dysfunction. Certainly drugs as well as extracorporeal therapies may contribute to controlling circulating signals that lead to organ dysfunction following AKI.

References

1 Uchino S, Kellum JA, Bellomo R, Doig GS, Morimatsu H, Morgera S, Schetz M, Tan I, Bouman C, Macedo E, Gibney N, Tolwani A, Ronco C: Acute renal failure in critically ill patients: a multinational, multicenter study. JAMA 2005;294:813–818.

2 Li X, Hassoun HT, Santora R, Rabb H: Organ crosstalk: the role of the kidney. Curr Opin Crit Care 2009;15:481–487.

3 Awad AS, Rouse M, Huang L, Vergis AL, Reutershan J, Cathro HP, Linden J, Okusa MD: Compartmentalization of neutrophils in the kidney and lung following acute ischemic kidney injury. Kidney Int 2009:689–698.

4 Hoke TS, Douglas IS, Klein CL, He Z, Fang W, Thurman JM, Tao Y, Dursun B, Voelkel NF, Edelstein CL, Faubel S: Acute renal failure after bilateral nephrectomy is associated with cytokine-mediated pulmonary injury. J Am Soc Nephrol 2007;18:155–164.

5 Klein CL, Hoke TS, Fang WF, Altmann CJ, Douglas IS, Faubel S: Interleukin-6 mediates lung injury following ischemic acute kidney injury or bilateral nephrectomy. Kidney Int 2008;74:901–909.

6 Koyner JL, Murray PT: Mechanical ventilation and lung-kidney interactions. Clin J Am Soc Nephrol 2008;3:562–570.

7 Ronco C, Haapio M, House AA, Anavekar N, Bellomo R: Cardiorenal syndrome. J Am Coll Cardiol 2008;52:1527–1539.

8 Kelly KJ: Distant effects of experimental renal ischemia/reperfusion injury. J Am Soc Nephrol 2003;14:1549–1558.

9 Golab F, Kadkhodaee M, Zahmatkesh M, Hedayati M, Arab H, Schuster R, Zahedi K, Lentsch AB, Soleimani M: Ischemic and non-ischemic acute kidney injury cause hepatic damage. Kidney Int 2009;75:783–792.

10 Li L, Okusa MD: Blocking the immune response in ischemic acute kidney injury: the role of adenosine 2A agonists. Nat Clin Pract Nephrol 2006;2:432–444.

Mark D. Okusa, MD
Division of Nephrology, Box 800133
Charlottesville, VA 22908 (USA)
Tel. +1 434 924 2187, Fax +1 434 924 5848
E-Mail mdo7y@virginia.edu

Organ Crosstalk in Multiple Organ Dysfunction Syndromes

Ronco C, Bellomo R, McCullough PA (eds): Cardiorenal Syndromes in Critical Care.
Contrib Nephrol. Basel, Karger, 2010, vol 165, pp 159–165

The Kidney in Respiratory Failure and Mechanical Ventilation

Patrick T. Murray

UCD School of Medicine and Medical Science, Catherine McAuley Centre, Mater Misericordiae University Hospital, Dublin, Ireland

Abstract

Acute lung Injury (ALI) and acute kidney injury (AKI) are complications often encountered in the setting of critical illness. Both forms of end-organ injury commonly occur in similar settings of systemic inflammatory response syndrome, shock, and evolving multiple organ dysfunction. Distant organ effects of apparently isolated injuries to the lungs, gut, and kidneys have all been discovered in recent years. In this review of the emerging evidence of deleterious bidirectional organ crosstalk between the lungs and kidney, we will focus on the role of ventilator-induced kidney injury in the pathogenesis of AKI in patients with ALI.

Acute Lung Injury and Acute Kidney Injury

It was not so long ago that the term 'pulmonary renal syndrome' was largely synonymous with the combination of immune alveolar hemorrhage and rapidly progressive glomerulonephritis. In contrast to the rarity of such autoimmune conditions, we are much more frequently faced with the clinical challenge of managing a harmful bidirectional interaction between acute lung injury (ALI) and acute kidney injury (AKI) caused by more common, non-autoimmune phenomena. This chapter will focus on the current understanding of one form of pulmonary renal syndrome: the effects of ALI and mechanical ventilation on renal function. In recent years, the implementation of lung-protective ventilatory strategies disclosed the role of inflammatory mediators of ALI and specifically ventilator-induced lung injury in the pathogenesis of AKI [1]. Animal models have demonstrated that in the setting of known ALI (e.g. acid aspiration), lung-protective ventilatory strategies serve to limit the proinflammatory cascade that has been implicated in the development of AKI [1]. Conversely, in models of AKI caused by unilateral renal ischemia, bilateral renal ischemia, or bilateral nephrectomy, it is clear that each of these distinct renal injuries results in a unique pattern of pulmonary inflammation and alveolar injury [1, 2]. Taken together, the interplay of renal injury inducing physiologic and inflammatory changes in the lung parenchyma, and pulmonary injury leading to distant effects on kidney function, leads to the existence of a vicious cycle of multiple organ failure.

Acute Lung Injury

ALI is a clinical syndrome characterized by acute (<7 days) onset of severe hypoxemia and bilateral pulmonary infiltrates in the absence of elevated left atrial pressures [3]. Alternatively, ALI has been defined as a subject who has a ratio of partial pressure of oxygen in the arterial blood (PaO_2) to the inspired fraction of oxygen (FIO_2) of <300. Acute respiratory distress syndrome (ARDS) is the most severe form of ALI where this ratio (PaO_2/FIO_2) is <200. ALI and ARDS occur in the setting of a variety of clinical scenarios including sepsis, trauma, blood product transfusion, pancreatitis, burns as well as pneumonia, and other forms of lung injury (inhalation, aspiration etc.). A mechanistic classification system of respiratory failure for patients requiring mechanical ventilation has been devised based upon the four pathophysiological patterns of respiratory failure [1]. Type I respiratory failure (acute hypoxemic respiratory failure) results from a failure of oxygenation due to increased intrapulmonary shunting of blood (characterized by an increased fraction of cardiac output passing through the pulmonary circulation without re-oxygenation, referred to as $\uparrow Q_{shunt}/Q_{t(cardiac\ output)}$), and is classically seen in the setting of pulmonary edema (both cardiogenic and non-cardiogenic ALI/ARDS) as well as pneumonia, ventilation-perfusion mismatch and pulmonary hemorrhage.

It is these concrete definitions of ALI and its severity that have led to the successful identification of patients for several large multicenter clinical trials in ALI [1, 3]. The results of some of these trials will be discussed in the review below, but it deserves noting that while these trials are not always positive, their protocols have led to standardization in the care of patients with ALI. This is in stark contrast to the AKI field, in which clinical investigation has been less successful.

Renal Effects of ALI and Mechanical Ventilation

AKI is an independent predictor of mortality in ICU patients. Unfortunately, AKI often develops as a component of multi-organ system dysfunction in critically ill patients and may lead to mortality rates in excess of 60%, depending on the setting. Severe AKI in critically ill patients is typically part of a triad with shock and respiratory failure requiring mechanical positive-pressure ventilation (PPV). The physiologic impact of PPV and its effects on renal perfusion and function are well documented, including hemodynamic and neurohormonal phenomena [1, 4]. Recent advances in critical care, including the implementation of lung-protective ventilatory strategies, have disclosed the role of inflammatory mediators of ALI and specifically ventilator-induced lung injury in the pathogenesis of AKI, which in some cases might appropriately be called ventilator-induced kidney injury. Mounting evidence points to the role of cytokines and chemokines in the pathogenesis of ARDS. Emerging data further suggest that the proinflammatory effects of PPV may be a source of AKI, especially in the setting of mechanical ventilation and lung-injurious ventilator strategies (higher tidal volumes and lower positive end expiratory pressure, PEEP) [1]. In summary, the decreased morbidity and mortality of patients with ALI achieved with lung-protective strategies for mechanical ventilation are likely mediated not only by amelioration of ventilator-induced lung injury and inflammation, but also by diminished injurious crosstalk with distant organs, including the kidneys.

Prevention and Treatment of AKI in Patients on Mechanical Ventilation

Low Tidal Volume Ventilation

It is recommended that patients with ALI are treated with the low tidal volume ventilation strategies that have been validated by the Acute Respiratory Distress Syndrome Network (ARDSNet) trial and subsequent studies by this research group [5]. The seminal trial of 861 subjects with ALI/ARDS (the most severe form of ALI) randomized patients to receive tidal volumes of 12 or 6 ml/kg of ideal body weight in order to determine the effect of lower tidal volumes on mortality. The trial was stopped early when mortality was found to be lower in the low tidal volume group (31 vs. 39.8%, $p = 0.007$). Additionally, those randomized to the low volume arm had more ventilator-free days during their first 4 weeks after study enrollment (mean ± SD, 12 ± 11 vs. 10 ± 11 days; $p = 0.007$). In this seminal trial, the investigators crudely defined AKI (renal failure) as a serum creatinine of >2 mg/dl (177 μmol/l), with no reference to the subjects' baseline values. Despite this rudimentary definition, they were still able to demonstrate that those treated with a lung-protective lower tidal volume incurred statistically fewer days with AKI compared to those exposed to the lung-injurious ventilation strategy (18 ± 11 vs. 20 ± 11 days, $p = 0.005$) [5]. This trial validated prior animal and human studies demonstrating the renal benefits of low tidal volume ventilation in the setting of ALI [6, 7].

Fluid and Volume Management

Following the low tidal volume trial, the ARDSNet group performed the Fluids and Catheters Treatment Trial (FACTT) study [8]. This trial used a 2 × 2 factorial design to combine two trials in 1,000 patients with ALI receiving low tidal volume mechanical ventilation. The second of these trials compared the effects of a fluid-conservative versus fluid-liberal algorithm [8]. This study was performed based on data which supported the premise that negative fluid balance (fluid restriction and diuretic use) was associated with shorter ICU stays as a result of decreased length of mechanical ventilation. The primary endpoint was 60-day survival, comparing a fluid-liberal strategy (target central venous pressure (CVP) <10 mm Hg or pulmonary capillary wedge pressure (PCWP) <14 mm Hg) versus a fluid-conservative arm (CVP <4 mm Hg, PCWP <8 mm Hg). Chronic kidney disease patients were not excluded (except those receiving chronic dialysis); baseline serum creatinine values in the groups were: fluid-conservative 1.24 ± 0.004 mg/dl; fluid-liberal 1.29 ± 0.04 mg/dl; $p = 0.39$).

Not surprisingly, the fluid-conservative arm received more furosemide and had a negative fluid balance over the first 7 study days (–136 ± 491 ml) compared with the substantial positive fluid balance of those in the liberal arm (6,992 ± 502 ml; $p < 0.001$). Additionally, the fluid-conservative arm had significantly more ICU-free days within the first week and month of the study, as well as an increased number of ventilator-free days (14.6 ± 0.5 conservative vs. 12.1 ± 0.5 liberal; $p < 0.001$). Importantly, this underpowered study was not able to demonstrate any difference in 60-day mortality (25.5 ± 1.9% conservative; 28.4 ±2.0% liberal; $p = 0.3$). Additionally, there was no difference in the occurrence of AKI (again defined crudely as a serum creatinine of >2 mg/dl) within the first 7 or 28 days of the study.

The major positive finding of this trial was the improvement in ventilator-free survival with fluid-conservative management. It is important to appreciate that this benefit was achieved by the prevention of positive fluid balance rather than the intended approach of achieving negative fluid balance. Specifically, although the conservative fluid arm protocol targeted a CVP of 4 mm Hg or a pulmonary artery occlusion pressure (PAOP) of 8 mm Hg compared to a CVP of 10–14 mm Hg or PAOP of 14–18 mm Hg in the fluid-liberal arm, the fluid-conservative targets were not generally achieved. Subjects with a central venous catheter began the trial with an average CVP of 12–13 mm Hg; those in the fluid-liberal arm only decreased slightly to 11–12 mm Hg over the first 7 days, while those in the fluid-conservative arm decreased to 8–9 mm Hg. In the pulmonary artery catheter arm, both groups started with a PAOP of 15–16 mm Hg, with those in the fluid-liberal arm finishing close to 16 mm Hg, which was higher than the 12–14 mm Hg for the fluid-conservative group. Regardless of the monitoring catheter being in place, while the fluid-conservative arms clearly achieved tighter fluid control, analysis of their intravascular fluid pressures demonstrates that the fluid-conservative group did not achieve the low level of filling pressure targeted by the study protocol, and this may have contributed to the lack of a significant difference in the overall mortality and in the AKI rates between the fluid-liberal and conservative arms.

Examination of the cumulative fluid balance data underscored this impression; the cumulative fluid balance in the fluid-conservative and fluid-liberal arms of this trial were compared with 2 patient populations which were recently studied in the setting of ALI. The first of these prior ARDSNet trials was the ALVEOLI study, which tested lower tidal volumes with higher PEEP and, after enrolling 549 patients, found no further improvement in survival with higher PEEP. The other group compared were those patients who received the lung-protective 6-ml/kg tidal volumes in the original ARDSNet tidal volume study. Of note, these data suggest that the fluid-liberal arm accurately serves as a 'real world' example of fluid balance in ALI patients, as the net fluid in (roughly 5.6 liters positive over the first 4 days and 7.1 liters positive for the first 7 days) is similar to the balance achieved in previous ARDSNet trials (which did not attempt to control fluid status). Conversely, the fluid-conservative arm was not able to obtain a net negative fluid balance over the first 4 days, but rather a net even fluid balance (0.2 liters net positive in 4 days; with 0.01 liters positive on day 7). As such, those clinicians looking to replicate the protocols set forth in this landmark trial in clinical practice need only to ensure that they achieve an even net fluid balance, rather than needing to achieve negative fluid balance. Similarly, one must recognize that this trial had a very intricate protocol to minimize hemodynamic instability. At any sign of shock (defined as a mean arterial pressure of <60 mm Hg or end-organ hypoperfusion), the protocol required the discontinuation of diuretics, as well as fluid boluses and the initiation and/or escalation of vasoactive medications (dobutamine, norepinephrine, epinephrine, vasopressin, or phenylephrine) as needed. This last point is yet another important caveat to consider when attempting to treat patients with ALI.

In summary, although in the FACTT study improved pulmonary function and lung mechanics did not translate into improved 60-day patient survival (the primary trial endpoint), improved lung function did impact the more restricted endpoint of ventilator-free survival. As discussed above, the failure to achieve actual negative fluid balance may have led to a less than maximal improvement in 60-day survival. Alternatively, unquantified adverse effects of fluid restriction and diuretics on non-pulmonary organs may have led to competing morbidities in the dry group. For example, although not statistically significant, there was a trend towards higher serum creatinine levels in the conservative group ($p = 0.07$), along with significantly higher blood urea nitrogen and bicarbonate levels. On the other hand, as previously noted there was a trend towards an increased need for renal replacement therapy (RRT) in the fluid-liberal group (14%) compared to the fluid-conservative group (10%; $p = 0.06$). However, it is important to note that the need for RRT is a subjective endpoint that is open to bias due to the unblended nature of the trial. Elevated serum creatinine and increased blood urea nitrogen were perhaps tolerated in the fluid-conservative arm, due to the trial's aim to run the subjects at lower intravascular volumes. Conversely, it is reasonable to suppose that positive fluid balance would be tolerated to a greater extent in the fluid-liberal group, but that eventually excess volume might lead to an increased rate of RRT. Further investigation of the role of fluid management, in an appropriately powered study, in the setting of ALI would be better served by the use of more precise definitions and biomarkers of AKI, among other measures of adequate tissue perfusion. Finally, it is also important to note that patients receiving RRT were excluded from participation in the FACTT trial, and the study did not include titration of fluid balance according to the protocol in AKI requiring RRT initiation developed following enrollment. Thus, although emerging data suggest an adverse effect of fluid overload on outcomes in AKI [9, 10], we cannot confidently extrapolate that use of RRT to achieve the approach to fluid management used in the FACTT trial will similarly improve ventilator-free survival.

Conclusion

In conclusion, it is evident that there is a deleterious bidirectional relationship between the AKI and ALI. Increasing evidence points to crosstalk between these two distant organs, and shows that injury to one organ may initiate and aggravate injury to the other. Recent data show that the kidneys play an important role in the production and elimination of mediators of inflammation and ALI. Conversely, exposure to the inflammatory milieu of ALI and mechanical ventilation may precipitate the onset of AKI. While there have been recent advances in approaches to limit ventilator-induced lung injury and decrease the duration of mechanical ventilatory support, the net effect of these advances on the incidence and severity of AKI in critically ill patients remains to be determined. Close collaboration between intensivists and nephrologists is required to optimize management and improve outcomes of patients with ALI and AKI.

References

1 Koyner JL, Murray PT: Mechanical ventilation and the kidney. Blood Purif 2009;29:52–68.

2 Scheel PJ, Liu M, Rabb H: Uremic lung: new insights into a forgotten condition. Kidney Int 2008;74:849–851.

3 Liu KD, Matthay MA: Advances in critical care for the nephrologist: acute lung injury/ARDS. Clin J Am Soc Nephrol 2008;3:578–586.

4 Pannu N, Mehta RL: Effect of mechanical ventilation on the kidney. Best Pract Res Clin Anaesthesiol 2004;18:189–203.

5 Ventilation with lower tidal volumes as compared with traditional tidal volumes for acute lung injury and the acute respiratory distress syndrome. The Acute Respiratory Distress Syndrome Network. N Engl J Med 2000;342:1301–1308.

6 Ranieri VM, Suter PM, Tortorella C, et al: Effect of mechanical ventilation on inflammatory mediators in patients with acute respiratory distress syndrome: a randomized controlled trial. JAMA 1999;282:54–61.

7 Ranieri VM, Giunta F, Suter PM, Slutsky AS: Mechanical ventilation as a mediator of multisystem organ failure in acute respiratory distress syndrome. JAMA 2000;284:43–44.
8 National Heart, Lung, and Blood Institute Acute Respiratory Distress Syndrome (ARDS) Clinical Trials Network, Wiedemann HP, Wheeler AP, Bernard GR, et al: Comparison of two fluid-management strategies in acute lung injury. N Engl J Med 2006;354:2564–2575.
9 Michael M, Kuehnle I, Goldstein SL: Fluid overload and acute renal failure in pediatric stem cell transplant patients. Pediatr Nephrol 2004;19:91–95.
10 Payen D, de Pont AC, Sakr Y, et al: A positive fluid balance is associated with a worse outcome in patients with acute renal failure. Crit Care 2008;12:R74.

Patrick T. Murray, MD
UCD School of Medicine and Medical Science, Catherine McAuley Centre
Mater Misericordiae University Hospital
Nelson Street, Dublin 7 (Ireland)
Tel. +353 1 7166319, Fax +353 1 7166355

Ronco C, Bellomo R, McCullough PA (eds): Cardiorenal Syndromes in Critical Care.
Contrib Nephrol. Basel, Karger, 2010, vol 165, pp 166–173

Pulmonary Renal Syndrome and Emergency Therapy

W. Kline Bolton

University of Virginia HS, Charlottesville, Va., USA

Abstract

The pulmonary renal syndrome is a symptom complex with simultaneous involvement of the kidneys and lungs. While this can be related to any concomitant kidney and lung dysfunction, the usual connotation is that of kidney dysfunction associated with pulmonary hemorrhage frequently manifesting as hemoptysis. Many causes of the pulmonary renal syndrome have been defined, and therapy is directed at the underlying etiologic factor for the syndrome. At times the presentation is life-threatening. In those situations emergency therapy must be initiated as soon as possible. A number of therapies are now available for this syndrome including pulse methylprednisolone and plasmapheresis. Emerging novel therapies show promise of augmenting the armamentarium of the clinician's therapeutic options. This review addresses the author's approach to the elaboration of the etiology of pulmonary renal syndrome and the therapeutic options available to the clinician.

The term *pulmonary renal syndrome* indicates that there is concomitant renal and pulmonary dysfunction. This may occur because of a primary renal abnormality with consequent pulmonary involvement, or vice versa with pulmonary involvement followed by renal dysfunction. This syndrome involves a very broad spectrum of clinical and laboratory presentations ranging from chronic kidney disease with bronchitis or volume overload to acute renal disease of multiple etiologies with pulmonary involvement [1]. The clinical presentation can be insignificant or can be a devastating life-threatening event requiring intensive care unit admission, monitoring and therapy. Technically, the pulmonary renal syndrome could consist of any type of pulmonary involvement with renal involvement but the usual connotation is that of kidney dysfunction associated with pulmonary hemorrhage. This may be manifest as hemoptysis although pulmonary hemorrhage can occur without overt hemoptysis. The pulmonary

Table 1. Renal disease with pulmonary hemorrhage

- Glomerulonephritis
  - Goodpasture's syndrome
  - Immune complex glomerulonephritis
  - Acute silicoproteinosis
  - Glomerulonephritis with CHF
  - Other types of glomerulonephritis
- Vasculitis
  - Polyarteritis nodosa
  - Microscopic polyangiitis
  - Wegener's granulomatosis
  - Henoch-Schönlein purpura
  - Mixed IgG-IgM cryoglobulinemia
  - Churg-Strauss
  - Lymphoid granulomatosis
  - Necrotizing sarcoid granulomatosis
  - Hypersensitivity
  - Behçet syndrome
  - Hughes-Stovin
  - Giant cell arteritis
  - Hypocomplementemic urticaria
  - Takayasu arteritis
- Collagen vascular disease
  - Systemic lupus erythematosus
  - Progressive systemic sclerosis
  - Rheumatoid arthritis
- Cardiovascular
  - Advanced uremia with superimposed CHF
  - Mitral stenosis
  - Pulmonary embolism with infarct and underlying renal vein thrombosis
- Infection
  - Legionnaire's disease
  - Necrotizing pneumonitis
  - Bronchitis
  - Tuberculosis
  - Fungus
  - Lung abscess
- Bronchiectasis
- Malignancy with associated glomerulopathy
- Blood dyscrasias
  - Thrombotic thrombocytopenic purpura
  - Hemolytic uremic syndrome
- Idiopathic hemosiderosis
- D-Penicillamine toxicity, trimellitic anhydride toxicity

renal syndrome is also considered to involve alveolar space disease rather than upper airway disease and results from inflammation and damage to the alveolar space or from alveolar hemorrhage, without inflammation [2]. Many causes of pulmonary renal syndrome are the result of systemic disease with incidental hemorrhaging into the alveolar space.

Renal disease with pulmonary involvement and pulmonary hemorrhage is classified most commonly according to the underlying etiologic factor (table 1). These include glomerulonephritis, different forms of systemic vasculitis, collagen vascular diseases, cardiovascular abnormalities, infection, primary pulmonary diseases such as bronchiectasis and malignancy as well as blood dysplasias and idiopathic hemosiderosis. Broadly speaking, this classification allows separation into inflammatory, mostly autoimmune processes, versus non-autoimmune etiologies. During my 30 plus years of caring for patients with rapidly progressive glomerulonephritis and pulmonary hemorrhage, I have seen most of the clinical presentations described in table 1, and have been able to be involved in successful treatment of many of these, although not all. Since the pathogenic process is different for these different entities, treatment is also different and needs to be tailored. Thus, it is critically important to make the diagnosis as quickly and accurately as possible. In some cases it is necessary to treat empirically before definitive diagnosis is made because of the clinical deterioration of the patient.

Clinical Presentation

As in other diseases, it is extremely important to obtain an accurate history and physical as well as relevant family and social history including employment activities such as silica exposure, exposure to possible toxic agents and the social use of agents that may be involved such as smoking, drugs of abuse, and covert risk factors. Patients with pulmonary renal syndrome generally have an abrupt onset of symptoms, frequently with hemoptysis, cough, fever, shortness of breath and may progress rapidly to the need for mechanical ventilation. Hemoptysis may be minor or absent on initial evaluation. In a number of patients the presence of a positive carbon monoxide diffusion test, falling hematocrit and pulmonary infiltrates associated with bronchoscopic findings may be required for the diagnosis. Both the physical examination and the radiologic examination are non-specific and consist of respiratory distress and pulmonary findings such as pulmonary infiltrates which may be seen with many different etiologies.

Laboratory Findings

Laboratory evaluation is essential and consists of obtaining the basic laboratory studies shown in table 2. Several caveats regarding the critical tests for anti-neutrophilic cytoplasmic antibody (ANCA) and anti-glomerular basement membrane (GBM) tests are worth emphasizing. Currently, p-ANCA and c-ANCA are performed using recombinant proteins and ELISA or immunoassays. Many laboratories utilize anti-IgG as the secondary antibody. A few patients can have other types of ANCA including most specifically IgM with more severe pulmonary involvement than IgG ANCA. Thus, it is important to ascertain that the diagnostic laboratory uses an anti-immunoglobulin second antibody recognizing all classes of immunoglobulins. Most anti-GBM assays now also involve recombinant proteins for the α3 non-collagenous domain of type IV collagen, but pulmonary hemorrhage has rarely been reported with autoimmune antibody responses to other α chains in patients with pulmonary hemorrhage [3]. This false negative result can be avoided by examination by indirect immunofluorescence on normal human kidney tissue by an experienced individual to detect antibodies to other constituents of the kidney. In those patients with

Table 2. Pulmonary renal syndrome: essential data for diagnosis and follow-up

1. Thorough history and physical examination, including rectal, pelvic and breast examinations
2. ANCA – IgG, IgM
3. Anti-GBM antibodies – recombinant protein, indirect immunofluorescence on kidney tissue
4. Collagen vascular disease serum profile
5. Hepatitis profile, HIV testing
6. Metabolic profile
7. CBC with differential
8. Urinalysis
9. Chest x-rays, stool guaiac
10. Sedimentation rate
11. Bronchoscopy with bronchoalveolar lavage
12. Carbon monoxide diffusion test
13. May need CT pulmonary angiogram
14. Echocardiogram
15. Renal, lung biopsy

pulmonary hemorrhages with anti-GBM kidney disease, the majority, >90%, will have a positive test for anti-GBM antibody if that is the etiology of the pulmonary renal syndrome.

Finally, the evaluation of the differential diagnosis of the pulmonary renal syndrome relies on some old-fashioned techniques as well as newer investigative procedures. First, a careful and critical examination of the urinary sediment, not just the dipstick, by an experienced clinician is essential in the differential diagnosis of pulmonary hemorrhage. This examination can roughly classify patients as those with inflammatory or non-inflammatory, benign sediments. In the latter case an inflammatory etiology of the kidney damage is unlikely to be involved in the syndrome. Patients with debris and muddy brown casts likely represent those who have acute kidney injury consequent to ischemic or toxic damage to the kidney, while patients who have dysmorphic red cells with RBC casts and proteinuria by dipstick represent those with glomerulonephritis. These etiologies consist of primary glomerulonephritis and vasculitis most commonly but also other types of glomerulonephritis. Finding leukocytes, especially eosinophils with or without bacteria in the urine, should raise the possibility of acute interstitial nephritis for the etiology of the acute kidney injury with secondary involvement of pulmonary structures. In all cases of acute pulmonary renal syndrome it is essential to obtain the renal failure parameters which indicate either a pre-renal or nephrotoxic component of renal failure. The use of the fractional excretion of sodium, urea and urate, in combination with the urinary sediment, should suggest the underlying relationship between the kidney and the lung.

Further elaboration of the pulmonary component of the pulmonary renal syndrome usually involves fiber optic bronchoscopic examination and should be performed with bronchoalveolar lavage focusing in areas of radiologic abnormalities (infiltrates). Three or more samples should be taken sequentially from the area. The presence of hemorrhage in these areas is characterized by continuing blood in sequential samples. There is frequently hemosiderin containing macrophages. Intra-alveolar bleeding is also indicated by carbon monoxide diffusion tests which will be enhanced if blood is present in alveolar spaces. If a definitive diagnosis has not been achieved by this point, then obtaining a histologic diagnosis from kidney is imperative to diagnose and direct therapy. Caveats include the fact that Wegener's granulomatosis is rarely diagnosed directly from kidney biopsies although segmental glomerular necrosis with glomerulonephritis suggestive of vasculitis may be seen. The presence or absence of anti-GBM antibody in the kidney is generally definitive for that diagnosis, although rarely, patients can have strong circulating anti-GBM antibody titers without GBM deposits, and are still presumed to have anti-GBM disease. Biopsy of the lung may suggest the type of glomerulonephritis. However, this is notoriously poor for defining anti-GBM disease as the lung basement membrane has a high degree of autofluorescence and is quite difficult to read without heavy deposits or a very experienced individual reading the sections. Finally, echocardiography is frequently obtained to assess the role of any cardiac dysfunction in the clinical syndrome and a computed tomographic pulmonary angiogram may be needed if pulmonary embolism is suspected.

Treatment

Therapy of the pulmonary renal syndrome obviously should be directed first at identified pathologic processes involving the kidney and the lung. This includes cessation of potential medications or injurious agents that could be involved, treatment of any underlying process such as systemic lupus erythematosus or other types of connective tissue disease, and treatment of underlying infections with appropriate antibiotics.

At times, especially in the intensive care setting, involvement of the lungs may be so critical that there may not be time to obtain the necessary tests to make a definitive diagnosis. In that case the mainstay of therapy is the infusion of intravenous methylprednisolone 1–2 g over 20 min every day for 3–5 days. Intravenous cyclophosphamide or rituximab have also been advocated but there is less information on these therapies [4]. It is critical in this setting to be certain that the patient does not have an underlying infection. I have seen patients die from disseminated aspergillosis even in the presence of a

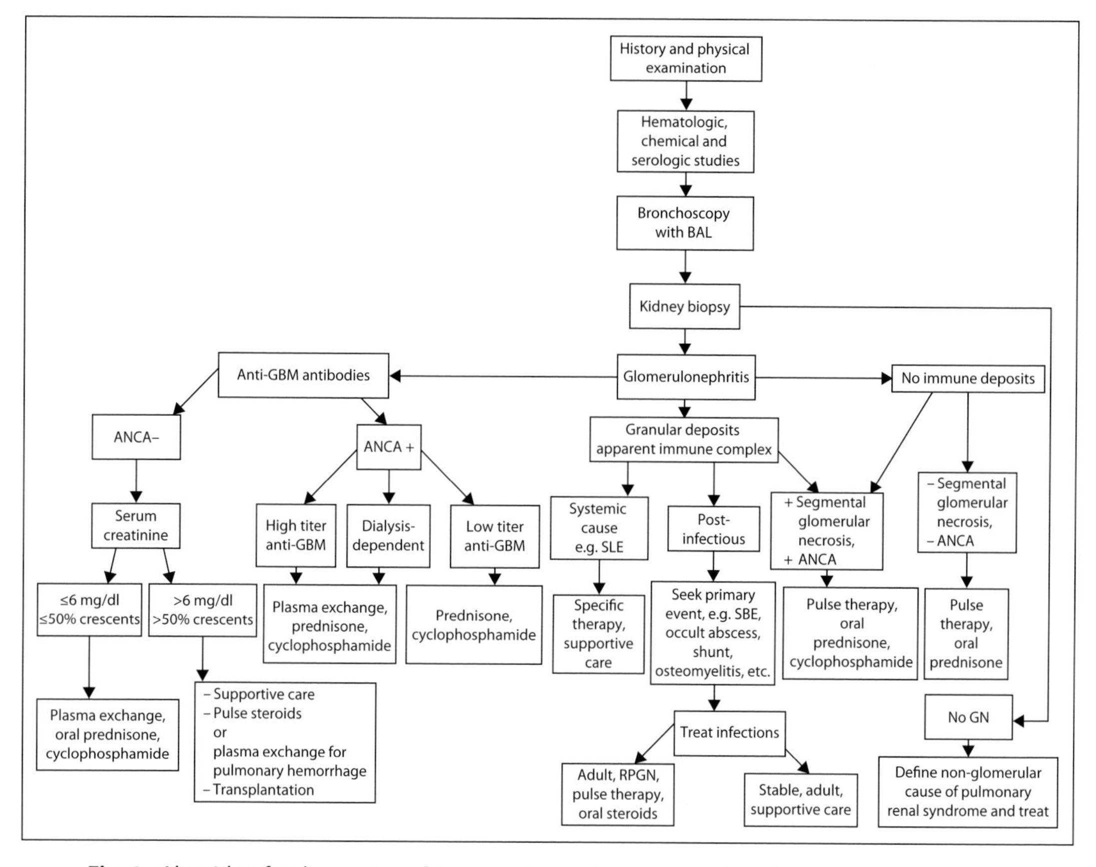

Fig. 1. Algorithm for diagnosis and treatment on pulmonary renal syndromes.

negative bronchoscopic examination. The institution of therapeutic apheresis, preferably with plasma, provides an added benefit for those patients with vasculitis and perhaps with anti-GBM disease, although the single randomized controlled prospective trial comparing these two modalities in Goodpasture's syndrome did not show a difference between these treatments [5–7]. The administration of intravenous γ-globin to those patients with refractory pulmonary hemorrhage, especially those with vasculitis, may be efficacious and should be considered [8]. Therapy with prednisone 1 mg/kg and cyclophosphamide starting at 2 mg/kg should also be instituted for Goodpasture's and vasculitis, following the leukocyte counts, with adjustment of the cyclophosphamide dose for leukopenia. Maintenance treatment with azathioprine or mycophenolate can be instituted when disease is quiescent. An algorithm for the diagnosis and treatment of the pulmonary renal syndrome is provided in figure 1 and will not only be dependent on the degree of renal involvement

Table 3. Other therapeutic modalities of the future based on animal models

General	Specific
Anti-cytokine/receptor agents	Targeted immunotherapy
1. Interleukin-1 receptor antagonist	1. Complementary peptides
2. TGF-β	2. Lysis of GBM-bound immunoglobulin – EndoS & IdeS
3. Soluble CR-1	3. Antigen-specific B-cell suicide ricin diphtheria toxin, etc.
4. TNF-α-soluble receptor protein	4. Antigen-specific T-cell suicide
5. CTLA4-Ig	5. Antigen-specific and non-specific T-regulatory cells
6. MCP-1	
Immunosuppressive	
1. Deoxyspergualin	
2. Anti-adhesion protein antibodies	
3. Peptides and mutated/substituted peptides	
4. Immunosorbent column/Goodpasture's epitope	
5. ALG – anti-lymphocyte globulin, monoclonal antibodies to T cells, ligands and receptors	
6. Anti-B-cell monoclonal antibodies, retaximab, others	

but also on the degree of pulmonary involvement. Although not discussed, the usual non-specific support modalities should be employed, such as dialysis, ventilation, nutrition, antibiotics, etc.

Future Therapies

Table 3 provides a list of potential future therapeutic interventions for pulmonary renal syndrome of autoimmune origin as demonstrated by empiric use and using animal models. These therapies may be directed at interference in the chemokine and cytokine environment, dissolution of antibodies bound to lung or kidney basement membrane, interference with epitope spreading, a process which augments damage in the target organ, B-cell- and T-cell-specific therapy and antigen-specific suicide of T cells and B cells responsible for antibody production as well as specific and non-specific upregulation of regulatory T cells. Until such time as more effective agents are developed, it behooves the clinician to move quickly to the current mainstays of diagnosis and therapy, since with these devastating diseases the longer the delay between diagnosis and treatment, the worst the outcomes.

References

1 Bolton WK: Goodpasture's syndrome. Nephrology Forum. Kidney Int 1996;50: 1753–1766.
2 Olson AL, Schwarz MI: Diffuse alveolar hemorrhage. Prog Respir Res 2007;36:250–263.
3 Kalluri R, Petrides S, Wilson CB, Tomaszewski JE, Palevsky HI, Grippi MA, Madaio MP, Neilson EG: Anti-α (IV) collagen autoantibodies associated with lung adenocarcinoma presenting as the Goodpasture syndrome. Ann Intern Med 1996;124:651–653.
4 Vincenti F, Cohen SD, Appel G: Novel B-cell therapeutic targets in transplantation and immune-mediated glomerular diseases. J Am Soc Nephrol 2010;5:142–151.
5 Cole E, Cattran D, Magil A, Greenwood C, Math M, Churchill D, Sutton D, Clark W, Morrin P, Posen G, Bernstein K, Dyck R, Canadian Apheresis Study Group: A prospective randomized trial of plasma exchange as additive therapy in idiopathic crescentic glomerulonephritis. Am J Kidney Dis 1992;20:261–269.
6 Jayne DRW, Gaskin G, Rasmussen N, Abramowicz D, Ferrario F, Guillevin L, Mirapeix E, Savage COS, Sinico RA, Stegeman CA, Westman KW, van der Woude FJ, de Lind van Wijngaarden RAF, Pusey CD: Randomized trial of plasma exchange or high-dosage methylprednisolone as adjunctive therapy for severe renal vasculitis. J Am Soc Nephrol 2007;18:2180–2188.
7 Johnson JP, Moore J Jr, Austin HA III, Balow JE, Antonovych TT, Wilson CB: Therapy of anti-glomerular basement membrane antibody disease: analysis of prognostic significance of clinical, pathologic and treatment factors. Medicine 1985;64:219–227.
8 Jayne DRW, Chapel H, Adu D, Misbah S, O'Donoghue D, Scott D, Lockwood CM: Intravenous immunoglobulin for ANCA-associated systemic vasculitis with persistent disease activity. Q J Med 2000;93:433–439.

Prof. W. Kline Bolton, MD, FASN
Box 800133, University of Virginia HS
Charlottesville, VA 22908 (USA)
Tel. +1 434 924 5125, Fax +1 434 924 5848
E-Mail wkb5s@virginia.edu

Ronco C, Bellomo R, McCullough PA (eds): Cardiorenal Syndromes in Critical Care.
Contrib Nephrol. Basel, Karger, 2010, vol 165, pp 174–184

Extracorporeal CO_2 Removal – A Way to Achieve Ultraprotective Mechanical Ventilation and Lung Support: The Missing Piece of Multiple Organ Support Therapy

Silvia Gramaticopolo · Alexandra Chronopoulos · Pasquale Piccinni · Federico Nalesso · Alessandra Brendolan · Monica Zanella · Dinna N. Cruz · Claudio Ronco

Departments of Nephrology, Dialysis & Transplantation, and Intensive Care, International Renal Research Institute, San Bortolo Hospital, Vicenza, Italy

Abstract

Extracorporeal therapies are able to sustain life through different mechanisms. This approach, called multiple organ support therapy, can in fact obtain blood purification by hemodialysis/hemofiltration to replace kidney function, temperature control, electrolyte and acid-base control to mimic homeostatic regulation of the kidney and circulation, fluid balance control to support the right hydration and cardiac performance, cardiac support removing cardiodepressant substances and equilibrating potassium levels, blood detoxification and liver support by coupled plasma filtration and adsorption or direct adsorption on blood (hemoperfusion), immunomodulation and endothelial support in the presence of sepsis by cutting the peaks of pro- and anti-inflammatory mediators, and immunoadsorption or adsorption of specific substances such as endotoxin. A missing piece of this group of therapies was the protective lung support. Today this is made possible by removal of CO_2 either by complete extracorporeal membrane oxygenation or by using decapneization in conjunction with hemofiltration in a system called DECAP/DECAPSMART. In conclusion, circulating blood outside the body and treating it with different filters or cartridges in a multiple organ support therapy may represent an important support for multiple organ dysfunction conditions induced by sepsis, acute respiratory distress syndrome and in recent times by complicated H1N1-related infections.

The incidence of the multiple organ dysfunction syndrome (MODS) is rapidly increasing in intensive care units (ICU). It is usually combined with sepsis and it is the most frequent cause of death in ICU patients. The ICU population displays a progressive increase of the average age and frequently presents complications from major surgery, trauma and sepsis with hemodynamic instability and evolution towards MODS. In the last decades, the only available and efficient therapy for organ failure was renal replacement therapy (RRT) in patients with acute renal failure [1]. The evolution of technology offers today the possibility to replace or at least sustain and support other failing organ systems through specific extracorporeal techniques [2, 3]. RRT and especially continuous renal replacement therapies (CRRT) allowed extracorporeal treatment in critically ill patients with hypercatabolism and fluid overload [4] with excellent hemodynamic stability. New techniques in CRRT, such as high volume hemofiltration or coupled plasma filtration adsorption, have been applied in septic patients with very promising results [5, 6]. But can extracorporeal therapies be applied to support different organ systems? A possible answer might come from the simple observation that all organs share one thing in common: contact with blood. All extracorporeal therapies also have one thing in common: treatment of blood. Based on these observations, a 'humoral' theory of MODS suggests the rationale for extracorporeal therapies as multiple organ support therapies (MOST) [7]. In recent years, the concept of MOST has been proven effective: blood purification and renal support is widely available and effective; temperature control can be achieved using the extracorporeal circuit as a heat exchange system; acid-base control can be obtained using specific buffer-containing replacement solutions; fluid balance control is one of the most important tasks performed by different ultrafiltration methods; cardiac support is achieved not only by an effective volume control but also by modulating electrolytes, acid base and removing myocardial depressant substances; brain protection is guaranteed by continuous treatments avoiding water shifts in the brain in response to sudden changes in osmolality and disequilibrium syndromes; bone marrow protection is made possible removing erythropoiesis inhibitory proteins; blood detoxification and liver support has been obtained with specific albumin dialysis techniques or selective adsorption techniques, and therapy of sepsis, immunomodulation and endothelial support has been made possible by the use of specific hemoperfusion techniques designed to remove endotoxin and inflammatory mediators.

Remarkable success has been obtained in all these efforts, but a missing piece was represented by the lack of effective protecting and supportive lung treatments. Today, the missing piece of lung support therapy is finally available; moving from extracorporeal membrane oxygenation (ECMO) that is a complex and sophisticated treatment not available in all clinical settings, we have today a simpler and easy-to-perform technique of extracorporeal CO_2 removal (DECAP). This does not represent a method for oxygenation and thus it does not represent 'lung replacement therapy' but rather a 'lung support or protective therapy'. In fact, it is well known that low tidal volume (T_V) techniques represent a protective mechanical ventilation, the use of DECAP may contribute to an additional CO_2 removal in the extracorporeal circuit used alone or in combination with CRRT such that 'ultraprotective ventilation' can be performed by applying ultra-low T_Vs (<6 ml/kg) [8–13].

Protective Lung Support and DECAP

Respiratory insufficiency is a common indication for ICU admission, and also frequently develops as a complication in patients initially admitted for other reasons. Patients with acute asthma, decompensated chronic obstructive respiratory syndrome, severe restrictive lung disease or hypoventilation of any cause may present with predominantly hypercapnic respiratory failure. Although the underlying medical problem may take several days to resolve, the initial severe hypercapnia is often rapidly corrected with medical therapy and invasive or non-invasive mechanical ventilation. The management of severe hypoxemic respiratory failure is however more problematic. Indeed, especially in patients with acute lung injury (ALI) or acute respiratory distress syndrome (ARDS), mechanical ventilation may be particularly challenging, leading to consequences such a volu- and barotrauma [14].

Historically, patients were usually ventilated with inspiratory pressure or V_T necessary to ventilate and oxygenate adequately. However, in the last two decades, several studies have shown that higher levels of volume or pressure may be detrimental to the patient, even if they help achieve seemingly adequate ventilation and oxygenation. When it was demonstrated that ventilation with smaller V_Ts and lower plateau pressures were associated with decreased mortality rates in ARDS, the concept of protective mechanical ventilation emerged, as did the notion of permissive hypercapnia [14]. If the hypercapnia is modest, it is often believed to be of no clinical importance. Actually, some experimental studies suggest that respiratory and metabolic acidosis can exert protective effects on tissue injury in several different organ systems, including the lung, and there is evidence indicating that acute hypercapnic acidosis may have anti-inflammatory effects and reduce lung injury [14–16]. Caution should be exerted however as more prolonged respiratory acidosis in animal models of infection-induced ALI has been shown to result in increased bacterial numbers, impaired neutrophil phagocytic response, and increased physiologic and pathologic lung damage. Moreover, buffering respiratory acidosis seems to worsen ALI.

However, when hypercapnia causes significant respiratory acidosis, it may have considerable adverse consequences. Indeed, severe academia (pH <7.20) results in impairment of myocardial contractility and reduction of cardiac output, arteriolar dilatation and reduction of arterial blood pressure, increased pulmonary vascular resistance, sensitization to re-entrant arrhythmias and reduction in threshold of ventricular fibrillation, attenuation of cardiovascular responsiveness to catecholamines, inhibition of cerebral metabolism and brain-cell volume regulation, obtundation and coma, and several adverse metabolic effects like increased metabolic demands, insulin resistance, reduction in ATP synthesis, inhibition of anaerobic glycolysis, increased protein degradation and hyperkalemia. Furthermore, even modest levels of hypercapnia are contraindicated in the presence of intracranial hypertension and in patients with uncontrolled seizure disorders.

The Acute Respiratory Distress Syndrome Network compared a traditional T_V (12 ml of predicted body weight) with a lower V_T (6 ml/kg of predicted body weight) in 861 patients: in the group receiving lower V_Ts, plateau pressure (airway pressure measured after a 0.5-s pause at the end of inspiration) did not exceed 30 cm of water and a detailed protocol was used to adjust the fraction of inspired oxygen and positive end-expiratory pressure (PEEP). The in-hospital mortality rate was 39.8% in the group treated with traditional V_Ts and 31.0% in the group treated with lower V_Ts ($p = 0.007$). Thus, mortality was reduced by 22% in the group treated with lower V_Ts. This large multicenter trial provides convincing evidence that a specific mechanical ventilation strategy for the ARDS can reduce mortality [14]. The fundamental concept in patients with ALI is to provide adequate gas exchange without causing further baro- or volutrauma to the lungs as well as decreasing extravascular lung water. Mechanical ventilation is typically needed in the treatment of ALI but is injurious to the lungs. High levels of PEEP can stabilize the lung units and prevent reocclusion, whereas low V_T ventilation reduces both baro- and volutrauma. Reducing the V_T lowers the plateau pressure, but at the cost of hypercapnia. In such conditions, the possibility of removing carbon dioxide (CO_2) from the circulating blood by means of extracorporeal methods may become extremely useful and beneficial for the patient. This approach has been previously explored with interest [10–13]. Today the idea of using a special CO_2-removing cartridge as a self-standing unit (DECAP) or in series with the hemofilter for CRRT might represent, in selected patients, a new chance for reducing the requirement for invasive mechanical ventilation leading to ultra-low V_Ts or even allowing for a non-invasive approach. Special membranes are under evaluation utilizing a dry/wet gas exchange process leading to significant values of CO_2 clearance in the extracorporeal circuit. Such systems might reduce the morbidity and mortality of ALI. Furthermore, by combining this with hemofiltration, we can optimize a patient's volume status offering additional support to the failing lung. Thus, such a system offers an important level of lung support. Studies have been performed to demonstrate the feasibility of the DECAP system [8]. New trials are being undertaken to demonstrate the possibility to further reduce V_Ts to a level <6 as it has been indicated that even protective strategies for mechanical ventilation are susceptible to induce lung injury.

Clinical Applications of Extracorporeal Carbon Dioxide Removal

CO_2 can be removed by an extracorporeal circuit using specialized filters. The use of such circuits can aid in life-threatening situations when maximal ventilator support does not suffice in decreasing CO_2 from dangerously high levels. A CO_2 removal device may be used to gain time as the underlying lung illness is treated or as a bridge to transplant, as has been demonstrated in a recently published case report. Another possibly beneficial application for DECAP is to permit the achievement of very low V_T when mechanically ventilating patients in whom a lung-protective strategy would be otherwise unattainable because of hypercapnia. Indeed, Terragni et al. [8] recently reported the results of a study of 32 patients with ARDS treated with low V_T strategy ($V_T \leq 6$ ml/kg and plateau pressure (P_{PLAT}) ≤ 30 cm H_2O) according to the ARDS Net protocol, in which P_{PLAT} 28–30 cm H_2O were treated with further V_T reduction. Hypercapnia was treated with bicarbonate infusion up to 20 mEq/h. If pH remained ≤ 7.25, DECAP was initiated by using a modified continuous veno-venous hemofiltration system equipped with a membrane lung. They concluded that $V_T < 6$ ml/kg enhances lung protection, and respiratory acidosis consequent to low V_T ventilation can be safely and efficiently managed by extracorporeal CO_2 removal.

Clearly, in each patient the benefits of CO_2 removal and lung-protective ventilation need to be weighed against the small but existent risks inherent to central venous catheter use, systemic anticoagulation with heparin, and the presence of an extracorporeal circuit.

Technical Aspects of Extracorporeal CO_2 Removal

Low-flow veno-venous DECAP is considered a mini-invasive procedure, as the risks are much smaller than those associated with the use of veno-venous ECMO. Indeed, the surgical insertion of large bore catheters for cannulation, and the need for very high blood flows in the range of 5 l/min, confer major risks of treatment complications during ECMO. The most worrisome are hematoma or aneurysm formation at cannulation site, cannula thrombosis, lower limb ischemia after cannulation, compartment syndrome in a lower limb, hemolysis, intracerebral hemorrhage, diffuse bleeding and shock syndrome. However, ECMO and DECAP are two different techniques: the first is a blood oxygenation procedure while the second is a CO_2 removal technique. Although low-flow veno-venous DECAP uses a membrane oxygenator, if functions solely as a decapneizator, and it does not effectively oxygenate. Hence it is not indicated in predominantly hypoxemic respiratory failure for the purpose of improving oxygenation.

Low-flow veno-venous DECAP can be obtained by several devices. Some circuits, such as the Decap® (Hemodec, Salerno, Italy) (fig. 1, 2) or the Decapsmart® (Medica, Medolla, Modena, Italy) (fig. 3, 4) are designed solely for CO_2 removal.

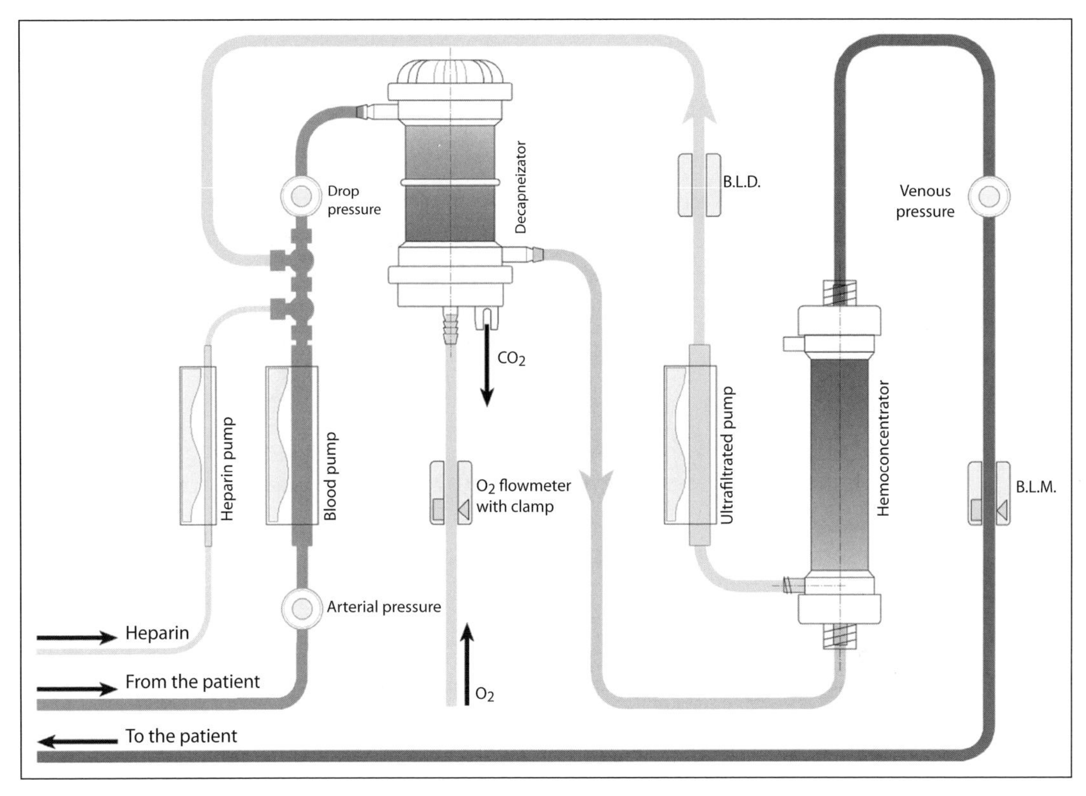

Fig. 1. Low-flow veno-venous DECAP circuit.

A membrane oxygenator for CO_2 removal can be used alone or in combination with a continuous RRT. In this case, the CO_2 removal is combined to RRT in the patient with multiple organ failure needing both respiratory and renal support as part of MOST. The DECAP/DECAPSMART system consists of a modified neonatal oxygenator (0.3 m^2 membrane surface polypropylene Neonatal Oxygenator (Polystan SAFE; Maquet, Rastatt, Germany), without heat exchanger, placed in series to a polysulfone hemofilter (1.35 m^2). The hemofilter, located downstream to the oxygenator, ensures adequate levels of pressure inside the latter, therefore preventing the formation of air bubbles in the blood. A roller pump (0–500 ml/min) drives patient blood from the access line to the oxygenator and the hemofilter. CO_2 is removed by diffusive gradient with a constant flow (10 l/min) of air/oxygen through the oxygenator. A second roller pump (0–150 ml/min) recirculates the ultrafiltrate (UF) to the inflow of the oxygenator. This UF acts as predilution, reducing the hematocrit, preventing blood packing, and ultimately optimizing the oxygenator function. In addition, the recirculation of CO_2 dissolved in the UF improves CO_2 removal [17, 18]. It is possible to regulate the

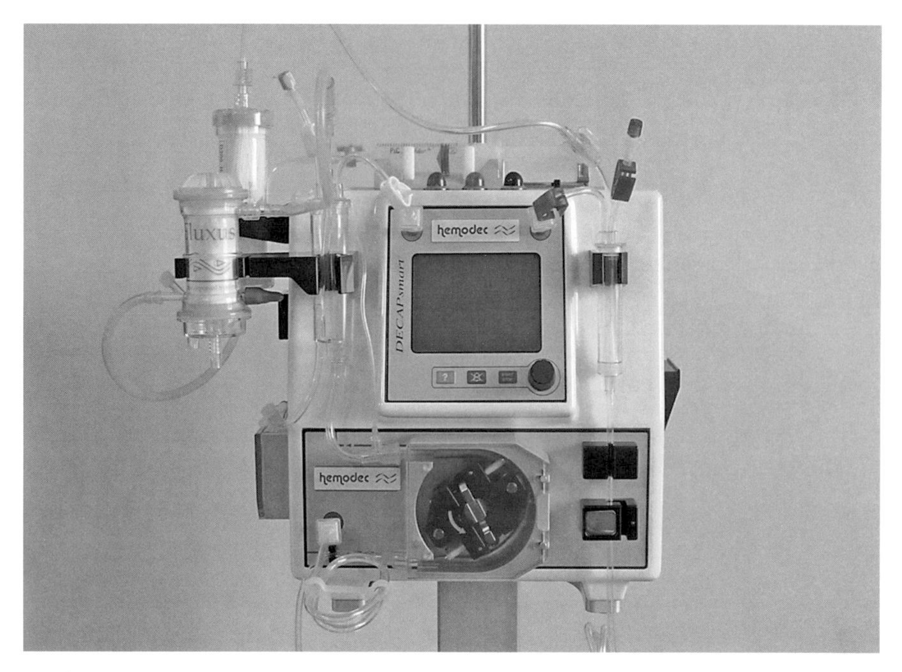

Fig. 2. Low-flow veno-venous DECAP circuit mounted on the machine.

blood flow (BF) (ml/min) and the UF as percentage of BF by acting, respectively, on BF and UF roller pumps. The suggested filtration fraction for adult patients is 20%. CO_2 removal is directly proportional to the BF. O_2 and CO_2 transfer rates are 25 ml O_2/l blood and 17 ml CO_2/l blood with 400 ml/min of BF and an equal flow of O_2. The following safety features are incorporated in the circuit: access, pre-membrane and return pressure sensors, an optical sensor for detecting blood in the UF, and an ultrasound sensor at the venous edge for the detection of microbubbles. Alarm warning and flow interruption occur in case of abnormal readings. The automated priming process incorporates several self-safety tests: pressure controls along the circuit (pressure sensors), leak test, functioning of blood optical sensor, and bubble ultrasound sensor. Priming requires 2 liters of normal saline with 10,000 IU of heparin (5 IU/ml). The circuit holds a maximum of 140–160 ml of fluid after priming.

Central venous access is ensured by a double-lumen 14-Fr catheter, and continuous anticoagulation with heparin aiming to maintain the activated partial thromboplastin time ratio to approximately 1.5 must be used in order to prevent circuit clotting. The maximum filter and circuit use time suggested by the manufacturer is 48 h, although CO_2 removal capacity decreases after several hours. In our experience it is best to change the filter approximately every 12–24 h for optimal performance. Filter efficacy assessment and decisions related to

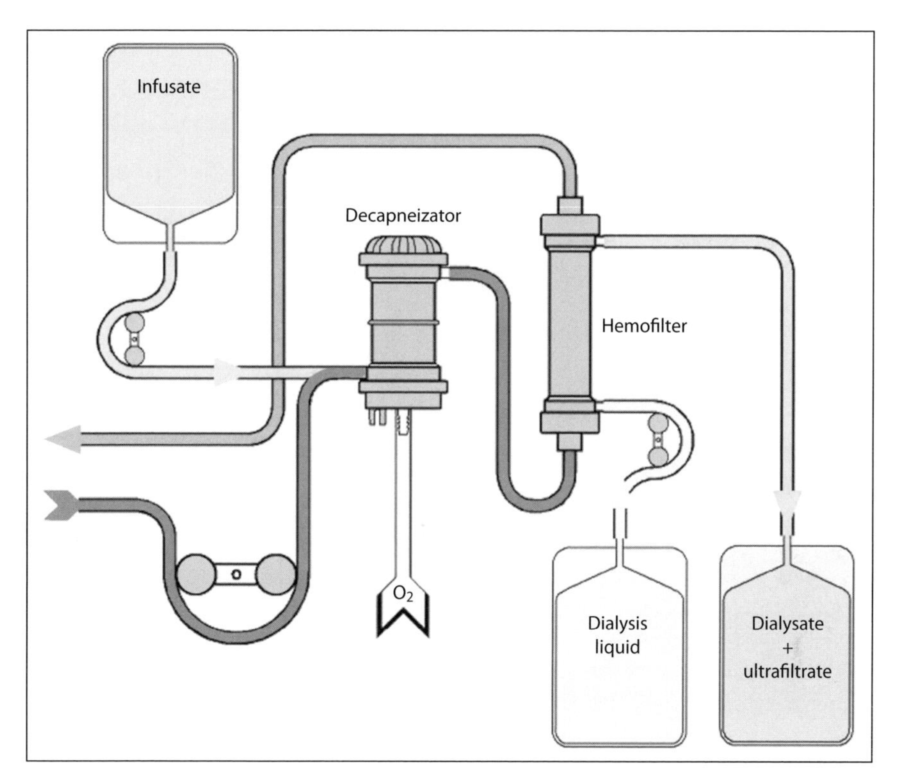

Fig. 3. Low-flow veno-venous DECAP circuit associated to a CRRT circuit (DECAPSMART).

filter change should be based on sequential blood gases, taken from the access, return and arterial lines.

Case Report

At San Bortolo Hospital in Vicenza, Italy, we have been using DECAP and DECAP+CRRT both as part of an ongoing research project, as well as off-protocol in some cases of severe refractory ARDS especially liked to H1N1 viral infection. One of our cases was a 41-year-old Caucasian man with no comorbidities who was admitted to our ICU in October 2009 with a presumptive diagnosis of septic shock due to leptospirosis. Adequate antibiotic therapy was instituted and leptospira infection was subsequently confirmed by a positive Galton test. The patient developed multiorgan failure with jaundice, acute kidney injury, and respiratory failure. Amine infusion was necessary to maintain adequate perfusion pressure. Ventilation was managed with pressure-control mode aiming for a P_{max} of <35 cm H_2O, high PEEP (16–18 cm H_2O) and T_V of about 6–7 ml/kg; prone position was employed as well to recruit lung parenchyma and enhance PaO_2/FiO_2 ratio; all these strategies resulted however on the third day after

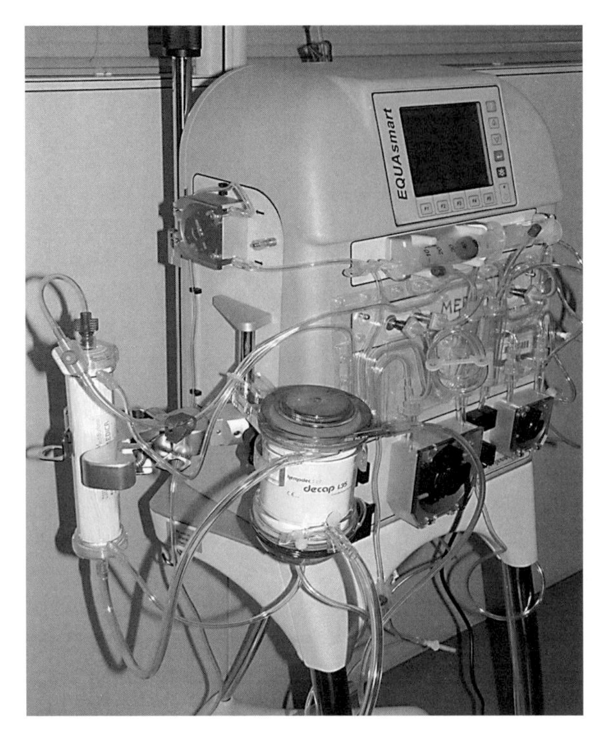

Fig. 4. Low-flow veno-venous DECAP circuit associated to a CRRT circuit (DECAPSMART) mounted on the machine.

admission in a moderately hypoxemic patient with high CO_2 values and ventilatory pressure unacceptably high. Moreover, fluid balance was positive, notwithstanding furosemide infusion. It was therefore decided to start CRRT therapy combined with DECAP. BF was held between 350 and 380 ml/min, the flow of O_2 through the oxygenator was maintained at 10 l/min; filtration fraction was maintained at 20%, and a continuous heparin infusion was used as anticoagulation system. Before DECAP, a moderate hypercapnic acidosis was present. Throughout the 12-hour treatment period, we were able to progressively decrease the P_{supp} from 12 to 10 cm H_2O and lower PEEP values in order to decrease P_{max} from 33 to 26 cm H_2O (fig. 5). pH during treatment ranged from 7.36 to 7.41. After a 12-hour cycle of treatment, renal function recovered and protective ventilation yielded acceptable values without extracorporeal support. In the subsequent 2 weeks the patient improved steadily, was weaned from mechanical ventilation and eventually dismissed home.

Conclusion

Respiratory failure is a frequent occurrence in the ICU. It can be isolated, but is often seen in the context of multiorgan failure. Some patients are predominantly

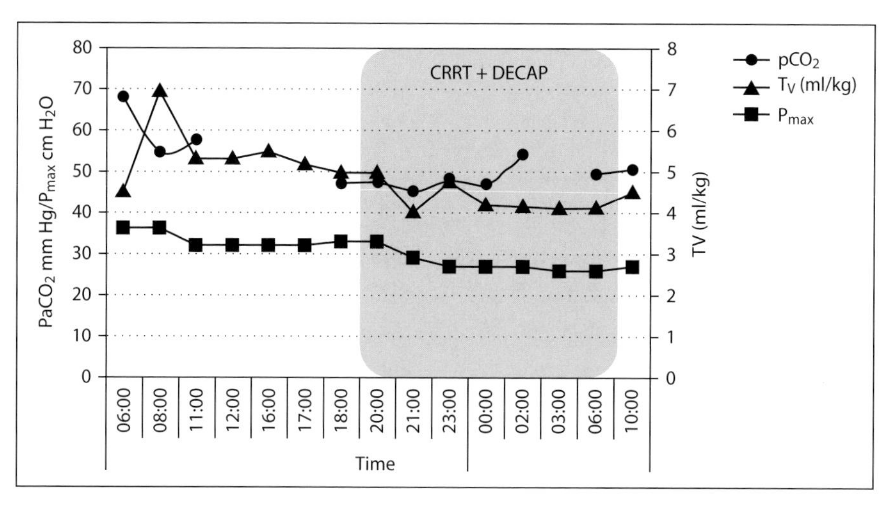

Fig. 5. Ventilatory parameters plotted versus time in a single ARDS patient before and during treatment with CRRT+DECAP: in the treatment period (shaded area) ventilatory pressure and V_T were reduced without significant CO_2 increase.

hypercapnic. Several problematic CO_2 elevations are seen in ALI or ARDS patients, as a consequence of protective mechanical ventilation. Although the clinical impact of modest degrees of hypercapnia is debated, it is clear that severe hypercapnic acidosis is detrimental and should be avoided at all costs. Moreover, the buffering of acidosis with bicarbonate may be deleterious in some conditions. Low-flow veno-venous DECAP can help gain time in patients with severe refractory respiratory insufficiency, either alone or as part of MOST. DECAP can also be used to achieve very low V_T when ventilating patients in whom a lung-protective strategy would be otherwise unattainable because of hypercapnia. Such a system may allow to achieve ultraprotective mechanical ventilation.

References

1 Clark WR, Letteri JJ, Uchino S, Bellomo R, Ronco C: Recent clinical advances in the management of critically ill patients with acute renal failure. Blood Purif 2006;24:487–498.

2 Ronco C, Ratanarat R, Bellomo R, Salvatori G, Petras D, De Cal M, Nalesso F, Bonello M, Brendolan A: Multiple organ support therapy for the critically ill patient in intensive care. J Organ Dysfunct 2005;1:57–68.

3 Bellomo R, Honore PM, Matson J, Ronco C, Winchester J: Extracorporeal blood treatment methods in SIRS/Sepsis. Int J Artif Organs 2005;28:450–458.

4 Brendolan A, D'Intini V, Ricci Z, Bonello M, Ratanarat R, Salvatori G, Bordoni V, De Cal M, Andrikos E, Ronco C: Pulse high volume hemofiltration. Int J Artif Organs 2004;27:398–403.

5 Ratanarat R, Brendolan A, Piccinni P, Dan M, Salvatori G, Ricci Z, Ronco C: Pulse high-volume haemofiltration for treatment of severe sepsis: effects on hemodynamics and survival. Crit Care 2005;9:R294–R302.
6 Ronco C, Brendolan A, Lonnemann G, Bellomo R, Piccinni P, Digito A, Dan M, Irone M, La Greca G, Inguaggiato P, Maggiore U, De Nitti C, Wratten ML, Ricci Z, Tetta C: A pilot study on coupled plasma filtration with adsorption in septic shock. Crit Care Med 2002;30:1250–1255.
7 Ronco C, Bellomo R: Acute renal failure and multiple organ dysfunction in the ICU: from renal replacement therapy to multiple organ support therapy. Int J Artif Organs 2002;25:733–747.
8 Terragni PP, Del Sorbo L, Mascia L, Urbino R, Martin EL, Birocco A, Faggiano C, Quintel M, Gattinoni L, Ranieri VM: Tidal volume lower than 6 ml/kg enhances lung protection: role of extracorporeal carbon dioxide removal. Anesthesiology 2009;111:826–835.
9 Ronco C, Chionh CY, Haapio M, Anavekar NS, House A, Bellomo R: The cardiorenal syndrome. Blood Purif 2009;27:114–126.
10 Ricci Z, Ronco C: Pulmonary/renal interaction. Curr Opin Crit Care 2010;16:13–18.
11 Heulitt MJ, Marshall J: Kidney function during extracorporeal lung assist techniques (ECMO/ECCO$_2$R); in Ronco C, Bellomo R (eds): Critical Care Nephrology. Dordrecht, Kluwer Academic, 1998, pp 1073–1079.
12 Ruberto F, Pugliese F, D'Alio A, Perrella S, D'Auria B, Ianni S, Anile M, Venuta F, Coloni GF, Pietropaoli P: Extracorporeal removal CO_2 using a venovenous, low-flow system (Decapsmart®) in a lung-transplanted patient: a case report. Transplant Proc 2009;41:1412–1414.
13 Livigni S, Maio M, Ferretti E, Longobardo A, Potenza R, Rivalta L, Selvaggi P, Vergano M, Bertolini G: Efficacy and safety of a low-flow veno-venous carbon dioxide removal device: results of an experimental study in adult sheep. Crit Care 2006;10:R151.
14 The Acute Respiratory Distress Syndrome Network: Ventilation with lower tidal volumes as compared with traditional tidal volumes for acute lung injury and the acute respiratory distress syndrome. N Engl J Med 2000;342:1301–1308.
15 Caironi P, Cressoni M, Chiumello D, Ranieri M, Quintel M, Russo SG, Cornejo R, Bugedo G, Carlesso E, Russo R, Caspani L, Gattinoni L: Lung opening and closing during ventilation of acute respiratory distress syndrome. Am J Respir Crit Care Med 2010;181:578–586.
16 Del Sorbo L, Slutsky AS: Ventilatory support for acute respiratory failure: new and ongoing pathophysiological, diagnostic and therapeutic developments. Curr Opin Crit Care 2010;16:1–7.
17 Cressoni M, Zanella A, Epp M, Corti I, Patroniti N, Kolobow T, Pesenti A: Decreasing pulmonary ventilation through bicarbonate ultrafiltration: an experimental study. Crit Care Med 2009;37:2612–2618.
18 Zanella A, Patroniti N, Isgrò S, Albertini M, Costanzi M, Pirrone F, Scaravilli V, Vergnano B, Pesenti A: Blood acidification enhances carbon dioxide removal of membrane lung: an experimental study. Intensive Care Med 2009;35:1484–1487.

Dr. Silvia Gramaticopolo
Departments of Nephrology and Intensive Care, San Bortolo Hospital
Viale Rodolfi 37, IT–36100 Vicenza (Italy)
Tel. +39 0 444 75 35 08, Fax +39 0 444 75 37 45
E-Mail silvia_gramaticopolo@iol.it

Ronco C, Bellomo R, McCullough PA (eds): Cardiorenal Syndromes in Critical Care.
Contrib Nephrol. Basel, Karger, 2010, vol 165, pp 185–196

Extracorporeal CO_2 Removal

Pier Paolo Terragni · Alberto Birocco · Chiara Faggiano ·
V. Marco Ranieri

Università di Torino, Dipartimento di Anestesia e di Medicina degli Stati Critici, Azienda Ospedaliera S. Giovanni Battista-Molinette, Torino, Italia

Abstract

The extracorporeal carbon dioxide removal ($ECCO_2R$) concept, used as an integrated tool with conventional ventilation, plays a role in adjusting respiratory acidosis consequent to tidal volume (Vt) reduction in a protective ventilation setting. This concept arises from the extracorporeal membrane oxygenation (ECMO) experience. Kolobow and Gattinoni were the first to introduce extracorporeal support, with the intent to separate carbon dioxide removal from oxygen uptake; they hypothesized that to allow the lung to 'rest' oxygenation via mechanical ventilation could be dissociated from decarboxylation via extracorporeal carbon dioxide removal. Carbon dioxide is removed by a pump-driven modified ECMO machine with veno-venous bypass, while oxygenation is accomplished by high levels of positive end-expiratory pressure, with a respiratory rate of 3–5 breaths/min. The focus was that, in case of acute respiratory failure, CO_2 extraction facilitates a reduction in ventilatory support and oxygenation is maintained by simple diffusion across the patient's alveoli, called 'apneic oxygenation'. Concerns have been raised regarding the standard use of extracorporeal support because of the high incidence of serious complications: hemorrhage; hemolysis, and neurological impairments. Due to the negative results of a clinical trial, the extensive resources required and the high incidence of side effects, low frequency positive pressure ventilation $ECCO_2R$ was restricted to a 'rescue' therapy for the most severe case of acute respiratory distress syndrome (ARDS). Technological improvement led to the implementation of two different CO_2 removal approaches: the iLA called 'pumpless arteriovenous ECMO' and the veno-venous $ECCO_2R$. They enable consideration of extracorporeal support as something more than mere rescue therapy; both of them are indicated in more protective ventilation settings in case of severe ARDS, and as a support to the spontaneous breathing/lung function in bridge to lung transplant. The future development of more and more efficient devices capable of removing a substantial amount of carbon dioxide production (30–100%) with blood flows of 250–500 ml/min is foreseeable. Moreover, in the future ARDS management should include a minimally invasive $ECCO_2R$ circuit associated with noninvasive ventilation. This would embody the modern mechanical ventilation philosophy: avoid tracheal tubes; minimize sedation, and prevent ventilator-induced acute lung injury and nosocomial infections.

Introduction

Since 1979 it is possible to markedly hypoventilate the lung at a rate of 2–4 breaths/min or allow spontaneous but insufficient ventilation to maintain normal arterial blood gases, while the metabolically produced CO_2 is removed by an extracorporeal membrane lung and the oxygen is fed through a tracheal tube or a helmet with continuous positive airways pressure (CPAP) [1, 2].

In 2001 the NIH published a randomized controlled trial that recommended ventilating patients with acute respiratory distress syndrome (ARDS) with a tidal volume (Vt) of 6 ml/kg (predicted body weight) and a maximum end-inspiratory plateau pressure of 30 cm H_2O, in order to prevent ventilation lung damage [3]. Recently published studies show that despite these limitations, a tidal hyperinflation may occur in about the 30% of ARDS patients; furthermore they may benefit from Vt reduction even if they already have a plateau pressure of <30 cm H_2O [4, 5].

Those results support the concept of extracorporeal carbon dioxide removal ($ECCO_2R$) as integrated to conventional ventilation to adjust respiratory acidosis consequent to very low Vt, and therefore allowing more protective ventilator settings [6]. This approach might also reduce ventilator-induced lung injury that is one of the most important and actual problems in diseased lungs and may allow a wider clinical implementation of the new concept 'less ventilation, less injury' [7].

Extracorporeal CO_2 Removal: The Concept

The basic concepts of the CO_2 removal technique can be extracted from the original description of extracorporeal membrane oxygenation (ECMO) that appeared in the clinical setting more than 30 years ago. At that time, Hill et al. [8] were the first to report the successful use of extracorporeal circulation to treat acute hypoxemic respiratory failure in an adult patient, and Bartlett [9] demonstrated for the first time the successful use of ECMO in a neonate.

In the same period, studies published by Gattinoni et al. [10] introduced extracorporeal support intended to separate carbon dioxide removal from oxygen uptake. Extracorporeal CO_2 removal ($ECCO_2R$) refers to extracorporeal support focused on the removal of blood CO_2 rather than on the improvement of the oxygenation. Extracorporeal oxygenation was initially designed as a heart-lung machine to render major cardiovascular surgery feasible and safe; this application was not that far from the one suggested by Gattinoni et al. [10]

for the acute respiratory failure. The blood flow levels required to obtain carbon dioxide removal are lower than the ones needed to achieve the oxygenation so Gattinoni et al. [10] exploited the concept that, if CO_2 is removed by a membrane lung through a low-flow veno-venous bypass, it is possible to reduce the ventilatory support in acute respiratory failure and severe ARDS, maintaining oxygenation simply with the patient's alveoli diffusion, also called 'apneic oxygenation', allowing the lung rest. Originally the veno-venous bypass was set out via the cannulation of the common femoral and jugular veins through a surgical cut; the larger lumen needed to be used for venous drainage, the smaller for blood return (toward the tricuspid valve minimizing the recirculation). Wounds and multiple cannulations determined continuous blood oozing, nursing care and patient mobility limitations, therefore the same authors developed a double-lumen femoral vein cannulation technique; with the introduction of newly designed percutaneous cannulas and the Seldinger technique, we arrive at the present technique. The first membrane lungs consisted of microporous polypropylene fibers and were associated with constant plasma leakage that determined the frequent need for membrane substitution and the circuit needed to be heparinized with 100 IU/kg at the cannula insertion. Heparin infusion was hence titrated on activated clotting time (150–200 s).

Gattinoni et al. [10], using this modified ECMO technique (using low frequency positive pressure LFPPV-$ECCO_2R$), reported ARDS survival of 49%, which was also attributed to patient selection, strict control of coagulation, and ventilator management directed to reach the 'lung rest' [11]. Anderson et al. in 1993, demonstrated 47% survival in adults with severe respiratory failure. In a retrospective review of 100 adult patients, Kolla et al. [12] reported a 54% overall survival. However, despite the later report by Brunet et al. [13] using LFPPV-$ECCO_2R$ to improve oxygenation, reduce pulmonary barotrauma in ARDS and achieve a mortality rate of 50%, in 1994 Morris et al. [14] presented the results of a randomized clinical trial where they used 'conventional' pressure-controlled inverse ratio ventilation compared to $ECCO_2R$ in ARDS patients. The study showed no significant difference in survival between the two interventions and reported several episodes of severe bleeding [14]. $ECCO_2R$ was hence restricted to the sickest patients in whom all other treatments had failed and limited only to centers with large expertise [15].

At the present time, ECMO (fig. 1) with the new technologies (centrifugal blood pump and new polymethylpentene low-resistance diffusion membrane oxygenators) still performs very well in maintaining oxygenation and eucapnia in the most severe ARDS patients with refractory hypoxemia [16]. In this application, survival of >50% can be achieved, whereas sepsis and multiple organ failure are the leading causes of unsuccessful use. Only a minority of patients suffer major complications related to the technique itself, and these serious complications are almost exclusively related to bleeding (particularly intracranial bleeding) [17].

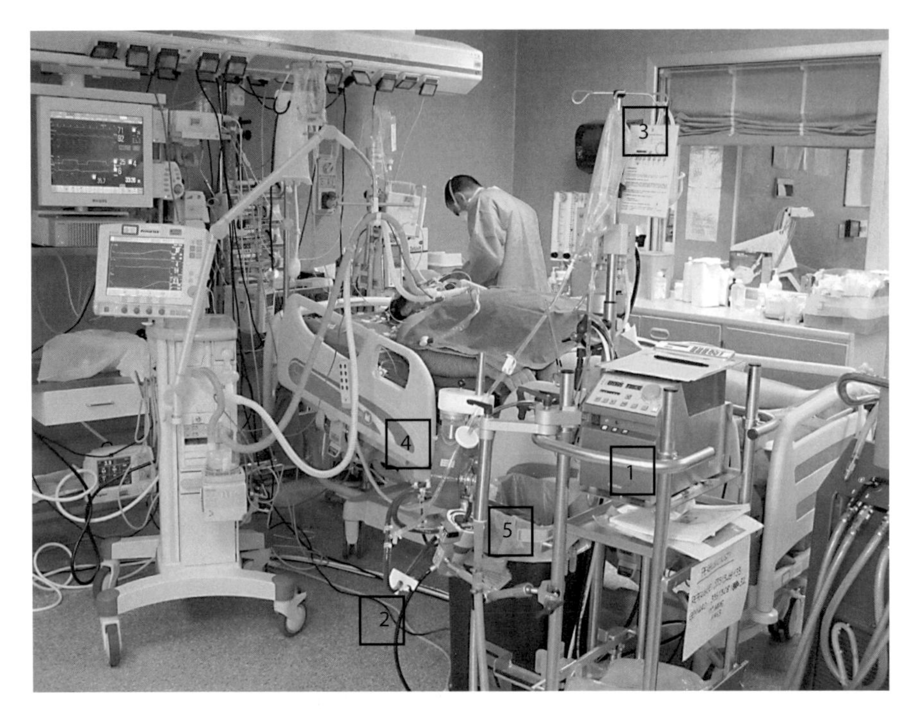

Fig. 1. Centrifugal pump (RotaflowTM, MAQUET Cardiopulmonary AG, Hechingen, Germany): (1) steering and control unit; (2) driveline; (3) pole for volume resuscitation; (4) membrane oxygenator, and (5) centrifugal pump.

Extracorporeal CO_2 Removal: The Clinical Data

The removal of 'only a portion of carbon dioxide production' was originally developed by Pesenti et al. [18], and has recently been implemented with new devices that may reduce side effects, complexity, and costs of $ECCO_2R$. In 1983, Ohtake et al. [19] described a simple method to remove carbon dioxide using arterial blood pressure in an arteriovenous setting including a hollow fiber oxygenator: the 'pumpless arteriovenous ECMO'.

The system was characterized by a new membrane gas exchange system based on heparin-coated hollow fiber technology that optimized blood flow reducing the resistances and was connected to the patient via arterial and venous cannulas inserted with the Seldinger technique.

The device did not required extended technical and staff support: blood flow is determined by the driving force given by cardiac output, and the mean arterial pressure: 20–25% of the cardiac output passes as a left to right shunt (an ultrasound flow meter might indicate the amount of blood passing per minute). Furthermore, the system used a 'low-dose' heparin infusion that did not exceed normal antithrombotic anticoagulation of the intensive care patient.

These findings lead to the newly designed interventional lung assist (iLA) device. Bein et al. [20] recently reported a retrospective analysis of 90 patients with critical hypoxemia/hypercapnia treated with the iLA device, who, despite ventilation with low Vt (320–470 ml), showed physiologic $PaCO_2$ (31–42 mm Hg) and pH (7.38–7.50) values. However, the authors reported a complication rate of 24%, including limb ischemia, compartment syndrome, and intracranial hemorrhage. In addition, continuous intravenous norepinephrine infusion was needed to maintain an arteriovenous pressure gradient.

In 2008 Fisher et al. [21] described the new iLA concept of protective ventilation as a bridge to lung transplant: in the Hannover experience 12 patients who developed severe ventilation-refractory hypercapnia and acidosis despite maximal conventional ventilatory support received iLA implantation, obtaining a $PaCO_2$ level reduction and a significant improvement in pH values. In a recent study, Zimmermann et al. [22] implemented an iLA system in 51 ARDS patients suffering from persistent hypoxemia and/or hypercapnia who were unresponsive to conventional therapy, and achieved de-escalation of invasive ventilatory variables preventing ventilator-induced lung injury.

Although iLA is a simple device that can be established quickly, and has easy monitoring, arterial cannulation is always required which cannot be performed in patients with serious peripheral arterial disease and has the potential risk of inducing limb ischemia. Additionally, the only system driving force is the patient's heart and frequently a continuous intravenous norepinephrine infusion is necessary in order to maintain an arteriovenous pressure gradient.

In the last years, a new concept of CO_2 removal was experimented to reduce complexity, side effects and expenses of extracorporeal lung assistance. Livigni et al. [23] described in an animal model the efficacy and safety of a veno-venous device ($ECCO_2R$) with a low-flow CO_2-removal system. In 2009 Terragni et al. [6] studied the effects of further decreasing Vt in a group of ARDS patients who developed plateau pressures of 28–30 cm H_2O. The Vt was decreased to 4 ml/kg of predicted body weight, and the predictable consequence of an increase in $PaCO_2$ was corrected through an extracorporeal circuit. The intervention was safe and produced notable physiologic improvements [6].

This new generation $ECCO_2R$ consists of a modified standard continuous veno-venous hemofiltration setup (Decap®, Hemodec, Salerno, Italy) that includes an oxygenator in series with the hemofilter. This system is less invasive since the veno-venous circuit is accessed via a double-lumen catheter through a femoral vein, and the blood flow is driven through the circuit by a roller non-occlusive low-flow pump through a membrane lung that is connected to a fresh gas flow source delivering 100% oxygen. Exiting the membrane lung, blood is driven to an hemofilter. The resulting plasmatic water is re-circulated through the membrane lung by a peristaltic pump. The membrane lung and the hemofilter are coupled in series in order to increase the pressure inside the membrane lung by adding the downstream resistance exerted by the hemofilter and therefore reducing the risk of air bubble formation, minimizing the need for heparin by diluting the blood entering the membrane lung by re-circulating the plasmatic water separated by the hemofilter, and enhancing the performance of the extracorporeal device extracting the carbon dioxide dissolved in the plasmatic water separated by the hemofilter and re-circulated through the membrane lung [6]. Ruberto et al. [24] described the use of this new generation $ECCO_2R$ to assist a patient affected by primary graft dysfunction after a single lung transplantation. Although this system should not be considered a replacement for traditional ECMO because the performances are not comparable in terms of CO_2 removal and especially oxygenation improvement, available data suggest that this 'mini-ECMO' optimized pH values, reduced partial pressure of CO_2 allowing minimization of ventilatory support and therefore minimizing ventilatory-induced lung injury with no adverse events in terms of bleeding, circuit clotting, severe hemodynamic instability, or venous embolism [24].

Extracorporeal CO_2 Removal: The Technological Development

Extracorporeal circulation can be achieved using an oxygenator for the CO_2 removal from its dry form (dissolved CO_2) or a hemodialyzer for CO_2 removal from its wet form [6, 23, 25, 26].

In the artificial lung, the real limiting factor of CO_2 elimination is physiological. The reaction speed of bicarbonate dehydration, and the consecutive rise in the CO_2 concentration in plasma, are very slow. This explains the need to bypass 25% of the cardiac output in order to eliminate metabolic CO_2 production. Oxygenator-associated acidification and hemodialyzer-associated alkalinization are methods that facilitate the shifting of the bicarbonate/dissolved CO_2 equilibrium in the respective senses of dissociation (CO_2 partial pressure raised) and hydration (bicarbonate raised). Even if the efficacy of CO_2 removal through hemodialysis (with or without NaOH dialysate alkalinization) was higher than the CO_2 elimination obtained through an oxygenator (with or without inlet HCl blood acidification) [25], the latter (without blood acidification) is today the most followed by clinicians in severe acute respiratory failure because of the reduction in the circuit complexity. At the present time, only preliminary data are available from Cressoni et al. [27] and Zanella et al. [28] about the effects of blood acidification to enhance carbon dioxide removal of membrane lung in a swine model. In these studies, the authors demonstrate that blood acidification at the inlet of a membrane lung can significantly increase the CO_2 removal by the membrane lung by converting the blood bicarbonate into physically dissolved CO_2 [27, 28].

For most adult patients with unresponsive severe respiratory failure, veno-venous support is the method of choice, including both $ECCO_2R$ and veno-venous ECMO. If the need of extracorporeal ventilatory support is partial, a new generation of $ECCO_2R$ devices that use arteriovenous pumpless bypass or low-flow venous-venous bypass that can remove only 20–30% of the CO_2 production are available in clinical practice [6, 24, 26].

1. Veno-Venous ECMO

$ECCO_2R$ refers to an extracorporeal support focused on the CO_2 leaching from blood rather than improving oxygenation [1, 29]. In cases of hypoxic/hypercapnic respiratory failure, but preserved cardiac function, a veno-venous ECMO support is preferred to carry over the pulmonary function.

Centrifugal pumps and surface-heparinized (bioline coating) hollow fiber membrane lungs both mounted on a specially designed multifunctional holder represent the state of the art. A flow meter and a bubble sensor are integrated into the pump unit. The tubing circuit consists of a pre-connected, heparin-coated closed-loop extracorporeal circulation system for rapid setup and priming. Total priming volume is 600 ml of normal saline. The centrifugal pump provides non-pulsatile flow rates of up to 4.5 liters/min (depending on the size of the cannula). The circuit needs to be heparinized with 100 IU/kg at the cannula insertion, heparin infusion is titrated on an activated clotting time of 150–200 s, but on a heparinized surface the circuit can work well without any systemic anticoagulation for at least 12–48 h [27]. Before cannulation, usually performed with a modified Seldinger technique, ultrasonic measurement of the femoral vessels is performed to assess the appropriate caliber of the cannula. Depending on the ultrasonic findings and the patient's biometric data, a 17- or 23-Fr cannula must be inserted for venous and a 15- or 17-Fr cannula for arterial vascular access. Outflow is achieved via the femoral vein, and inflow is gained by cannulation of the internal jugular vein or femoral vein and thereafter into the superior vena cava.

Pumpless Arteriovenous Interventional Lung Assist

The iLA (NovaLung® GmbH, Hechingen, Germany) is a single-use compact extrapulmonary gas exchange system perfused by a passive femoral artery-femoral vein shunt, generated by the arterial blood pressure (60–80 mm Hg femoral artery-femoral vein) through a lung-assist device; a blood flow rate of approximately 1.0–2.5 liters/min produces effective CO_2 extraction and an improvement in arterial oxygenation. Apart from an oxygen supply (10–12 liters/min), the system does not require additional energy or substrate sources [20] (fig. 2). A polymethylpentene diffusion membrane resistant to plasma leakage is used as a separation layer between phases (blood/gas). Due to the molecular structure of this layer, the passage of air bubbles from gas to the blood path in the event of negative pressure on the blood side is impossible.

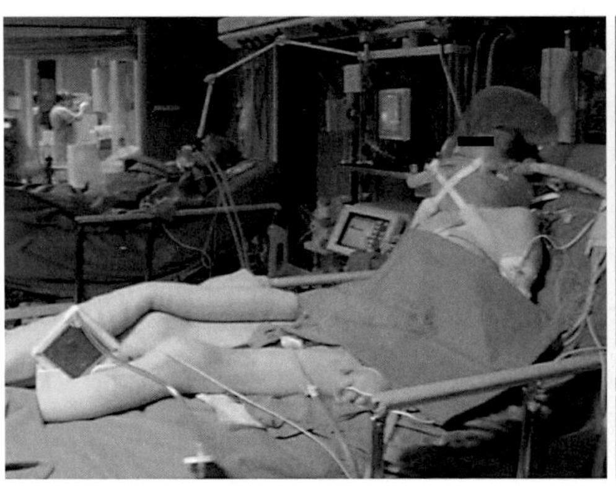

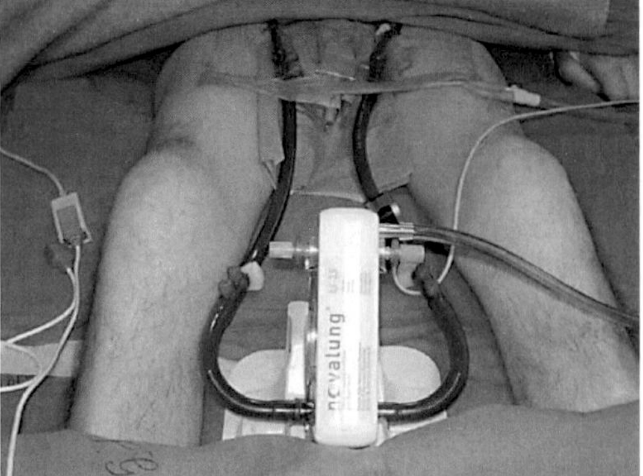

Fig. 2. Interventional lung assist device in place in a patient with CPAP.

The entire effective gas exchange surface area amounts to 1.3 m^2, integration of a heat exchanger is not necessary as temperature loss due to convection is negligible. To optimize hemocompatibility, the system is entirely (tip to tip) homogeneously treated with the coating method (Novalung Coating, NovaLung GmbH).

To connect the iLA to the patient, a special percutaneous cannulation system has to meet the following conditions: implantable with the Seldinger technique, cannula walls must be extremely thin to minimize resistance to flow and should be available in various diameters (13–21 Fr). In every individual case, the cannula size used is determined by the diameter of the vessel to be cannulated and the required shunt flow, diameter should be measured by ultrasound. Functional control is achieved through a monitoring Doppler device. Weaning from iLA is attempted by a reduction in mechanical ventilation and gas supply to approximately 1 liter/min performed for 30 min.

The frequency of complications reported is actually very high: about 25% serious complications were observed; episodes of ischemia of a lower limb after arterial cannulation were major problems; in other cases of ischemia, the cannulas were removed and normal perfusion of the limb was restored. Cannula thrombosis was only observed in the early period without specially designed cannulas before 2001. The main contraindication for the application of the system is hemodynamic depression. Fischer et al. [21] also reported the application of iLA as a bridge to lung transplant.

Low-Flow ECCO$_2$R Technique

The Decap® (Hemodec, Salerno, Italy) ECCO$_2$R device is a modified renal replacement circuit incorporating a neonatal membrane lung coupled in series with a hemofilter (fig. 3). Vascular access is granted by the femoral vein and via a

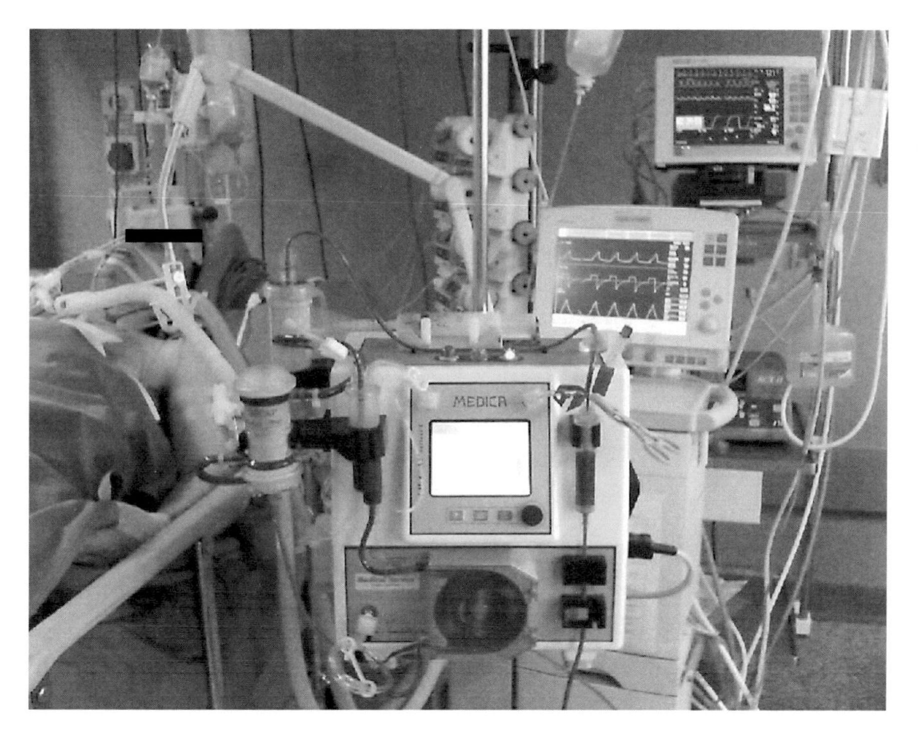

Fig. 3. Extracorporeal carbon dioxide-removal device (Decap® Hemodec, Salerno, Italy). The system consists of a standard continuous veno-venous hemofiltration system equipped with a membrane lung with a total membrane surface of 0.33 m^2.

double-lumen catheter with a 14-Fr diameter inserted using the Seldinger technique. Blood flow is driven by a roller non-occlusive low-flow pump (maximum flow 450 ml/min) through a membrane lung connected to a fresh gas flow source delivering 100% oxygen at a constant rate of 6 liters/min. Exiting the membrane lung, blood is driven to an hemofilter and the resulting plasmatic water is re-circulated through the membrane lung by a peristaltic pump (0–155 ml/min). Detectors of leaks and bubbles are inserted within the circuit. The circuit, including the membrane lung, is primed with saline at a volume that ranges between 140 and 160 ml. The new concept introduced by this newly designed technique is that the membrane lung and the hemofilter are coupled in series. This characteristic of the circuit increases the pressure inside the membrane lung by adding the downstream resistance exerted by the hemofilter and therefore reduces the risk of air bubble formation; minimizes the need for heparin by diluting the blood entering the membrane lung by re-circulating the plasmatic water separated by the hemofilter; produces a performance enhancement of the extracorporeal device extracting the carbon dioxide dissolved in the plasmatic water separated by the hemofilter and re-circulated through the membrane lung [6, 23, 24, 26].

Extracorporeal CO_2 Removal: The Future

The NIH protocol represents the standard for mechanical ventilation of ARDS patients, recommending the use of a low Vt of 6 ml/kg (predicted body weight) and an end-inspiratory plateau pressure of a maximum of 30 cm H_2O [3]. Despite these ventilatory limitations, tidal hyperinflation may occur in up to 30% of ARDS patients who could benefit from an additional Vt reduction [4]. In this scenario, extracorporeal lung support may play a role integrating conventional care and allowing the use of more protective ventilator settings. The concept of removing 'only a portion of carbon dioxide production', originally developed by Kolobow et al. [29], has recently been implemented in new devices that may reduce side effects, complexity, and costs of $ECCO_2R$.

Terragni et al. [6] effectively and safely managed respiratory acidosis consequent to a Vt of <6 liters/kg predicted body weight and reestablished normal arterial pH through an $ECCO_2R$ technique. The system, at 380 ml/min blood flow, could allow a $PaCO_2$ reduction of approximately 20% at constant ventilation [6]. Therefore, the key for a revolutionary approach to ARDS ventilatory management, is shifting from invasive mechanical ventilation to the application of low extracorporeal blood flow combined with high efficiency $ECCO_2R$ as lung support.

Venous blood contains large amounts of carbon dioxide, most carried as bicarbonate ion (approximately 500 ml/l of carbon dioxide under normocapnic conditions) and with a blood flow through the extracorporeal circuit of 500 ml/min, the Vt could theoretically be reduced to zero. From these preliminary clinical data (waiting for additional studies to further confirm these results), with the development of very efficient devices capable of removing a substantial amount of carbon dioxide production (30–100%) with blood flows of 250–500 ml/min we could assume the possibility of avoiding endotracheal intubation, with related complications such as pulmonary infections and need of sedation. In this way, severe ARDS patients could be managed without any form of mechanical ventilation, simply providing enough positive airway pressure to keep the lung open and high FiO_2 to avoid hypoxemia, as a bridge to recovery from pulmonary disease.

Conclusions

With improved technology and experience, low extracorporeal blood flow with high performance $ECCO_2R$ may be the key to management of severe ARDS with a new respiratory support, shifting from invasive mechanical ventilation to the application of extracorporeal lung support, similar to renal support. Lung-protective ventilatory strategies with new solutions to remove CO_2, might make clinicians rethink the role of extracorporeal lung support procedures in the treatment algorithm of ARDS.

References

1 Gattinoni L, Kolobow T, Damia G, Agostoni A, Pesenti A: Extracorporeal carbon dioxide removal ($ECCO_2R$): a new form of respiratory assistance. Int J Artif Organs 1979;2: 183–185.

2 Gattinoni L, Kolobow T, Agostoni A, Damia G, Pelizzola A, Rossi GP, Langer M, Solca M, Citterio R, Pesenti A, Fox U, Uziel L: Clinical application of low frequency positive pressure ventilation with extracorporeal CO_2 removal (LFPPV-$ECCO_2R$) in treatment of adult respiratory distress syndrome (ARDS). Int J Artif Organs 1979;2:282–283.

3 Ventilation with lower tidal volumes as compared with traditional tidal volumes for acute lung injury and the acute respiratory distress syndrome. The Acute Respiratory Distress Syndrome Network. N Engl J Med 2000;342: 1301–1308.

4 Terragni PP, Rosboch G, Tealdi A, Corno E, Menaldo E, Davini O, Gandini G, Herrmann P, Mascia L, Quintel M, Slutsky AS, Gattinoni L, Ranieri VM: Tidal hyperinflation during low tidal volume ventilation in acute respiratory distress syndrome. Am J Respir Crit Care Med 2007;175:160–166.

5 Hager DN, Krishnan JA, Hayden DL, Brower RG: Tidal volume reduction in patients with acute lung injury when plateau pressures are not high. Am J Respir Crit Care Med 2005; 172:1241–1245.

6 Terragni PP, Del Sorbo L, Mascia L, Urbino R, Martin EL, Birocco A, Faggiano C, Quintel M, Gattinoni L, Ranieri VM: Tidal volume lower than 6 ml/kg enhances lung protection: role of extracorporeal carbon dioxide removal. Anesthesiology 2009;111: 826–835.

7 Bigatello LM, Pesenti A: Ventilator-induced lung injury: less ventilation, less injury. Anesthesiology 2009;111:699–700.

8 Hill JD, O'Brien TG, Murray JJ, Dontigny L, Bramson ML, Osborn JJ, Gerbode F: Prolonged extracorporeal oxygenation for acute post-traumatic respiratory failure (shock-lung syndrome). Use of the Bramson membrane lung. N Engl J Med 1972;286: 629–634.

9 Bartlett RH: Esperanza. Presidential address. Trans Am Soc Artif Intern Organs 1985; 31:723–726.

10 Gattinoni L, Pesenti A, Mascheroni D, Marcolin R, Fumagalli R, Rossi F, Iapichino G, Romagnoli G, Uziel L, Agostoni A, et al: Low-frequency positive-pressure ventilation with extracorporeal CO_2 removal in severe acute respiratory failure. JAMA 1986;256: 881–886.

11 Zwischenberger JB, Alpard SK: Artificial lungs: a new inspiration. Perfusion 2002;17: 253–268.

12 Kolla S, Awad SS, Rich PB, Schreiner RJ, Hirschl RB, Bartlett RH: Extracorporeal life support for 100 adult patients with severe respiratory failure. Ann Surg 1997;226:544–566.

13 Brunet F, Mira JP, Belghith M, Monchi M, Renaud B, Fierobe L, Hamy I, Dhainaut JF, Dall'ava-Santucci J: Extracorporeal carbon dioxide removal technique improves oxygenation without causing overinflation. Am J Respir Crit Care Med 1994;149:1557–1562.

14 Morris AH, Wallace CJ, Menlove RL, Clemmer TP, Orme JF Jr, Weaver LK, Dean NC, Thomas F, East TD, Pace NL, et al: Randomized clinical trial of pressure-controlled inverse ratio ventilation and extracorporeal CO_2 removal for adult respiratory distress syndrome. Am J Respir Crit Care Med 1994;149:295–305.

15 Bartlett RH, Roloff DW, Custer JR, Younger JG, Hirschl RB: Extracorporeal life support: the University of Michigan experience. JAMA 2000;283:904–908.

16 Grasselli G, Foti G, Patroniti N, Pesenti A: A case of ARDS associated with influenza A-H1N1 infection treated with extracorporeal respiratory support. Minerva Anestesiol 2009;75:741–745.

17 Tobin MJ: Principles and practice of mechanical ventilation; in Bartlett R, Kolobow T (eds): Extracorporeal Membrane Oxygenation and Extracorporeal Life Support. MacGraw-Hill 2006, pp 493–500.

18 Pesenti A, Rossi GP, Pelosi P, Brazzi L, Gattinoni L: Percutaneous extracorporeal CO2 removal in a patient with bullous emphysema with recurrent bilateral pneumothoraces and respiratory failure. Anesthesiology 1990;72:571–573.

19 Ohtake S, Kawashima Y, Hirose H, Matsuda H, Nakano S, Kaku K, Okuda A: Experimental evaluation of pumpless arteriovenous ECMO with polypropylene hollow fiber membrane oxygenator for partial respiratory support. Trans Am Soc Artif Intern Organs 1983;29:237–241.
20 Bein T, Weber F, Philipp A, Prasser C, Pfeifer M, Schmid FX, Butz B, Birnbaum D, Taeger K, Schlitt HJ: A new pumpless extracorporeal interventional lung assist in critical hypoxemia/hypercapnia. Crit Care Med 2006;34: 1372–1377.
21 Fischer S, Hoeper MM, Bein T, Simon AR, Gottlieb J, Wisser W, Frey L, Van Raemdonck D, Welte T, Haverich A, Strueber M: Interventional lung assist: a new concept of protective ventilation in bridge to lung transplantation. ASAIO J 2008;54:3–10.
22 Zimmermann M, Bein T, Arlt M, Philipp A, Rupprecht L, Mueller T, Lubnow M, Graf BM, Schlitt HJ: Pumpless extracorporeal interventional lung assist in patients with acute respiratory distress syndrome: a prospective pilot study. Crit Care 2009;13:R10.
23 Livigni S, Maio M, Ferretti E, Longobardo A, Potenza R, Rivalta L, Selvaggi P, Vergano M, Bertolini G: Efficacy and safety of a low-flow veno-venous carbon dioxide removal device: results of an experimental study in adult sheep. Crit Care 2006;10:R151.
24 Ruberto F, Pugliese F, D'Alio A, Perrella S, D'Auria B, Ianni S, Anile M, Venuta F, Coloni GF, Pietropaoli P: Extracorporeal removal CO_2 using a venovenous, low-flow system (Decapsmart) in a lung transplanted patient: a case report. Transplant Proc 2009;41:1412–1414.
25 Gille JP, Lautier A, Tousseul B: EC CO_2R: oxygenator or hemodialyzer? An in vitro study. Int J Artif Organs 1992;15:229–233.
26 Terragni PP ML, Urbino R, Rosboch G, Basso M, Birocco A, Cochelli L, Degiovanni C, Nicosia M, Ranieri VM: Protective ventilation with CO_2-removal technique in patients with ARDS. Intensive Care Med 2007;33(suppl 2):S235.
27 Cressoni M, Zanella A, Epp M, Corti I, Patroniti N, Kolobow T, Pesenti A: Decreasing pulmonary ventilation through bicarbonate ultrafiltration: an experimental study. Crit Care Med 2009;37:2612–2618.
28 Zanella A, Patroniti N, Isgrò S, Albertini M, Costanzi M, Pirrone F, Scaravilli V, Vergnano B, Pesenti A: Blood acidification enhances carbon dioxide removal of membrane lung: an experimental stud. Intensive Care Med 2009;35:1484–1487.
29 Kolobow T, Gattinoni L, Tomlinson TA, Pierce JE: Control of breathing using an extracorporeal membrane lung. Anesthesiology 1977;46:138–141.
30 Lewandowski K, Rossaint R, Pappert D, Gerlach H, Slama KJ, Weidemann H, Frey DJ, Hoffmann O, Keske U, Falke KJ: High survival rate in 122 ARDS patients managed according to a clinical algorithm including extracorporeal membrane oxygenation. Intensive Care Med 1997;23:819–835.

V. Marco Ranieri, MD
Università di Torino, Dipartimento di Anestesia e di Medicina degli Stati Critici,
Azienda Ospedaliera S. Giovanni Battista-Molinette
Corso Dogliotti 14
IT–10126 Torino (Italy)
Tel. +39 011 633 4001, Fax +39 011 696 0448, E-Mail marco.ranieri@unito.it

Organ Crosstalk in Multiple Organ Dysfunction Syndromes

Ronco C, Bellomo R, McCullough PA (eds): Cardiorenal Syndromes in Critical Care.
Contrib Nephrol. Basel, Karger, 2010, vol 165, pp 197–205

Management of Acute Kidney Injury in Liver Disease

A. Davenport

UCL Centre for Nephrology, Royal Free Campus, University College London Medical School, London, UK

Abstract

Acute deterioration in kidney function often occurs in patients with acute and chronic liver disease admitted to hospital. The true incidence of acute kidney injury (AKI) is unknown due to the difficulty in accurately measuring kidney function in patients with liver failure, and the lack of a consensus definition of AKI. There is a current consensus definition of hepatorenal syndrome (HRS), but in current clinical practice, volume-unresponsive AKI or acute tubular necrosis are much more common causes of AKI in patients with liver disease. Due to arterial vasodilatation and compensatory neuroendocrine activation, these patients are more prone to AKI due to a combination of changes in renal autoregulation and a sudden reduction in effective plasma volume or hypotension. The key management strategy is to prevent the development of HRS, by avoiding nephrotoxins, and preventing infective complications, and maintaining an effective circulating volume. In patients with HRS, treatment centres around daily plasma volume expansion, typically with albumin, in combination with vasopressors, typically vasopressin analogues, such as terlipressin, or noradrenaline. If renal function does not recover, then renal support with dialysis may be appropriate.

As patients develop progressive liver disease, whether acute or chronic, renal sodium retention occurs. Initially this occurs without significant compensatory increased neuroendocrine activity, but as arterial vasodilatation increases, this is associated with increased renin-angiotensin-aldosterone activity, hypothalamic vasopressin release, circulating catecholamines and sympathetic nervous system activity [1]. Within the kidney, intrarenal vasoconstriction is enhanced by local endothelin production, secondary to the hepatorenal reflex, and raised renal venous pressure due to increased intra-abdominal pressure, and renal perfusion-dependent upon increased circulating natriuretic peptides, and increased synthesis and/or release of vasodilatory prostanoids, nitric oxide and kallikrein-kinin system. Hyponatraemia, due to reduced renal water excretion, only tends to occur as a late phenomenon, as initially the increased non-osmotic vasopressin release [2] is counteracted by intra-renal prostaglandin E_2 synthesis.

Not surprisingly, acute kidney injury (AKI), of varying severity, is relatively common in patients with liver disease and has important prognostic significance. The incidence of AKI and in particular the incidence in the different subgroups of liver disease (acute liver failure – rapidly progressive encephalopathy, coagulopathy and jaundice in the absence of preexisting liver disease; chronic liver disease – advanced liver disease with or without portal hypertension; acute on chronic liver disease – acute decompensation of chronic liver disease from multiple acute insults), is generally unknown, due to the lack of consensus agreement of defining AKI in liver failure patients.

Serum creatinine measurements are affected not only by the aetiology of underlying liver disease and time course of liver failure, but also by reduced dietary protein intake, loss of muscle mass, sex, ethnic origin as well as by interference with the standard Jaffé colorimetric method of creatinine determination, by bilirubin and other compounds which accumulate in liver failure [3]. Similarly, other biomarkers of kidney function, such as cystatin C, are not as accurate in patients with liver failure, and there is no universally accepted definition of AKI in this group of patients.

Hepatorenal Syndrome

Although it is customary to classify AKI into volume-responsive and volume-unresponsive AKI, in reality, a number of overlapping insults typically occur on a background of underlying liver disease (fig. 1). Whilst hepatorenal syndrome (HRS) [4] remains the stereotypical cause of renal failure in the context of liver disease, it only occurs in up to 10% patients with cirrhosis and even fewer patients with acute liver failure. The definition of HRS was originally agreed by a consensus meeting of the International Ascites Club (table 1) [5], and further defined as type 1: progressive renal impairment with a doubling of serum creatinine to >250 mg/dl (220 μmol/l) within a 2-week period, and type 2, with a stable or slow deterioration in renal function, and those who did not fulfil the criteria of type 1. The original consensus definition was subsequently simplified (table 2), specifying volume replacement and rationalisation of the minor criteria, as the fractional excretion of urea was a better marker of HRS than sodium.

HRS represents functional renal disease without significant histological damage, and is therefore potentially reversible, however if left untreated may progress to volume-unresponsive AKI. Deterioration of renal function typically parallels the progressive decline in liver function, as patients progress through

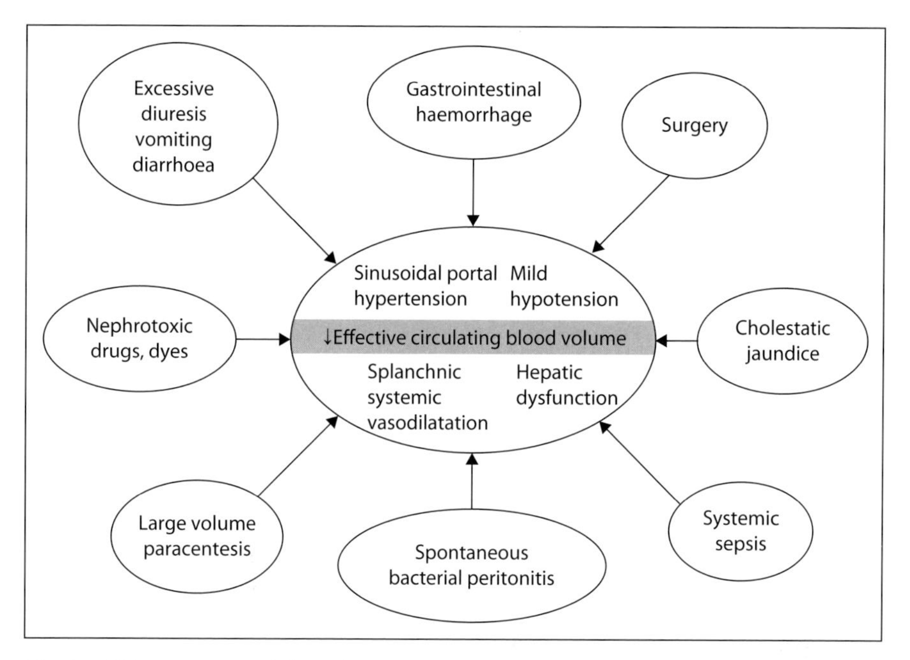

Fig. 1. AKI in patients with liver disease is typically secondary to one or more intercurrent events which further compromise renal perfusion, on a background of a relative reduction in renal perfusion.

stages of cirrhosis without ascites, to ascites, which then becomes less diuretic-responsive and finally to HRS (fig. 2). As renal autoregulation is impaired, patients are then vulnerable to acute deterioration in renal function occurring with episodes of acute decompensation of liver disease (fig. 1) [6], and volume-unresponsive AKI (acute tubular necrosis) accounts for around a third of AKI in advanced cirrhotics [1]. Sepsis, prolonged hypovolaemia and the use of nephrotoxic drugs and radiocontrast agents lead to ischaemic tubular damage which may initially be volume-responsive (fig. 1).

Other causes of AKI in patients with liver failure include drugs and toxins causing acute tubulointerstitial AKI, whilst vasculitis, haemolytic uraemic syndrome, thrombotic thrombocytopenic purpura, acute sickle cell crisis, host-versus-graft disease and toxaemia of pregnancy can affect the glomerular capillaries causing thrombosis and reduced glomerular blood flow.

Medical Therapy for Hepatorenal Syndrome

The key management strategy for patients admitted with acute or chronic liver failure is to avoid the development of HRS wherever possible, by preventing

Table 1. Diagnostic consensus criteria for HRS devised by the International Ascites Club, 1996 [5]. Only major criteria are required for the diagnosis

Major criteria	cirrhosis with ascites or acute liver failure
	serum creatinine >1.5 mg/dl (132 μmol/l) or 24-hour creatinine clearance <40 ml/min
	absence of shock, ongoing bacterial sepsis, fluid losses, and current treatment with nephrotoxic drugs
	no sustained improvement in renal function (decrease in serum creatinine to ≤1.5 mg/dl, or increase in creatinine clearance to ≥40 ml/min after diuretic withdrawal and expansion with 1.5 l of a plasma expander
	proteinuria <500 mg/day, and no ultrasound evidence of obstructive uropathy or parenchymal renal disease
Minor criteria	urine volume <500 ml/day
	urine sodium <10 mEq/l
	urine osmolality > plasma osmolality
	urine RBC <50 per high-power field
	serum Na concentration <130 mEq/l

Table 2. Revised criteria for diagnosis of hepatorenal failure agreed by International Ascites Club, 2007 [4]. Only major criteria are required for the diagnosis

Major criteria	cirrhosis with ascites or acute liver failure
	serum creatinine >1.5 mg/dl (130 μmol/l)
	absence of shock
	no improvement in serum creatinine (decrease to <1.5 mg/dl after 2 days with diuretic withdrawal and volume expansion with albumin. 1 g albumin/kg body weight/day up to a maximum of 100 g/day)
	absence of parenchymal renal disease (proteinuria <500 mg/day, microhaematuria <50 RBC/high-power field) and/or abnormal renal ultrasound
Minor criteria	no current or recent treatment with nephrotoxic drugs

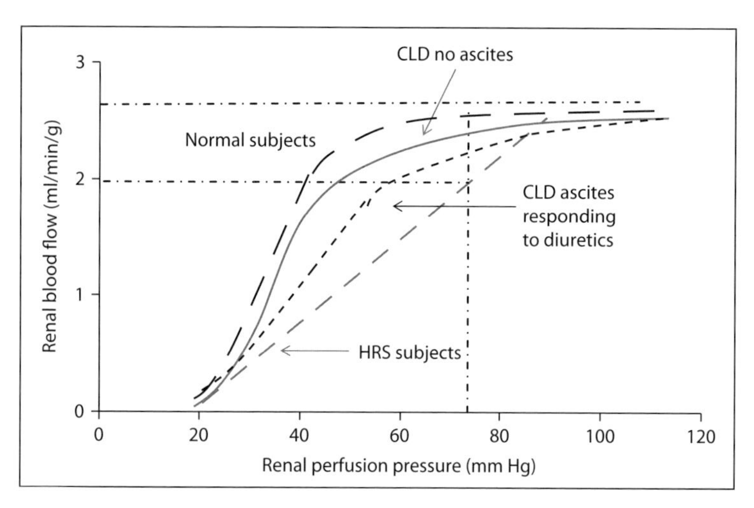

Fig. 2. Change in renal autoregulation as patients develop progressive liver disease, resulting in HRS [6].

relative renal hypoperfusion, by maintaining an effective circulating volume, and renal perfusion pressure. This includes the administration of prophylactic antibiotics in patients with cirrhosis and ascites to reduce the risk of spontaneous bacterial peritonitis and variceal haemorrhage. Some patients with HRS may be relatively hypoadrenal, and should therefore be supported with hydrocortisone. Volume expansion with sodium-containing fluids may exacerbate ascites, and increased intra-abdominal pressure may then compromise renal perfusion by increasing intrarenal endothelin levels.

Initial trials of vasoconstrictors failed to reverse HRS, as vasoconstrictors were typically only given for a short period and without intravascular volume expansion. More recently, medical therapy from patients with type 2 HRS has been more successful by administering subcutaneous terlipressin, a vasopressin analogue, initially at a low dose of 0.5 mg qds, and increasing the dose in a stepwise fashion every few days up to a maximum of 12 mg/day, in addition to plasma volume expansion with albumin, initially with a loading dose of 1 g/kg (up to a maximum of 100 g), followed by 20–40 g/day [7]. Other vasoconstrictors such as intravenous noradrenaline, or oral midodrine in combination with octreotide appear to be also effective for HRS, and similarly other plasma volume expanders such as gelatins and low molecular weight starches may be equally effective as albumin. Other experimental treatments for HRS have included external pressure suits, designed to increase the effective circulating volume. Endothelin antagonists have been developed, but initial trials were not effective in reversing HRS, due to systemic hypotension.

Hyponatraemia is commonly found in patients with chronic liver failure, in particular those with HRS. Vasopressin receptor blockers are now available, and have been used to increase serum sodium in patients with chronic liver disease and ascites, and at higher doses (200 mg of tolvaptan) can induce weight loss, but have not been proven to improve renal function or reverse HRS.

Surgical Management of Hepatorenal Syndrome

Historically portosystemic shunting was used for the management of refractory ascites, and this has now been replaced by the transjugular placement of portosystemic shunts (TIPSS) within the liver. Insertion of TIPSS leads to a reduction in intrarenal endothelin, and improved renal sodium excretion, but reduces hepatic plasma flow and increases the risk of encephalopathy, particularly in the elderly.

Renal Replacement Therapy

The long-term survival of patients with HRS remains dismal, and therefore some healthcare systems restrict access to renal support for only those patients thought to be able to spontaneously recover sufficient liver function, or those deemed suitable for liver transplantation.

Standard renal dialysis systems are designed to predominantly remove small water-soluble solutes by diffusion, such as ammonia. The addition of haemofiltration allows greater removal of water-soluble middle-sized molecules, typically ≤20 kDa, but is less efficient in clearing small solutes, such as ammonia. As patients with liver failure have arterial vasodilatation, with high levels of nitric oxide, treatment with intermittent forms of haemodialysis is often associated with increased frequency of intradialytic hypotension [8]. This can be reduced by pretreating patients with oral midodrine, or increasing noradrenaline or vasopressin infusion rates at the start of treatment. Peritoneal dialysis is used by some centres, but there is the risk of peritonitis, which may potentially result in portal vein thrombosis, making subsequent liver transplantation more technically challenging. Continuous modes of renal replacement therapy lead to fewer hypotensive episodes [9], although higher volume exchanges may be required to control acidosis in patients with acute liver failure and reduced muscle perfusion. Thus continuous modes of renal replacement therapy are used for the acute treatment of inpatients with HRS or established chronic kidney disease, and intermittent haemodialysis or peritoneal dialysis for outpatient management of HRS or end-stage kidney disease.

Renal Support Designed to Increase Hepatic Toxin Removal

Many of the toxins which accumulate in liver failure are lipid-soluble, or protein-bound, so are not readily cleared by standard dialysis or haemofiltration systems. This has led to the development of several extracorporeal therapies designed to remove lipid-soluble and protein-bound toxins, including charcoal adsorption, charcoal perfusion, single-pass albumin dialysis, albumin-based dialysis, plasma separation with adsorption and reinfusion and plasma exchange. In addition there are several centres developing bioartificial support systems. Earlier studies showed the importance of albumin infusions in treating patients with HRS, and albumin is now recognised as having a key detoxifying role, due to its numerous binding sites. Three therapies based on albumin are currently in clinical practice: single-pass albumin dialysis, in which albumin is added to the dialysate during continuous dialysis, albumin dialysis (molecular adsorbent recirculation system (MARS); Gambro AB, Lund, Sweden) which uses a fixed volume of an albumin-based dialysate, which is then regenerated by passage through charcoal and ion exchange cartridges, and fractionated plasma separation, adsorption and dialysis (Prometheus, Fresenius, Bad Homberg, Germany), which uses a highly permeable membrane allowing the passage of plasma albumin, which is then pumped through a neutral resin cartridge followed by ion exchanger before returning to the patient. These treatments have typically been used in cases of hyperacute liver failure with good potential for hepatic regeneration, as a bridge for liver transplantation, primary non-function whilst awaiting urgent retransplantation, cases of acute decompensation on a background of chronic liver disease, and small for size liver post-hepatic resection, but also type 2 HRS. Although the MARS treatment has been used in clinical practice for some 10 years [10], there are very few published randomised controlled trials. Evidence to date has shown that although encephalopathy and cardiovascular stability is improved by MARS, neither treatment (MARS or Prometheus) has currently been proven to improve either patient survival, or extended survival to allow successful liver transplantation compared to standard supportive medical management with continuous renal support.

Conclusion

Changes in renal function, particularly sodium retention, occur in patients with liver disease, well before arterial vasodilatation is detectable and the development of ascites. As liver damage progresses, further dilatation occurs with corresponding compensatory increase in neuroendocrine activation, and changes in renal autoregulation, making the kidney more susceptible to acute reduction in effective plasma volume or hypotension. Although HRS is the most recognised form of AKI in patients with liver disease, it only accounts for less than 10% of cases in current clinical practice, as volume-unresponsive AKI or acute tubular necrosis is now much more common. The key management strategy for patients with liver disease is to prevent HRS developing by avoiding nephrotoxins, including radiocontrast dyes, relative hypovolaemia and hypotension. Treatment centres on treating potentially precipitating factors, with daily plasma volume expansion, typically albumin, in combination with vasoconstrictors, such as terlipressin, a vasopressin analogue, or noradrenaline. In those patients who do not spontaneously recover renal function, then dialysis may be appropriate for those suitable for liver transplantation. Continuous therapies are more suited to critically ill patients with liver failure, whereas intermittent haemodialysis and/or haemofiltration can be used for outpatient management, but may be complicated by intradialytic hypotension. Although other treatments, using albumin-based dialysate or plasma filtration with haemodialysis/haemofiltration have been advocated, evidence of their superiority over standard supportive medical therapy remains awaited.

Acknowledgement

I wish to thank the support of my colleagues from the Sherlock Hepatobiliary-Pancreatic and Liver Transplantation Unit, Royal Free Hospital, London, UK.

References

1 Slack AJ, Wendon J: The liver and the kidney in critically ill patients. Blood Purif 2009;28: 124–134.
2 Davenport A: The brain and the kidney-organ cross talk and interactions. Blood Purif 2008;26:526–536.
3 Cholongitas E, Marelli L, Kerry A, Senzolo M, Goodier DW, Nair D, Thomas M, Patch D, Burroughs AK: Different methods of creatinine measurement significantly affect MELD scores. Liver Transpl 2007;13:523–529.
4 Arroyo V, Gines P, Gerbes AL, Dudley FS, Gentili P, Laffi G, Reynolds TB, Ring-Larsen H, Scholmerich J: Definition and diagnostic criteria of refractory ascites and hepatorenal syndrome in cirrhosis. International Ascites Club. Hepatology 1996;23:164–176.
5 Salerno F, Gerbes A, Ginès P, Wong F, Arroyo V: Diagnosis, prevention and treatment of hepatorenal syndrome in cirrhosis. Postgrad Med J. 2008;84:662–670.
6 Dagher L, Moore K: The hepatorenal syndrome. Gut 2001;49:729–737.
7 Uriz J, Gines P, Cardenas A, Sort P, Jiminez W, Salmeron JM, Bataller R, Mas A, Navasa M, Arroya V, Rodes JR: Terlipressin plus albumin infusion; an effective and safe therapy of hepatorenal syndrome. J Hepatol 2000;33: 43–48.
8 Davenport A: Continuous renal replacement therapy in patients with hepatic and acute renal failure. Am J Kidney Dis 1996; 28(suppl 3):S62–S66.

9 Davenport A, Will EJ, Davison AM: Improved cardiovascular stability during continuous modes of renal replacement therapy in critically ill patients with acute hepatic and renal failure. Crit Care Med 1993;21:328–338.

10 Stange J, Mitzner SR, Risler T, Erley CM, Lauchart W, Goehl H, Klammt S, Peszynski P, Freytag J, Hickstein H, Löhr M, Liebe S, Schareck W, Hopt UT, Schmidt R: Molecular adsorbent recycling system (MARS): clinical results of a new membrane-based blood purification system for bioartificial liver support. J Artif Organs 1999;23:319–330.

Dr. Andrew Davenport
UCL Centre for Nephrology
Royal Free & University College Medical School, Hampstead Campus
Rowland Hill Street, London NW3 2QG (UK)
Tel. +44 20 77 940 500, Fax +44 20 78 302 125, E-Mail Andrew.davenport@royalfree.nhs.uk

Ronco C, Bellomo R, McCullough PA (eds): Cardiorenal Syndromes in Critical Care.
Contrib Nephrol. Basel, Karger, 2010, vol 165, pp 206–218

Fluid Management in Septic Acute Kidney Injury and Cardiorenal Syndromes

Rinaldo Bellomo[a] · John R. Prowle[a] · Jorge E. Echeverri[b] · Valentina Ligabo[b] · Claudio Ronco[b]

[a]Department of Intensive Care, Austin Health, Heidelberg, Melbourne, Vic., Australia, and [b]Department of Nephrology, Dialysis & Transplantation, International Renal Research Institute, San Bortolo Hospital, Vicenza, Italy

Abstract

Intravenous fluids are commonly administered to patients with developing septic acute kidney injury (AKI). Conversely, fluids are just as commonly removed with diuretics or renal replacement therapy (RRT) techniques or ultrafiltration in patients with cardiorenal syndromes (CRS). In both groups, there is controversy regarding fluid management. However, in patients with septic AKI, the deleterious consequences of overzealous fluid therapy are increasingly being recognized, while concerns exist both about the possible adverse effects of excessive and/or insufficient fluid removal with diuretics or ultrafiltration in CRS. In this article, we discuss how interstitial edema can further delay renal recovery and why conservative fluid strategies are now being advocated in septic AKI. In patients with septic AKI, this strategy might require RRT to be given earlier to assist with fluid removal. However, in patients with either septic AKI or CRS, hypovolemia and renal hypoperfusion can occur if excessive fluid removal is pursued with diuretics or extracorporeal therapy. Thus, accurate assessment of fluid status and careful definition of targets are needed to improve clinical outcomes. Controlled studies of conservative versus liberal fluid management in patients with AKI or CRS seem justified.

Appropriate management of intravenous fluid therapy is a key aspect of the treatment of septic acute kidney injury (AKI) [1] and supplemental intravenous fluids often given to mitigate the effects of hemodynamic renal insults. On the other hand, the adverse effects of fluid overload may be most pronounced in situations such as systemic sepsis [2]. In patients with cardiorenal syndrome (CRS) (especially types 1 and 2), fluid management is similarly important, because many patients present with fluid overload and require fluid removal and because fluid removal itself may induce excessive reductions in intravascular volume and/or cardiac output, which may induce further renal injury. In this brief review, we will focus on the major issues surrounding fluid management in septic AKI and type 1 and type 2 CRS. We will argue that an approach which avoids progressively positive fluid balances with extracellular volume expansion and organ edema but also shuns overzealous fluid removal may serve these two different types of patients best.

Rationale for Giving Fluids in Septic AKI

Much of the rationale for fluid therapy in septic AKI is rooted in the notion that AKI is the consequence of reduced cardiac output, systemic hypotension and triggered neuroendocrine reflexes. Although initially reversible, persistent renal ischemia is then considered responsible for so-called acute tubular necrosis. Vigorous fluid administration in this setting is therefore aimed at reversing renal ischemia to avert the onset of acute tubular necrosis or prevent recurrent injury that might compromise renal recovery. Within this framework, oliguria or hypotension prompt intravenous fluid challenges, and maintenance fluids are prescribed to promote diuresis, maintain cardiac output, and keep the patient 'well filled' [3, 4].

Challenges to the Need for Fluids in Septic AKI

The presumed clinical benefit of the above approach, however, is being challenged by increasing evidence that positive fluid balances in the order of 5–10% of body weight are associated with worsening organ dysfunction in the critically ill [5] and with worse postoperative outcomes after routine surgery [6] with no evidence of any beneficial effects on renal function. Given this dichotomy between traditional teaching and evolving evidence, wide variations in clinical fluid management continue to exist.

Pathophysiology of Fluid Therapy in Septic AKI

From a renal standpoint, fluid therapy is used to restore glomerular filtration and thus increase urine output. Fluid therapy is aimed at restoring blood pressure (a major determinant of renal perfusion pressure) and cardiac output (a prerequisite for adequate renal blood flow). Restoration of these parameters might relax the neuroendocrine reflexes responsible for increasing renal vascular resistance and diminishing glomerular filtration rate (GFR) (fig. 1).

Fluid administration aimed at restoring systemic blood pressure should 'work', mechanistically speaking, by increasing preload and stroke volume. Unfortunately, conventional goals of fluid resuscitation (restoration of blood pressure, central venous pressure and/or urine output), are only indirect

<table>
<tr><th colspan="3">Loss of glomerular ultrafiltration in septic AKI</th></tr>
<tr><th>Abnormality</th><th>Physiological effect</th><th>Consequence</th></tr>
<tr><td>Low systemic blood pressure</td><td rowspan="3">Decreased intraglomerular pressure</td><td rowspan="8">Decreased glomerular filtration rate</td></tr>
<tr><td>Afferent arteriolar vasoconstriction</td></tr>
<tr><td>Efferent arteriolar vasodilatation</td></tr>
<tr><td>Renal interstitial edema</td><td rowspan="4">High intracapsular pressure</td></tr>
<tr><td>Extrinsic compression</td></tr>
<tr><td>Tubular obstruction</td></tr>
<tr><td>Increased intra-abdominal pressure</td></tr>
<tr><td>Hyperoncotic fluids</td><td>Rise in oncotic pressure</td></tr>
</table>

Fig. 1. Pathological sequelae of fluid overload in organ systems.

measures of cardiac output and much less indicative of restoration of adequate organ blood flow [7, 8]. Similarly, the effects of critical illness, preexisting chronic disease and pharmacotherapy can unpredictably alter the determinants of fluid responsiveness such as myocardial compliance and contractility, systemic vascular resistance, regional blood flow distribution, venous capacitance and capillary permeability. These changes make the effects of volume resuscitation variable in extent and duration, and make the assessment of adequacy of volume replacement challenging. In such situations, fluid therapy is either insufficient to restore hemodynamic stability, or excessive volumes are required. Importantly, fluids do not correct vasodilatation. Invasive monitoring may, therefore, be required to guide treatment, not only to ensure adequate volume expansion, but also to prevent excessive fluid administration. Irrespective of how it is guided, however, indiscriminate use of intravenous fluid to maximize cardiac output might not be beneficial. Increasing cardiac output from a normal or elevated value with fluids might be effective in transiently increasing blood pressure, reducing vasopressor requirement and inducing diuresis, but such responses are not sustained and may not improve organ function In particular, when one considers the short-lived hemodynamic effect of exogenous fluids, continuing such an approach beyond the initial few hours of illness would require regular, repeated volume challenges and, inevitably, lead to a markedly positive fluid balance with its associated adverse consequences.

Colloid solutions are frequently used to avoid excessive fluid therapy, despite limited evidence to justify their use [9]. Most such fluids are also likely to leak into the extravascular compartment, have a limited theoretical advantage over crystalloids in preventing tissue edema and lead to only small decreases in the total quantity of fluid administered [9]. Moreover, higher molecular weight (>200 kDa) hyperoncotic starches are associated with an increased risk of AKI [10].

Adverse Effects of Fluid Therapy

Crystalloid solutions expand the extracellular compartment, and, over time, leave the circulation and distribute in the extracellular volume, particularly in critically ill patients with increased capillary leakiness. Renal excretion of exogenous sodium is generally slow and is further impaired in acute illness.

In the intensive care unit (ICU), 0.9% saline is still widely used as an intravenous crystalloid, and many colloids and drugs that are suspended in saline. Saline contains a supranormal chloride concentration and its administration will result in relative or absolute hyperchloremia. Hyperchloremia has been shown to reduce renal blood flow and to impair sodium excretion in humans.

Fluid overload also contributes to visceral edema, a risk factor for intra-abdominal hypertension (IAH). IAH increases renal venous pressure, reduces blood flow and increases pressure in Bowman's space [11]. In an ICU population, positive fluid balances have been associated with an increased risk of IAH, which, in turn, is strongly associated with the development of AKI [19].

In the absence of overt IAH, renal interstitial edema alone might impair renal function. As an encapsulated organ, the kidney is affected by fluid congestion and raised venous pressures with a disproportionate elevation in intracapsular pressure leading to a decrease in renal blood flow and GFR [11]. Observational studies found an association between positive fluid balance and an increased risk of AKI or renal non-recovery [5]. In both, a positive fluid balance was associated with increased mortality in AKI patients. Elevated tubular pressure may play a role in the continued loss of renal function during the maintenance phase of AKI, yet another mechanism by which persistent fluid overload might increase the duration and severity of AKI.

Interstitial Edema and Organ Dysfunction

Physiologically, fluid overload results in tissue edema. Edema impairs oxygen and metabolite diffusion, distorts tissue architecture, obstructs capillary blood flow and lymphatic drainage, and disturbs cell-cell interactions and may contribute to progressive organ dysfunction. These effects are more pronounced in encapsulated organs – such as the liver and kidneys – which lack the capacity to accommodate additional volume without an increase in interstitial pressure and compromised organ blood flow. Myocardial edema can worsen ventricular function and oxygen delivery Recovery of gastrointestinal function, wound healing, and coagulation are also all adversely effected by edema [6].

The adverse effects of fluid overload are perhaps most evident in the lungs, where overzealous fluid resuscitation can lead to acute pulmonary edema. In patients with established acute lung injury, prospective, multicenter, randomized controlled trials [12], have provided evidence associating more positive fluid balances with poorer pulmonary outcomes.

Indeed, the Fluid and Catheter Treatment Trial (FACTT) study [12], by far the largest multicenter randomized controlled trial examining fluid balance in patients with acute lung injury, reported an almost statistically significant decrease in the requirement for renal replacement therapy (RRT) in the *conservative* fluid management group (10 vs. 14% in the liberal fluid management group, $p = 0.06$). This finding is important because it occurred in patients with acute lung injury (of whom the majority had sepsis as the trigger) who were on mechanical ventilation at high levels of positive end-expiratory pressure. In these conditions, patients would be expected to be at particular risk of AKI secondary to hypoperfusion. If, despite this high risk, fluid restriction resulted in a near significant *decrease* in the need for RRT, it seems unlikely that similar fluid restriction in less severe acute illness would contribute to AKI. It is important to note, however, that fluid removal in the FACTT study was pursued in a clinically judicious way. Namely, fluid was removed only when hemodynamically safe. Specifically, no fluid removal took place in either group during periods of hemodynamic instability (mean arterial pressure <60 mm Hg or the need for vasopressor infusion). Thus, contrary to received opinion, maintaining a positive fluid balance may worsen renal function and impede recovery from septic AKI.

Possible Approaches to Fluid Management

Although correction of the relative volume deficit is crucial, restoration of preload can fail to normalize blood pressure or cardiac output in many situations, including sepsis and postoperative vasodilatory states. Invasive cardiovascular monitoring may be required to recognize such situations and to allow the timely institution of vasopressor and/or inotropic support. Numerous methods for preload assessment are available, but any measurement is only as useful as its clinical interpretation. Rather than focusing on fluid responsiveness and the maximization of cardiac output to combat hypotension (which is a path towards a relentlessly positive fluid balance), it might be better to merely ensure that preload is sufficient to generate an adequate cardiac output, so that the appropriate point to stop fluid resuscitation can be determined. Such an approach may require earlier or greater use of vasopressor therapy, which is in fact associated with increased renal blood flow and restoration of urine output [13].

Despite guided fluid resuscitation and early vasopressor support, initial resuscitation of acute severe illness almost always results in a positive fluid balance and tissue edema. Thus, in the subsequent plateau phase of sepsis, treatment focus should shift towards the prevention of further fluid overload and the removal of accumulated excess salt and water.

Septic patients generally have high mandatory water and sodium intake in drugs and nutrition, thus 'maintenance' intravenous fluids are rarely required. It is much easier not to give fluid than to remove edema once accumulated. Furthermore, as soon as it can be hemodynamically tolerated, sodium and water balance should be neutral, or even negative, as in the FACTT trial.

Achieving a neutral or negative fluid balance spontaneously during acute illness can be difficult. Loop diuretics are frequently employed [12], but their use can be complicated by electrolyte abnormalities, worsening of renal function and progressive diuretic resistance. Resistance can be overcome by the additional administration of diuretics that target the distal tubule and collecting ducts. Even with such treatment, however, sufficient and sustained natriuresis can be difficult to achieve. Hyperoncotic colloids can be used to encourage vascular refilling but, as already noted, they may be nephrotoxic, particularly high molecular weight starches.

The prevention of edema is particularly difficult in patients with AKI as excessive diuretic therapy may worsen renal function, induce hypernatremia and/or be unable to induce a sufficient diuresis. Indeed, diuretics have not been shown to beneficially affect the clinical course of AKI or to speed up recovery after RRT [14]. Initiation of RRT may thus be required to successfully maintain neutral fluid balance. However, intermittent ultrafiltration during conventional hemodialysis might be associated with intradialytic hypotension and an increased risk of recurrent renal injury. Continuous renal replacement therapy (CRRT) has many theoretical advantages in this setting because a constant slow rate of ultrafiltration allows time for vascular refilling and effective control of fluid balance while maintaining hemodynamic stability [15]. In the treatment of AKI in the ICU, use of intermittent hemodialysis correlates with progressively positive fluid balances whereas CRRT enables net fluid removal, despite the fact that CRRT is preferred for sicker, more hemodynamically unstable patients. Similarly, use of CRRT for initial treatment has been associated with higher rates of renal recovery than immediate use of intermittent hemodialysis in critically ill individuals. These observations have been strengthened by the findings of a large randomized controlled trial of CRRT for AKI in the ICU [16]. In our opinion, consideration should be given to the early initiation of CRRT, if fluid balance cannot be adequately controlled with diuretic therapy. This approach anticipates and limits the extent of fluid overload rather than treating its consequences. In addition, it permits adequate nutritional support without worsening fluid balance. Therefore, in a large proportion – perhaps the majority – of critically ill septic AKI patients, we believe that CRRT should be initiated within the first 24 h of ICU admission. Such earlier intervention with CRRT is now commonly practiced worldwide and has been associated with improved survival.

In light of the above data, we advocate a two-stage approach of directed resuscitation in acute illness, with an early transition to neutral and then negative fluid balances, to best limit the adverse consequences of fluid overload. This philosophy is in harmony with recent consensus guidelines for the management of perioperative patients Nonetheless, further clinical trials are clearly warranted to examine optimal fluid balances and to better document renal outcomes, yet the evidence available so far favors more restrictive fluid management strategies than have been employed historically. These strategies should be implemented with appropriate monitoring to avoid iatrogenic hypovolemia and further functional renal impairment, particularly during active fluid removal. Medical education, practice patterns and clinical guidelines should increasingly reflect these new paradigms.

Fluid Management in Cardiorenal Syndrome Types 1 and 2

Similar to septic AKI, there are increasing concerns that inadequately treated or simply tolerated fluid overload may contribute to worse renal and patient outcomes in patients with type 1 and type 2 CRS. For example, right atrial pressures are a more important predictor of renal dysfunction in patients with heart failure than measures of left-sided performance such as cardiac output or left-sided filling pressure [17–19]. Recently, Damman et al. [18] and collaborators published a retrospective series of 2,557 patients who had right-side cardiac catheterization because of variable causes of heart failure with documentation of central venous pressure (CVP). Multivariate analysis showed an independent and inversely proportional relationship between CVP and GFR. Higher CVP values correlated with greater renal deterioration ($p < 0.0001$). This relationship was maintained even after excluding patients with heart transplants and heart failure. In support of these observations, Mullens et al. [19] concomitantly published a retrospective study using a cohort of 145 acute heart failure (AHF) patients who required intensive medical care. Pulmonary artery catheter measurements were used to monitor progress towards therapeutic targets: cardiac indices >2.4 l/min/m^2, PCWP ≤18 mm Hg and CVP ≤8 mm Hg. In total, 40% of patients developed AKI despite optimization of heart function and reduction in indices of hypervolemia. Patients who developed AKI, compared to those with stable renal function, showed higher CVP values at the time of admission (18 ± 7 vs. 12 ± 6 mm Hg, $p < 0.001$) and after medical therapy (11 ± 8 vs. 8 ± 5 mm Hg, $p < 0.001$), despite having a better cardiac index both at baseline (2 vs. 1.8 l/m^2/min, $p < 0.008$) and after medical therapy (2.7 vs. 2.4 l/m^2/min, $p < 0.001$). The CVP value was as an independent predictor of renal dysfunction, with increasing progressive risk, particularly with CVP >24 mm Hg. No difference was found between groups when left-side pressure (PCWP), pulmonary systolic pressure and arterial systolic pressure were evaluated. Deterioration of renal function in AHF patients is thus not totally dependent on cardiac output or mean arterial pressure. Hypervolemia, venous congestion and tissue edema might also explain acute renal deterioration. Thus, renal functional impairment in the setting of AHF should not lead to the acceptance or undertreatment of hypervolemia. Indeed, its prevention or rapid and diligent treatment may lead to better renal outcomes.

How can hypervolemia and edema cause kidney injury in either chronic heart failure or AHF? Elevated right heart pressures are likely to decrease renal perfusion pressure through an increase in backpressure and the formation of renal edema. As renal perfusion pressure is equal to mean arterial pressure minus intrarenal tissue pressure, increased backpressure may induce renal hypoperfusion and activate the renin-angiotensin-aldosterone system. In addition, as the kidney is surrounded by a capsule, organ edema may generate further backpressure and a degree of intracapsular 'tamponade' with additionally decreased renal perfusion, decreased urine output, more fluid retention and further edema. This vicious cycle can easily contribute to diuretic resistance. In such patients, even large doses of loop diuretics may fail to achieve a neutral or negative fluid balance. This situation of deteriorating kidney function is further exacerbated by the effect of fluid accumulation and myocardial dilatation on cardiac output and systemic blood pressure, both of which decrease in this setting. When all of the above cardiovascular and fluid retention states exist then diuretic therapy may fail and require ultrafiltration [20, 21]. However, diuretic resistance is often a relative state which can be overcome if sufficient doses of loop diuretics are given (up to 50 mg/h) by continuous infusion and if loop diuretic therapy is aided by the concomitant administration of diuretics which target other segment of the tubule (e.g. thiazides, carbonic anhydrase inhibitors and aldosterone antagonists).

Additional Therapeutic Challenges

Having overcome diuretic resistance, further issues related to diuretic use must be considered: What is a safe and yet sufficient urine output in this setting? What are the consequences of such urine output on electrolytes like sodium, chloride, potassium and magnesium? How much fluid should be removed? How can one know when an optimal fluid state has been achieved for a specific patient?

Although only empirical observations can help address these issues in a specific patient, some general principles should be taken into account in all cases. First, a urine output of 3–4 ml/kg/h rarely causes intravascular volume depletion as capillary refill can meet such rates in almost all patients. Thus, in an 80-kg man, a urine output to 250 ml/h almost never causes hypotension or decreases in cardiac output. Second, diuretic infusion is clearly superior to boluses of diuretics in ensuring a steady, predictable and smooth urinary output. In addition, greater diuresis is typically achieved and a lesser dose of loop diuretic given. Finally, if large amounts of diuretics are necessary, the avoidance of rapidly given boluses lessens the risk of temporary deafness. Third, in the absence of pulmonary edema, tissue edema does not require correction over minutes or even hours. The fluid that has accumulated over days is more logically removed at a similar rate. This means that large volumes of hourly output are unnecessary and potentially dangerous. Fourth, a loop diuretic infusion will typically result in a marked kaliuresis. This requires that steps be taken to avoid hypokalemia. The administration of oral potassium preparations seems logical and easy. It is however important to estimate the likely requirements. This can be done by measuring the urinary potassium concentration and calculating the daily losses of potassium which require replacement. For example, at 300 ml/h of diuresis and a potassium concentration of 50 mmol/l, a patient will required approximately 100 mmol of potassium every 6 h. This may not be achievable with oral therapy alone. A continuous potassium infusion may be necessary. An alternative may be to add potassium-sparing diuretics, which are indicated to help improve outcome in patients with heart failure such as spironolactone.

Another concern relates to magnesium losses. Although the body has a large reservoir of magnesium in bones, magnesium replacement therapy in patients experiencing a large diuresis seems wise and can be achieve either intravenously or orally, typically with 20–30 mmol/day.

In some patients, chloride losses exceed sodium losses and hypochloremic metabolic alkalosis develops. This is usually corrected with potassium chloride and magnesium chloride as described above and is typically not a major source of concern. However, in some patients, clinicians may feel that intervention is warranted. In these patients, the addition of acetazolamide and a decrease in loop diuretic use is often sufficient to maintain diuresis and achieve normalization of chloride levels and metabolic acid-base status.

Finally, in some patients, water losses exceed sodium losses leading to hypernatremia. In some patients, this can be an indication for ultrafiltration. In others, cautious administration of water and the addition of a thiazide diuretic can achieve improved natriuresis and maintain normal serum sodium levels. Importantly, no matter what the chosen interventions are, assiduous monitoring of renal function and electrolytes is mandatory under these circumstances.

At this point it must be noted that some authors have questioned the use of loop diuretics in heart failure. They comment that loop diuretics have not been shown to change the outcome of either heart failure or AKI, their impact on heart failure mortality is questionable and the use of large doses is associated with an increased mortality and risk of AKI. However, loop diuretics are widely used in most types of CRS because they achieve specific relief of symptoms and resolution of the congestive state in most patients. Moreover, their association with unfavorable outcome when given in large doses probably represents the fact that diuretic resistance is a powerful marker of disease severity.

Other approaches to diuresis in these patients have also recently been promoted. Nesiritide, a recombinant analog to brain natriuretic peptide (BNP), has been proposed for the management of AHF as it counteracts renin-angiotensin-aldosterone system hyperactivity and optimizes fluid management. However, results have not been optimal and on occasion, renal function deterioration has

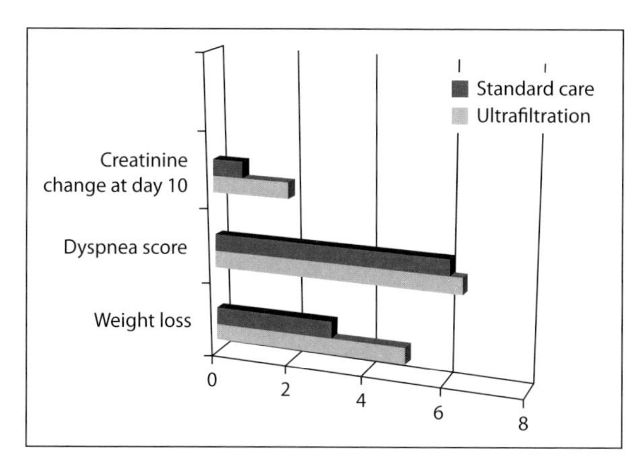

Fig. 2. Changes in weight loss, dyspnea score and serum creatinine in patients randomized to standard care or ultrafiltration for the treatment of type 1 CRS.

been observed. Thus, the role of nesiritide in the treatment of heart failure and fluid overload remains unclear [22].

If diuretics fail, the treating clinician needs to investigate the reasons for diuretic resistance and correct them and/or resort to other techniques (e.g. ultrafiltration) to relieve the symptoms of fluid overload and/or restore a desirable and safe physiological state.

Ultrafiltration

Diuretic resistance has led the development of new technologies based on slow ultrafiltration. For example, slow ultrafiltration in patients with AHF refractory to diuretics can remove up to remove 5 l/day without hemodynamic instability. Edema is thus reduced, with an improvement in cardiac and respiratory function, increased diuretic responsiveness and hemodynamic stability.

For example, the RAPID [20] (Relief for Acute Fluid Overload Patients with Decompensated Chronic Heart Failure) study randomly assigned patients to either receive diuretic management or ultrafiltration for 8 h. Fluid removal during the first 24 h was higher in the ultrafiltration group (4.6 vs. 2.8 l, $p = 0.001$) and occurred without renal functional deterioration or hemodynamic instability. In the larger UNLOAD trial [21], 200 patients were randomly assigned to receive either diuretic therapy or ultrafiltration. All patients with signs of fluid overload, independent of the ejection fraction, were included. Weight loss was greater in the ultrafiltration group (5.0 ± 3.1 vs. 3.1 ± 3.5 kg, $p = 0.001$) (fig. 2). The requirement for inotropic support was also decreased in the ultrafiltration group (3 vs. 12%, $p = 0.015$). Follow-up to 90 days showed a lower rehospitalization rate for HF in the ultrafiltration group (18 vs. 32%, $p = 0.022$), without renal functional deterioration. However, of note, the mean dose of frusemide during the study period (48 h) was 180 mg. This is relatively low and suggests that more aggressive diuretic therapy may have achieved equivalent results. Finally, changes in serum creatinine during hospital admission showed no correlation with the amount of fluid removed.

In some patients, deterioration of renal function can be substantial and dialytic therapy may become necessary. In these patients, CRRT and peritoneal dialysis may be more appropriate that intermittent dialysis. Prompt referral of such patients to the intensive care and/or nephrology team is then a priority.

A Judicious Approach

Diuretics can be called 'two-headed swords' in the context of chronic heart failure, pointing out their potent efficacy in relieving symptoms of fluid overload, but also the risk of potential worsening of renal function. However, allowing patients to have persistent severe edema in order to 'protect' the kidney is physiologically irrational and probably clinically undesirable. Similarly, the pursuit of a patient without any trace of edema in the setting of right heart failure is likely to induce marked renal dysfunction with its attendant risks and consequences without appreciably improving the patient's quality of life. A more careful approach which controls edema to a state which balances risks and benefits as physiologically and clinically seen in an individual patient seems safer. This assessment of adequacy of fluid removal may be assisted by the use of biomarkers, which can now be used to guide therapy and optimize fluid management [23]. In particular, BNP was used to guide therapy in randomized study and compared with symptom-guided treatment in study of 499 patients with systolic heart failure. In this study, BNP-guided therapy resulted in similar rates of survival free of all cause hospitalizations. However, it improved the secondary endpoint of survival free of heart failure-related hospitalization (hazard ratio of 0.68). In addition, aldosterone-blocking diuretics were prescribed more frequently in the BNP-guided therapy group. These observations reinforce the view that biomarkers like BNP may assist clinicians in titrating diuretic therapy.

Conclusions

Fluid removal with diuretics (especially loop diuretics) remains key to the relief of symptoms and the improvement of pathophysiological states in type 1 and 2 CRS. The administration of diuretics requires careful titration, occasional use of continuous infusion, attention to electrolyte and fluid balance, measured and slow removal of fluid, avoidance of excessive overall fluid removal, combination of different diuretic medications targeted to different aspects of tubular function and appropriate dosage. In cases of resistance, hemodynamic factors contributing to such resistance must be addressed. In selected patients, ultrafiltration may be necessary to achieve the desired physiological and clinical goals. If clinicians adhere to the above principles, they will find that fluid removal can be achieved safely and effectively in essentially all CRS patients.

References

1 Leblanc M, Kellum JA, Gibney RT, et al: Risk factors for acute renal failure: inherent and modifiable risks. Curr Opin Crit Care 2005; 11:533–536.

2 Payen D, de Pont AC, Sakr Y, et al: Sepsis Occurrence in Acutely Ill Patients (SOAP) Investigators: A positive fluid balance is associated with a worse outcome in patients with acute renal failure. Crit Care 2008;12:R74.

3 Bellomo R, Li W, May CN: Managing septic acute renal failure: 'fill and spill'? 'squeeze and diurese'? or 'block Bax to the max'? Crit Care Resusc 2004;6:12–16.

4 Duke GJ: Renal protective agents: a review. Crit Care Resusc 1999;1:265–275.

5 Bouchard J, Soroko SB, Chertow GM, et al: Program to Improve Care in Acute Renal Disease (PICARD) Study Group: Fluid accumulation, survival and recovery of kidney function in critically ill patients with acute kidney injury. Kidney Int 2009;76:422–427.

6 Brandstrup B, Tonnesen H, Beier-Holgersen R, et al: Effects of intravenous fluid restriction on postoperative complications: comparison of two perioperative fluid regimens: a randomized assessor-blinded multicenter trial. Ann Surg 2003;238;641–648.

7 Michard F, Teboul JL: Predicting fluid responsiveness in ICU patients: a critical analysis of the evidence. Chest 2002;121: 2000–2008.

8 Wan L, Bellomo R, May CN: The effect of normal saline resuscitation on vital organ blood flow in septic sheep. Intensive Care Med 2006;32:1238–1242.

9 Finfer S, Bellomo R, Boyce N, et al: SAFE Study Investigators: A comparison of albumin and saline for fluid resuscitation in the intensive care unit. N Engl J Med 2004; 350:2247–2256.

10 Brunkhorst FM, Engel C, Bloos F, et al: German Competence Network Sepsis (SepNet): Intensive insulin therapy and pentastarch resuscitation in severe sepsis. N Engl J Med 2008;358:125–139.

11 Herrier T, Tischer A, Meyer A, et al: The intrinsic renal compartment syndrome: new perspectives in kidney transplantation. Transplantation 2010;89:40–46.

12 The National Heart, Lung, and Blood Institute Acute Respiratory Distress Syndrome (ARDS) Clinical Trials Network: Comparison of two fluid-management strategies in acute lung injury. N Engl J Med 2006;354:2564–2575.

13 Bellomo R, Wan L, May C: Vasoactive drugs and acute kidney injury. Crit Care Med 2008; 36:S179–S186.

14 Bagshaw SM, Delaney A, Haase M, et al: Loop diuretics in the management of acute renal failure: a systematic review and meta-analysis. Crit Care Resusc 2007;9:60–68.

15 Bellomo R, Ronco C: Renal replacement therapy in the intensive care unit. Crit Care Resusc 1999;1:13–24.

16 The RENAL Replacement Therapy Study Investigators: Intensity of continuous renal-replacement therapy in critically ill patients. N Engl J Med 2009;361:1627–1638.

17 Androne AS, Hryniewicz K, Hudaihed A, et al: Relation of unrecognized hypervolemia in chronic heart failure to clinical status, hemodynamics, and patient outcomes. Am J Cardiol 2004;93:1254–1259.

18 Damman K, van Deursen VM, Navis G, et al: Increased central venous pressure is associated with impaired renal function and mortality in a broad spectrum of patients with cardiovascular disease. J Am Coll Cardiol 2009;53:582–588.

19 Mullens W, Abrahams Z, Francis GS, et al: Importance of venous congestion for worsening of renal function in advanced decompensated heart failure. J Am Coll Cardiol 2009;53:589–596.
20 Bart BA, Boyle A, Bank AJ, et al: Ultrafiltration versus usual care for hospitalized patients with heart failure. The Relief for Acutely Fluid overloaded Patients with Decompensated Congestive Heart Failure (RAPID-CHF) trial. J Am Coll Cardiol 2005; 46:2043–2046.
21 Costanzo MR, Guglin ME, Saltzberg MT, et al: Ultrafiltration versus intravenous diuretics for patients hospitalized for acute decompensated heart failure. J Am Coll Cardiol 2007;49:675–683.
22 Sackner-Bernstein JD, Kowalski M, Fox M, et al: Short-term risk of death after treatment with nesiretide for decompensated heart failure: a pooled analysis of randomized controlled trials. JAMA 2005;293:1900–1905.
23 Pfisterer M, Buser P, Rickli H, et al: BNP-guided vs. symptom-guided heart failure therapy: the trial of intensified vs. standard medical therapy in elderly patients with congestive heart failure randomized trial. JAMA 2009;301:383–392.

Prof. Rinaldo Bellomo
Department of Intensive Care, Austin Health
Heidelberg, Melbourne, Vic 3084 (Australia)
Tel. +61 3 9496 5992, Fax +61 3 9496 3932
E-Mail rinaldo.bellomo@austin.org.au

Ronco C, Bellomo R, McCullough PA (eds): Cardiorenal Syndromes in Critical Care.
Contrib Nephrol. Basel, Karger, 2010, vol 165, pp 219–225

Use of Loop Diuretics in the Critically Ill

Thomas Rimmelé · Vijay Karajala · Raghavan Murugan · John A. Kellum

The CRISMA (Clinical Research, Investigation, and Systems Modeling of Acute Illness) Laboratory, Department of Critical Care Medicine, University of Pittsburgh Medical Center, Pittsburgh, Pa., USA

Abstract

Diuretics are commonly used in the intensive care unit, especially for patients with oliguric acute kidney injury. This practice is controversial since there is a lack of evidence regarding any beneficial effects of diuretics either on prevention or treatment of acute kidney injury. Some data even suggest harm when diuretics are used with the goal to influence renal function. However, diuretics can minimize fluid overload, making patient management easier and potentially avoiding many cardiopulmonary and non-cardiopulmonary complications. We will briefly review the available evidence for and against the use of diuretics in the critically ill, including cardiorenal syndromes.

Administration of diuretics for patients in the intensive care unit (ICU) remains controversial. We have to acknowledge that, regarding diuretics, there is still an important gap between current knowledge that can be extracted from evidence-based medical literature and daily medical practice. The aim of this review is to provide a critique of the current literature concerning diuretics in critically ill patients. Management of this drug in the subgroup of patients with a cardiorenal syndrome (CRS) will be described in the context of recently published classification/definition for this disease [1].

Diuretics in the Critically Ill

Diuretics represent one of the most commonly used drugs in critically ill patients with acute kidney injury (AKI). This widespread use can partially be explained by the fact that non-oliguric AKI is reported to have a better prognosis compared to oliguric AKI [2]. It was once common to attribute AKI in the ICU to acute tubular necrosis where tubules are obstructed with denuded epithelium resulting in a back leak of glomerular filtrate into the renal interstitium, which is thought to further perpetuate renal injury [3]. In such a scenario, clinicians seek to preserve urine flow by using loop diuretics in order to flush out debris to limit tubular obstruction and consequently the injury [3, 4]. Furthermore, evidence from experimental studies in animals suggests that diuretics may also reduce metabolic demand of the renal tubular cells and consequently their oxygen consumption leading to a possible protection of the renal tubular cells from ischemia [3].

Unfortunately, despite these theoretical benefits of loop diuretics, there is no evidence that these agents can improve clinical outcomes. In 2002, Mehta et al. [5] looked at 552 patients with severe AKI in a cohort study aimed to determine if diuretics influence outcomes. After adjustment for numerous potential confounding factors, a multivariable analysis found that diuretic use was associated with significant increases for in-hospital mortality and non-recovery of renal function. In a large, subsequent prospective, multinational epidemiologic study that included 1,743 critically ill patients with AKI from 23 countries, Uchino et al. [6] found that the use of diuretics did not increase the risk of death. However, there was also no evidence of any therapeutic benefit and this study pointed out the need for randomized controlled trials.

Small, randomized controlled trials assessing the effect of loop diuretics in AKI patients have demonstrated an increase of urine output without modifications in outcomes such as mortality, renal recovery or need for dialysis. One study conducted by Shilliday et al. [7] randomized 92 patients with AKI and stated that loop diuretics increased urine output but without any significant effect on renal recovery, need for dialysis, or mortality at day 21. Patients remaining oliguric had a worse outcome, but they were also sicker at randomization. Cantarovich et al. [8] randomized 330 patients requiring renal replacement therapy (mean serum creatinine >4.5 mg/dl). The daily administration of 25 mg/kg of intravenous furosemide shortened the time to achieve a diuresis of 2 l/day, however mortality, time on dialysis, number of dialysis sessions, and time to achieve a creatinine level <2.6 mg/dl were not affected. In 2009, Van der Voort et al. [9] randomized 71 patients who had been treated with continuous venovenous hemofiltration. After the end of the need for renal replacement therapy, in the recovery phase of AKI, one group received a continuous infusion of furosemide and the other group received a placebo. The authors found that furosemide increased urinary volume and sodium excretion but did not lead to a shorter duration of renal failure or more frequent renal recovery.

Several meta-analyses have confirmed these findings. Ho and Sheridan [10] reported that loop diuretics used for AKI did not lead to a reduction of hospital mortality (relative risk = 1.09 [range: 0.9–1.31]; n = 474) or a reduction of requirement for renal replacement therapy (relative risk = 0.94 [range

0.71–1.26]; n = 204). The number of dialysis sessions and proportion of patients with persistent oliguria were also unaffected. More recently, Bagshaw et al. [11] reviewed 62 studies and found that loop diuretics were not associated with improved mortality or an accelerated rate of independence from RRT. However, the diuretic treatment was associated with a shorter duration of RRT, shorter time to spontaneous decline in serum creatinine levels, and increased urine output. Study heterogeneity and limited reporting of some outcomes across studies make firm conclusions regarding these secondary outcomes impossible.

Numerous hypotheses have been proposed to explain this absence of beneficial effects of loop diuretics on renal function in humans despite therapeutic potential from experimental data [3]. First, loop diuretics decrease the effective circulating volume, potentially resulting in reduced renal blood flow and therefore decreased glomerular filtration rate. Second, diuresis obtained with diuretics may deceive physicians, hiding the real problem and thus delaying the appropriate treatment. Third, reduction of the effective arterial volume provokes stimulation of the adrenergic and renin-angiotensin systems both leading to vasoconstriction in the renal cortex [12]. This leads to a corticomedullary redistribution of the renal blood flow that can abolish the beneficial tubuloglomerular feedback system (protective adaptative mechanism which autoregulates the medullary oxygen balance), thereby worsening renal injury [13]. Fourth, if diuretics are supposed to prevent tubular obstruction by flushing out debris such as denuded epithelium, they can also promote the aggregation of Tamm-Horsfall proteins in the tubular lumen and consequently worsen the tubular obstruction which makes diuretics unable to reach their site of action [14]. Diuretics cannot reach their site of action into the tubular fluid for several other reasons that can be seen in the critically ill patient such as the reduction of the renal function by itself which decreases the proximal tubular secretion, the increase concentration of endogenous and exogenous organic acids that compete for the organic anion transporters in the proximal tubule, and hypoalbuminemia which increases the volume distribution of furosemide and consequently the extrarenal clearance that makes urinary secretion lower [3].

Thus, there is lack of evidence regarding any beneficial effects on renal function (prevention or treatment of AKI). It is therefore unlikely that use of diuretics in the setting of oliguric AKI affords any benefit to the kidney and any other outcomes. Given this, there is still some evidence and strong clinical rationale that diuretics may be of benefit in the treatment of volume overload.

Fluid overload is indeed an obvious complication of the impaired sodium and water excretion observed in oliguric AKI. Critically ill patients with oliguric AKI are also at additional increased risk for fluid overload due to widespread systemic inflammation, reduced plasma oncotic pressure and increased capillary leak [15]. Finally, current trends in ICU practice worldwide with early goal-directed therapy, meaning administration of large volumes of fluid therapy to patients with septic shock, have the potential to compound fluid overload in these critically ill patients [16]. Fluid overload leads to cardiopulmonary complications such as congestive heart failure, pulmonary edema requiring mechanical ventilation, pulmonary-restrictive defects and reduced pulmonary compliance, lung function and oxygenation [15]. Besides, fluid overload is implicated (at least indirectly) in the occurrence of other complications such as leaking of surgical anastomoses, bleeding, wound infection or dehiscence [15]. Some studies have even reported that a high positive cumulative fluid balance is an independent risk for hospital mortality [17]. Finally, fluid overload may directly worsen kidney function if it results in abdominal compartment syndrome or renal venous congestion. There is therefore evidence that positive fluid accumulation in ICU patients can adversely impact outcomes and diuretics can be used to treat this fluid overload state.

In conclusion, the use of diuretics in a critically ill setting such as oliguric AKI, while still fairly common, is not supported by the evidence and emerging data suggest this attitude could even be harmful if the goal is to prevent or to treat AKI. On the other hand, diuretics are often necessary in the treatment of fluid overload, which is a state commonly observed in ICU patients with AKI. Given the available evidence, we recommend that loop diuretics should therefore be restricted to this indication for patients in the ICU.

Diuretics in Cardiorenal Syndromes

A convenient classification of the CRS has recently been released, including five subtypes reflecting the primary and secondary pathology and the time frame over which the syndrome evolves (table 1) [1, 18]. Along with providing a clear definition of this disorder, this classification also allows an easier identification of patients and a better characterization of the pathophysiological mechanisms underlying each subtype. Consequently, a management strategy can be proposed for each subtype, supported by its pathophysiological rationale. However, loop diuretic administration remains an obvious difficult paradigm in this disorder since diuretic use can be beneficial while still carrying important risks including harm to the kidney [19]. Importantly, our discussion is limited to the role of loop diuretics – other diuretics agents (e.g. nesiritide) are beyond the scope of this review.

In type I CRS, acute worsening of cardiac function leads to AKI through multiple and complex mechanisms implying inadequate renal perfusion [1]. A marked increased in venous pressure leading to kidney congestion can explain why diuretic responsiveness is sometimes impaired in these patients. Besides, diuretics may also worsen renal function in this situation so they may best be given with evidence of fluid overload with the goal of achieving a gradual diuresis. If given, diuretics have to be titrated according to renal function, and systolic blood pressure. High doses are not recommended but continuous infusion of furosemide may be preferred [20]. Furthermore, diuretic management should

Table 1. Classification of the CRS [adapted from 1]

General definition	Pathophysiologic disorder of the heart and kidneys, whereby acute or chronic dysfunction in one organ induces acute or chronic dysfunction in the other
Type I CRS	Acute worsening of cardiac function leads to AKI
Type II CRS	Chronic abnormalities in cardiac function such as chronic congestive heart failure are responsible for progressive and permanent impairment of kidney function
Type III CRS	Acute and primary worsening of renal function contributes to acute cardiac dysfunction
Type IV CRS	Primary chronic kidney disease contributes to chronic heart disease
Type V CRS	Combined cardiac and renal dysfunction is due to acute or chronic systemic disorders

always be guided by fluid status assessment, and often, by cardiac output measurement. Since worsening renal function in the setting of acute heart failure is not simply a marker of illness severity but it is rather a powerful and independent risk factor for mortality, attention should be paid to preserve it. Diuretics should therefore be managed with extreme caution.

In type II CRS, chronic abnormalities in cardiac function such as chronic congestive heart failure are responsible of progressive and permanent impairment of kidney function. Patients with type II CRS are more likely to receive loop diuretics at higher doses than patients with stable renal function [21] and such treatment may precipitate further renal injury. As for the acute setting, reduction in renal function in the context of heart failure is associated with adverse outcomes and diuretics may be at least partially involved in the occurrence of this chronic kidney disease. The prevalence of renal dysfunction in chronic heart failure is estimated to be around 25% [22].

In type III CRS, an abrupt and primary worsening of renal function contributes to acute cardiac dysfunction. It is now very well established that AKI is a very common condition with a growing incidence in hospital and ICU patients. Type III CRS is therefore likely to be a more common condition than is currently recognized. AKI can affect the heart through several pathways such as fluid overload (pulmonary edema), hyperkalemia (arrhythmias), uremia levels (reduction of myocardial contractility and pericarditis), acidemia (pulmonary vasoconstriction and negative inotropic effects) and renal ischemia (activation of inflammation and apotosis at cardiac level) [18]. The use of loop diuretics in type III CRS, as in other forms of AKI, should be limited to the treatment of fluid overload.

In type IV CRS, primary chronic kidney disease contributes to chronic heart disease. Therapeutic choices such as diuretic administration are particularly challenging in these patients due to the fact that a drug can be beneficial on one hand but deleterious on the other. Unfortunately, large randomized controlled trials that have shaped the treatment of chronic heart failure in the last two decades have typically excluded patients with significant renal disease [19]. Thus, no firm recommendations for diuretic use can be offered other than to individualize therapy and monitor renal function closely.

In type V CRS, also called secondary CRS, the combined cardiac and renal dysfunction is due to acute or chronic systemic disorders. Severe sepsis is one common cause. There is no reason to recommend diuretic administration in a type V CRS unless it leads to the constitution of a deleterious fluid overload state.

Thus, the complexity of the interactions between heart and kidney in the different subtypes of CRS requires a multidisciplinary approach combining cardiology, nephrology and critical care expertise in order to administrate the best therapy. This appears to be especially true for diuretics since we have seen they can be harmful and worsen outcomes if not appropriately administered. The recent interest for this disease and the release of the CRS classification should help to identify groups of patients for prospective studies, in which specific therapies such as diuretics can be better assessed and validated.

Conclusion

Loop diuretics are widely administered in critically ill patients despite the lack of evidence regarding any beneficial effects on renal function (prevention or treatment of AKI) and the fact that some authors even report worse outcomes when diuretics are used for AKI. However, these agents appear to be useful for the treatment of fluid overload. The recent CRS classification should be useful for the design of future studies which are necessary to better assess the role of diuretics in the treatment of this disease, which commonly forces physicians to make conflicting therapeutic choices.

References

1 Ronco C, Haapio M, House AA, et al: Cardiorenal syndrome. J Am Coll Cardiol 2008;52:1527–1539.

2 Venkataraman R, Kellum JA: Prevention of acute renal failure. Chest 2007;131:300–308.

3 Karajala V, Mansour W, Kellum JA: Diuretics in acute kidney injury. Minerva Anestesiol 2009;75:251–257.

4 Klahr S, Miller SB: Acute oliguria. N Engl J Med 1998;338:671–675.

5 Mehta RL, Pascual MT, Soroko S, Chertow GM: Diuretics, mortality, and non-recovery of renal function in acute renal failure. JAMA 2002;288:2547–2553.

6 Uchino S, Doig GS, Bellomo R, et al: Diuretics and mortality in acute renal failure. Crit Care Med 2004;32:1669–1677.
7 Shilliday IR, Quinn KJ, Allison ME: Loop diuretics in the management of acute renal failure: a prospective, double-blind, placebo-controlled, randomized study. Nephrol Dial Transplant 1997;12:2592–2596.
8 Cantarovich F, Rangoonwala B, Lorenz H, et al: High-dose furosemide for established ARF: a prospective, randomized, double-blind, placebo-controlled, multicenter trial. Am J Kidney Dis 2004;44:402–409.
9 Van der Voort PH, Boerma EC, Koopmans M, et al: Furosemide does not improve renal recovery after hemofiltration for acute renal failure in critically ill patients: a double-blind randomized controlled trial. Crit Care Med 2009;37:533–538.
10 Ho KM, Sheridan DJ: Meta-analysis of frusemide to prevent or treat acute renal failure. BMJ 2006;333:420.
11 Bagshaw SM, Delaney A, Haase M, et al: Loop diuretics in the management of acute renal failure: a systematic review and meta-analysis. Crit Care Resusc 2007;9:60–68.
12 Guild SJ, Eppel GA, Malpas SC, et al: Regional responsiveness of renal perfusion to activation of the renal nerves. Am J Physiol 2002;283:R1177–R1186.
13 Mason J, Kain H, Welsch J, Schnermann J: The early phase of experimental acute renal failure. VI. The influence of furosemide. Pflügers Arch 1981;392:125–133.
14 Sanders PW, Booker BB: Pathobiology of cast nephropathy from human Bence-Jones proteins. J Clin Invest 1992;89:630–639.
15 Bagshaw SM, Bellomo R, Kellum JA: Oliguria, volume overload, and loop diuretics. Crit Care Med 2008;36:S172–S178.
16 Rivers E, Nguyen B, Havstad S, et al: Early goal-directed therapy in the treatment of severe sepsis and septic shock. N Engl J Med 2001;345:1368–1377.
17 Sakr Y, Vincent JL, Reinhart K, et al: High tidal volume and positive fluid balance are associated with worse outcome in acute lung injury. Chest 2005;128:3098–3108.
18 Ronco C, Cruz DN, Ronco F: Cardiorenal syndromes. Curr Opin Crit Care 2009;15: 384–391.
19 Ronco C, Chionh CY, Haapio M, et al: The cardiorenal syndrome. Blood Purif 2009; 27:114–126.
20 Howard PA, Dunn MI: Aggressive diuresis for severe heart failure in the elderly. Chest 2001;119:807–810.
21 Butler J, Forman DE, Abraham WT, et al: Relationship between heart failure treatment and development of worsening renal function among hospitalized patients. Am Heart J 2004;147:331–338.
22 Hillege HL, Nitsch D, Pfeffer MA, et al: Renal function as a predictor of outcome in a broad spectrum of patients with heart failure. Circulation 2006;113:671–678.

John A. Kellum, MD
604 Scaife Hall, The CRISMA Laboratory
Critical Care Medicine, University of Pittsburgh
3550 Terrace Street, Pittsburgh, PA 15261 (USA)
Tel. +1 412 647 7125, Fax +1 412 647 8060, E-Mail Kellumja@ccm.upmc.edu

Ronco C, Bellomo R, McCullough PA (eds): Cardiorenal Syndromes in Critical Care.
Contrib Nephrol. Basel, Karger, 2010, vol 165, pp 226–235

Use of Bioimpedance Vector Analysis in Critically Ill and Cardiorenal Patients

W. Frank Peacock, IV

Department of Emergency Medicine, The Cleveland Clinic, Cleveland, Ohio, USA

Abstract

Prospective outcome prediction and volume status assessment are difficult tasks in the acute care environment. Rapidly available, non-invasive, bioimpedance vector analysis (BIVA) may offer objective measures to improve clinical decision-making and predict outcomes. Performed by the placement of bipolar electrodes at the wrist and ankle, data is graphically displayed such that short-term morality risk and volume status can be accurately quantified. BIVA is able to provide indices of general cellular health, which has significant prognostic implications, as well as total body volume. Knowledge of these parameters can provide insight as to the short-term prognosis, as well as the presenting volume status.

Accurate volume assessment is necessary for optimal clinical decision-making as mortality is increased when inappropriate therapy is administered to critical patients. Not only is accurate volume assessment necessary, speed is important as delayed therapy is also associated with increased mortality. Unfortunately, volume assessment is challenging. And because the history and physical examination can be inaccurate, there is a significant need for an objective measure. Few testing modalities satisfy the requirement for both accuracy and speed. The chest x-ray is insensitive for volume overload, B-type natriuretic peptide (BNP) is inaccurate in critically ill patients, and the volume assessment gold standard (radioisotopic measurement) is neither rapid nor inexpensive. Bioimpedance vector analysis (BIVA) may fulfill the requirements of an inexpensive, rapid, and accurate tool for evaluating critically ill patients.

Bioimpedance

Determination of hemodynamic status has historically required the insertion of a catheter directly into the heart. Because of the significant morbidity and mortality associated with this invasive procedure, non-invasive techniques are now promoted as an acceptable alternative [1]. One of the more researched non-invasive measurements is impedance cardiography (ICG).

First published in 1826, Ohm's law [2] states that the flow of electrical current (I) is equal to the voltage drop (E) between two ends of a circuit divided by the impedance (Z) to current flow, represented as E = IZ. If the current remains constant, then changes in voltage across the circuit are equal to changes in the impedance to current flow. Furthermore, if Z is dependent upon the cross-sectional area (A), length (L), and resistivity (ρ) of the conducting material, then its changes can be related to changes in volume (V) of the conductor by $Z = \rho(L^2/V)$. These suppositions are the fundamental principle that supports the fact that impedance changes over time are proportional to the volume of moving fluid and are related to cardiac output.

Using ICG to evaluate hemodynamics is based on the concept that the human thorax is an inhomogeneous electrical conductor [3, 4]. When a high-frequency current is injected across the thorax, impedance can be measured by electrode pairs located at the edge of the chest. Measurements of the average thoracic impedance reflect the static volume of all the combined thoracic compartments. Dynamic voltage changes that result from volumetric and velocity changes occurring within the aorta during cardiac systole are then proportional to cardiac output. By integrating these changes with ECG-derived timing measures, hemodynamic parameters are approximated. The accuracy of the cardiac output measurements obtained with ICG has been compared to invasive measures acquired by thermodilution. A recent meta-analysis of over 200 studies found a correlation of 0.81 for ICG determined stroke volume and cardiac output, when compared to traditional measurements. Besides being non-invasive, the major advantage of ICG technology is that it can also be utilized for continuous monitoring and trend identification.

Limitations to obtaining accurate impedance measures include: (1) erroneous electrode position, e.g. operator error or physiologic complications such as central lines or cutaneous alterations; (2) poor skin-electrode interface, e.g. diaphoresis, excessive hair; (3) uncooperative patient, e.g. dementia, psychiatric disease; (4) contact with an electrical ground (e.g. metal bed frame) or electrical interference, and (5) abnormal body habitus, e.g. severe obesity.

Bioimpedance Vector Analysis

A newer assessment technique is BIVA. BIVA combines bioimpedance with capacitance measures (the time required to charge a circuit). Whole-body

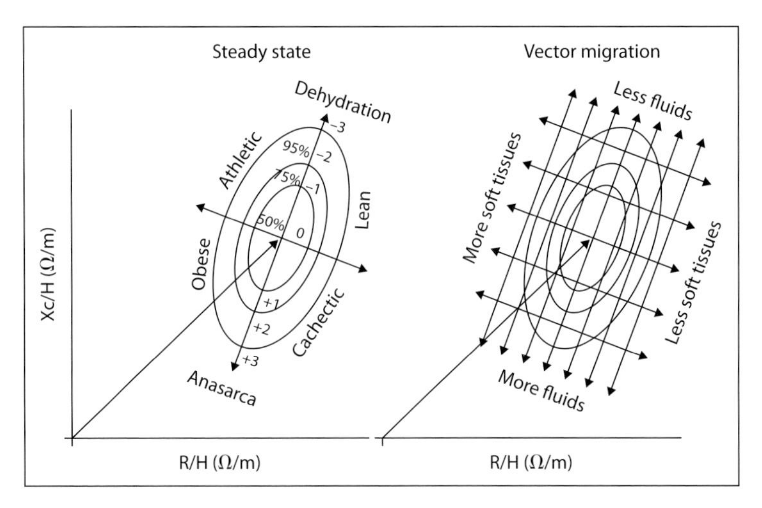

Fig. 1. BIVA results displayed graphically comparing R/H to Xc/H. BIVA patterns: major axis → tissue hydration, minor axis → soft tissue mass.

impedance may be considered a combination of resistance, R (the opposition to flow of an alternating current through intra- and extracellular electrolytic solutions), and reactance, Xc (the capacitance produced by tissue interfaces and cell membranes). The arc tangent of Xc/R is termed the phase angle, and represents the phase difference between voltage and current, determined by the reactive component of R. In conductors without cells (e.g. saline), no capacitance exists and thus Xc cannot be measured. By including Xc, the accuracy of volume assessment is improved over conventional bioimpedance measurements.

Standardized by height (H), BIVA results can be displayed graphically, comparing R/H to Xc/H, and generates output that simultaneously reflects hydration abnormalities and alterations of soft tissue mass (fig. 1). When obtained as a single measure, BIVA data can be compared to the normal population, represented on the graph by confidence ellipses, with data expected to fall within the reference 75% tolerance ellipse. When presented as a vector, shorter or longer lengths are associated with more or less hydration, respectively. The height of the vector (the arc tangent Xc/R), measured in degrees of elevation from the x-axis and termed the phase angle (PA), has been described as a prognostic tool in many clinical situations.

Although the biological meaning of the PA is not completely understood, it reflects not only body cell mass, but is one of the best indicators of cell membrane function. The PA indicates the distribution of water between the intra- and extracellular spaces, and corresponds to a low ratio of extra- to intracellular water [5, 6]. An upward or downward displacement of the PA

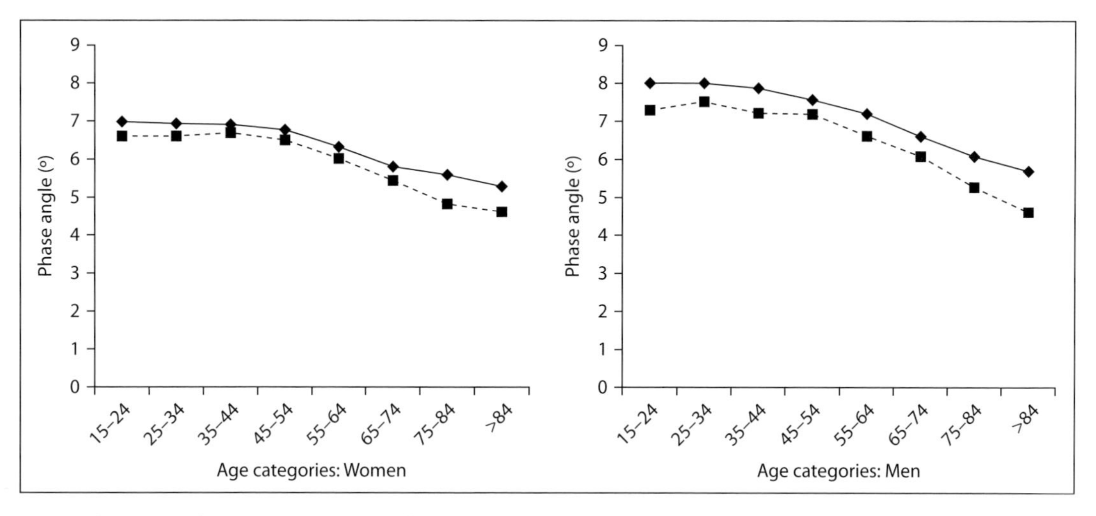

Fig. 2. PA for American (—) and Swiss (– –) populations stratified by gender and age [13].

is associated with alterations in soft tissue mass. While there are some controversies about considering it as a nutritional marker, studies in burn victims and sickle cell disease corroborate its ability to evaluate cell membrane function.

Serial BIVA measures are also clinically useful. A shortening or lengthening of the vector over the confidence ellipses is indicative of increasing fluid overload (edema) or the occurrence of dehydration, respectively. Conversely, if the vector of the PA increases or decreases, it is representative of increasing or decreasing, respectively, cell mass.

When interpreting the PA, results must be considered in the context of sex and age. In a study of 1967 healthy aged 18- to 94-year-old adults, the PA was smaller in women (6.53 vs. 7.488) and decreased with increasing age (from 7.438 in the youngest to 5.888 in the oldest) [5]. In another study of 653 20- to 90-year-old volunteers [6], age was reported as the strongest predictor of PA in a stepwise multiple regression analysis (fig. 2).

Most importantly, PA is predictive of mortality, although the absolute cutpoints are variable when comparing studies due to a lack of standardization. In a study of HIV patients, a PA <5.38 was the most important survival predictor, with performance exceeding the CD4 count [7]. Another HIV study found an increase of the PA of only 1° represented a 29% increase in survival [8]. PA measures have also been found to be prognostic in cancer. In lung cancer, values <4.58 were associated with a 25% higher mortality (OR = 1.25) [9], and in other studies of advanced colorectal [10] and pancreatic cancer [11], a PA >5.58 was a strong predictor of survival.

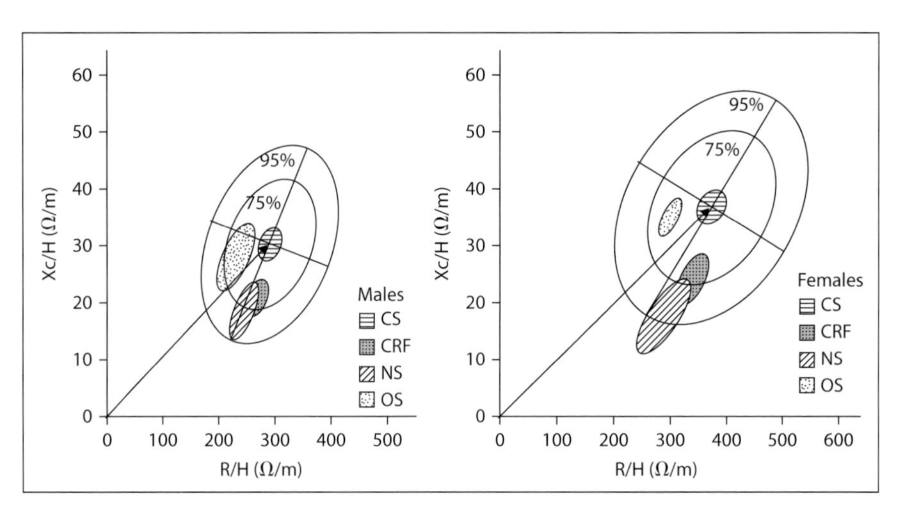

Fig. 3. BIVA accurately defines volume status. CS = Controls, CRF = chronic renal failure, NS = nephrotic syndrome, OS = obese subjects, H = height, Xc = reactance, R = resistance.

PA measurement may also be prognostic of mortality in other conditions, including sepsis. In 30 patients meeting two or more SIRS criteria [12], an initial PA >4° was correlated with survival. This compared to all patients who died where the PA was <4° ($p = 0.038$). Finally, PA measurements may predict postsurgical outcomes. In 225 adult patients undergoing gastrointestinal surgery [13], nutritional status was assessed by several methods. They found that only the PA predicted postoperative complications (relative risk = 4.3 for a low PA).

BIVA is also useful for prediction of hydration status as the vector length is proportional to whole-body fluid volume [14, 15], has a correlation coefficient of 0.996, and a measurement error that ranges from 2 to 4%. BIVA has been validated in kidney [16], liver, and heart disease for the prediction of volume status [17–19]. In one study, using BIVA as a proxy for the adequacy of ultrafiltration in >3,000 hemodialysis patients, vector length indices corresponding to greater soft tissue hydration (less adequate ultrafiltration) were associated with a significant mortality increase [20]. In another study of 217 patients consisting of 86 healthy controls, 55 mild-to-terminal chronic renal failure in conservative treatment (15% with apparent edema), 36 idiopathic nephrotic proteinuria (58% with apparent edema), and 40 obese subjects, BIVA accurately identified edematous versus normovolemic populations [21] (fig. 3). Finally, in a population of 2,092 patients (1,116 hemodialysis, 726 healthy controls, 200 CAPD, and 50 nephrotic patients), total body water was estimated by anthropometry and BIVA measures [22]. They reported that while anthropometry measures were misleading, indicating the same hydration for edematous and non-edematous states, BIVA was very sensitive to fluid overload, indicating a 10% excess of fluid in volume overloaded patients. They also noted that BIVA vectors from patients with edema were displaced downward on the RXc graph, out of the 75% ellipse (88% sensitivity, 87% specificity) and similar to vectors from nephritic patients.

BIVA has also been evaluated in heart failure (HF). In a study of 22 HF patients, using deuterium dilution as a gold standard for total body water evaluation, BIVA demonstrated excellent correlation with total body water measures ($r = 0.93$, $p = 0.01$) [23]. In another analysis of 243 HF patients, BIVA measurements were an objective measure of New York Heart Association Class [24]. The authors concluded that BIVA allows an easier and more objective evaluation of body composition and might be particularly useful to stratify the severity of HF.

The use of BIVA to evaluate HF has also been examined in the context of existing myocardial stress measures. In a prospective study combining BIVA and BNP measurement in 292 dyspneic patients [25], 58.9% who had acute decompensated heart failure (ADHF) and the remainder were non-ADHF, ADHF patients showed significantly ($p < 0.001$) higher BNP values (591.8 ± 501 vs. 69.5 ± 42 pg/ml). Regression analysis revealed that whole-body BIVA was a strong predictor of ADHF alone or in combination with BNP. In fact, the combination of BNP and BIVA measures resulted in the most accurate volume status determination (fig. 4). As both BNP and BIVA are available as 'point of care' testing, providing results within minutes of contact with the medical system, this strategy may have a role in the early evaluation of patients presenting with dyspnea of unknown etiology.

BIVA has also been used to assist in guiding acute HF therapy. In 186 hospitalized HF patients undergoing BNP-guided therapy [26], the definition of stability and eligibility for discharge included no orthopnea, stable hemodynamics, diuresis >1 liter daily, and improvement in body hydration status as defined by BIVA. They found that a discharge BNP <250 pg/ml predicted successful management and that BIVA was able to monitor diuretic therapy, assess changes in body fluids, and increase the usefulness of BNP as a guide to therapy. When these objective discharge criteria were validated in a separate study of 166 hospitalized HF patients and compared 149 patients discharged based on clinical acumen [27], the objective group had 12% fewer 6-month readmissions (23 vs. 35%, $p = 0.02$) and a lower overall cost of care (EUR 2,978 vs. 2,781, $p < 0.01$). This suggests that the combination of clinical acumen and objective measures improve outcomes over clinical impression alone.

BIVA can be utilized to evaluate volume status in patients undergoing hemodialysis. Piccoli [28] reported on 1,367 patients stratified as 726 healthy controls and 641 receiving weekly hemodialysis. Of the hemodialysis cohort, 251 were considered unstable due to symptomatic hypotension, or a systolic blood

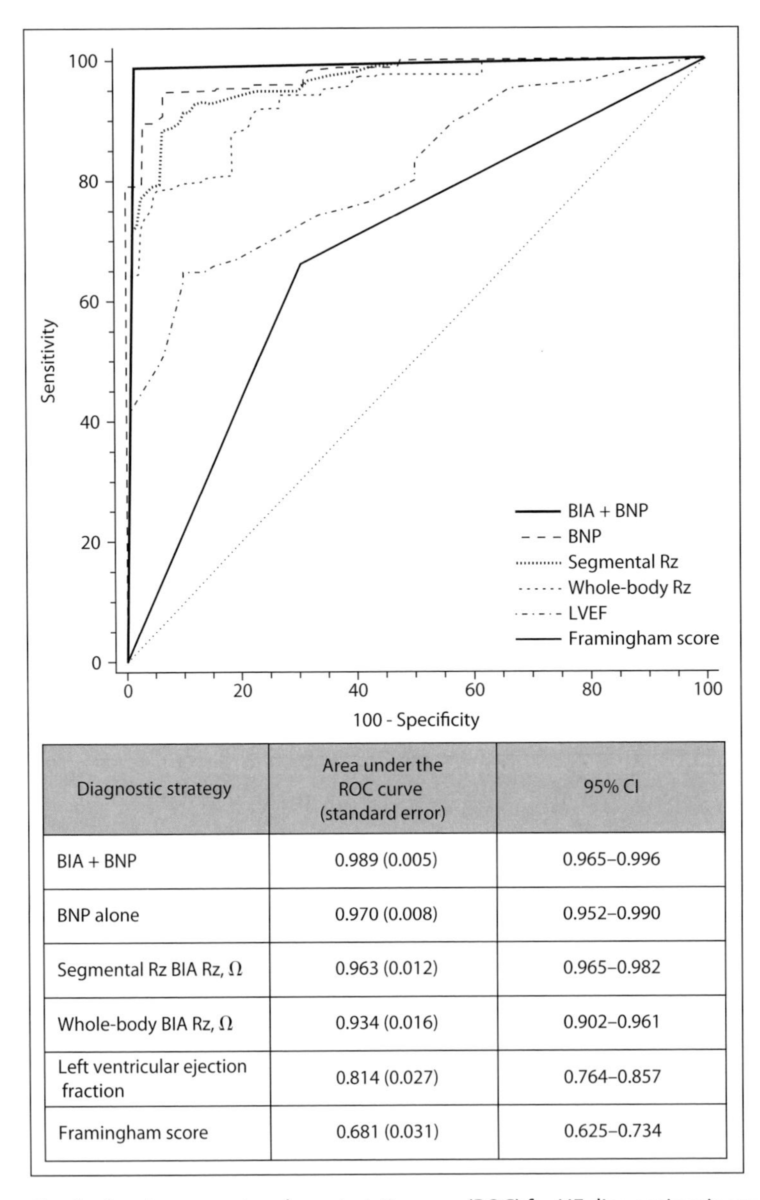

Diagnostic strategy	Area under the ROC curve (standard error)	95% CI
BIA + BNP	0.989 (0.005)	0.965–0.996
BNP alone	0.970 (0.008)	0.952–0.990
Segmental Rz BIA Rz, Ω	0.963 (0.012)	0.965–0.982
Whole-body BIA Rz, Ω	0.934 (0.016)	0.902–0.961
Left ventricular ejection fraction	0.814 (0.027)	0.764–0.857
Framingham score	0.681 (0.031)	0.625–0.734

Fig. 4. Receiver operator characteristic curve (ROC) for HF diagnosis using various parameters [25].

pressure <90 mm Hg during >30% of dialysis sessions. BIVA was performed before and after dialysis. They found significantly shorter impedance vectors in either hemodialysis population compared to controls. Further, in both men and women, unstable hemodialysis patients had longer vectors (303 vs. 294 Ω/m and 373 vs. 355 Ω/m, respectively) and smaller PAs (4.5 vs. 5.1 and 4.2 vs. 4.7, respectively) compared to the stable hemodialysis cohort. After dialysis, the vector displacement caused by fluid removal was shorter and less steep in the unstable patients, as compared to the stable cohort, despite having similar volumes removed.

Outcomes associated with BIVA measures in hemodialysis have also been studied. Pillon et al. [20] reported on 3,009 patients with BIVA measured before each three times weekly dialysis session. They found vector length was longer in women than men (mean 340 vs. 270 Ω/m), African-Americans, and non-diabetics, and the relative risk of death associated with each 100 Ω/m vector length increase was 0.75 (95% CI 0.57–0.88). Defining 300–350 Ω/m as the normal vector length, the relative risks of death were 1.54 (95% CI 1.08–2.21) and 2.83 (95% CI 1.55–5.14) for vector lengths of 200–250 and <200 Ω/m, respectively. They concluded that shorter predialysis bioimpedance vectors, indicating greater soft tissue hydration, were associated with diminished survival in hemodialysis patients and validated clinical observations linking longevity to maintenance of dry body weight. They suggested BIVA may be used to evaluate the adequacy of ultrafiltration in hemodialysis.

BIVA is not without limitations that should be considered in the interpretation of its results. Because total body resistance is calculated with BIVA, accurate body position is required (limb abduction with arms separated from trunk by about 30° and legs separated by about 45°). Another limitation of BIVA is its failure to distinguish compartmentalized edema such as pericardial, pleural, or abdominal effusion [29]. Lastly, BIVA is standardized for Western Europeans and normal values for African heritage patients are not yet established [30].

Alternatively, BIVA does offer advantages over simple bioimpedance measurements as it requires measurement on only one side of the body. Thus, where bioimpedance may be limited in patients with unilateral abnormalities, BIVA can simply be measured on the contralateral side. Finally, BIVA does not demonstrate the same sensitivity to errors in accurate electrode placement location as occurs with bioimpedance.

Conclusion

BIVA is a simple, rapid, reproducible, non-invasive, and objective tool to better understand prognosis and hemodynamic status such that optimized evaluation and treatment may occur.

References

1 Sandham JD, Hull RD, Brant RF, et al: A randomized, controlled trial of the use of pulmonary-artery catheters in high-risk surgical patients. N Engl J Med 2003;348:5–14.

2 Winter UJ, Klocke RK, Kubicek WG, Niederlag W (eds): Thoracic Impedance Measurements in Clinical Cardiology. New York, Thieme Medical, 1994.

3 Summers RL, Shoemaker W, Peacock WF, Ander DS: Bench to bedside: electrophysiologic and clinical principles of noninvasive hemodynamic monitoring using impedance cardiography. Acad Emerg Med 2003;10: 669–680.

4 Sramek BB: Thoracic electrical bioimpedance: basic principles and physiologic relationship. Noninvas Cardiol 1994;3:83–88.

5 Kyle UG, Genton L, Karsegard VL, et al: Percentiles (10, 25, 75 and 90th) for phase angle, determined by bioelectrical impedance, in 2,740 healthy adults aged 20–75 years. Clin Nutr 2004;23:758.

6 Dittmar M: Reliability and variability of bioimpedance measures in normal adults: effects of age, gender, and body mass. Am J Phys Anthropol 2003;122:361–370.

7 Ott M, Fischer H, Polat H, et al: Bioelectrical impedance analysis as a predictor of survival in patients with human immunodeficiency virus infection. J Acquir Immune Defic Syndr Hum Retrovirol 1995;9:20–25.

8 Schwenk A, Beisenherz A, Römer K, et al: Phase angle from bioelectrical impedance analysis remains an independent predictive marker in HIV-infected patients in the era of highly active antiretroviral treatment. Am J Clin Nutr 2000;72:496–501.

9 Toso S, Piccoli A, Gusella M, et al: Altered tissue electric properties in lung cancer patients as detected by bioelectric impedance vector analysis. Nutrition 2000;16:120–124.

10 Gupta D, Lammersfeld CA, Burrows JL, et al: Bioelectrical impedance phase angle in clinical practice: implications for prognosis in advanced colorectal cancer. Am J Clin Nutr 2004;80:1634–1638.

11 Gupta D, Lis CG, Dahlk SL, et al: Bioelectrical impedance phase angle as a prognostic indicator in advanced pancreatic cancer. Br J Nutr 2004;92:957–962.

12 Swaraj S, Marx G, Masterson G, Leuwer M: Bioelectrical impedance analysis as a predictor for survival in patients with systemic inflammatory response syndrome. Critical Care 2003;7(suppl 2):185.

13 Barbosa-Silva MC, Barros AJ: Bioelectric impedance and individual characteristics as prognostic factors for post-operative complications. Clin Nutr 2005;24:830–838.

14 Kushner RF, Schoeller DA, Fjeld CR, et al: Is the impedance index (ht2/R) significant in predicting total body water? Am J Clin Nutr 1992;56:835–839.

15 Sergy G, Bussolotto M, Perini P, et al: Accuracy of bioelectrical impedance analysis in estimation of extracellular space in healthy subject and in fluid retention states. Ann Nutr Metab 1994;38:158–165.

16 Piccoli A, Rossi B, Pillon L, et al: A new method for monitoring body fluid variation by bioimpedance analysis: the RXc graph. Kidney Int 1994;46:534–539.

17 Panella C, Guglielmi FW, Mastronuzzi T, et al: Whole-body and segmental bioelectrical parameters in chronic liver disease: effect of gender and disease stages. Hepatology 1995; 21:352–358.

18 Woodrow G, Oldroyd B, Turney JH, et al: Segmental bioelectrical impedance in patient with chronic renal failure. Clin Nutr 1996; 15:275–279.

19 Paterna S, Di Pasquale P, Parrinello G, et al: Changes in brain natriuretic peptide levels and bioelectrical impedance measurements after treatment with high-dose furosemide and hypertonic saline solution versus high-dose furosemide alone in refractory congestive heart failure. A double-blind study. J Am Coll Cardiol 2005;45:1997–2003.

20 Pillon L, Piccoli A, Lowrie EG, Lazarus JM, Chertow GM: Vector length as a proxy for the adequacy of ultrafiltration in hemodialysis. Kidney Int 2004;66:1266–1271.

21 Piccoli A, Ross, Pillon LJ, Bucciante G: Body fluid overload and bioelectrical impedance analysis in renal patients. Miner Electrolyte Metab 1996;22:76–78.

22 Piccoli A for the Italian CAPD-BIA Study Group: Bioelectric impedance vector distribution in peritoneal dialysis patients with different hydration status. Kidney Int 2004; 65:1050–1063.
23 Uszko-Lencer NH, Bothmer F, van Pol PE, Schols AM: Measuring body composition in chronic heart failure: a comparison of methods. Eur J Heart Fail 2006;8:208–214.
24 Martínez LC, Ramírez EC, Tejeda AO, Lafuente EA, Rosales LPB, González VR, et al: Bioelectrical impedance and strength measurements in patients with heart failure: comparison with functional class. Nutrition 2007;23:412–418.
25 Parrinello G, Paterna S, DiPasquale P, et al: The usefulness of bioelectrical impedance analysis in differentiating dyspnea due to decompensated heart failure. J Card Fail 2008;14:676–686.
26 Valle R, Aspromonte N, Giovinazzo P, Carbonieri E, Chiatto M, Di Tano G, et al: B-type natriuretic peptide-guided treatment for predicting outcome in patients hospitalized in sub-intensive care unit with acute heart failure. J Card Fail 2008;14:219–224.
27 Valle R, Apromonte N, Carbonieri E, D'Eri A, Feola M, Giovinazzo P, et al: Fall in readmission rate for heart failure after implementation of B-type natriuretic peptide testing for discharge decision: a retrospective study. Int J Cardiol 2008;126:400–406.
28 Piccoli A for the Italian Hemodialysis-Bioelectrical Impedance Analysis (HD-BIA) Study Group: Identification of operational clues to dry weight prescription in hemodialysis using bioimpedance vector analysis. Kidney Int 1998;53:1036–1043.
29 Lichtenstein D: General Ultrasound in the Critically Ill. New York, Springer, 2004.
30 Piccoli A, Pillon L, Dumler F: Impedance vector distribution by sex, race, body mass index, and age in the United States: standard reference intervals as bivariate Z-scores. Nutrition 2002;18:153–167.

W. Frank Peacock IV, MD, Vice Chief
Desk E-19, Department of Emergency Medicine
9500 Euclid Ave, Cleveland, OH 44195 (USA)
Tel. +1 216 445 4546, Fax +1 216 445 4552
E-Mail peacocw@ccf.org

Ronco C, Bellomo R, McCullough PA (eds): Cardiorenal Syndromes in Critical Care.
Contrib Nephrol. Basel, Karger, 2010, vol 165, pp 236–243

Extracorporeal Fluid Removal in Heart Failure Patients

Maria Rosa Costanzo[a] · Claudio Ronco[b]

[a]Midwest Heart Foundation, Naperville, Ill., USA, and [b]Department of Nephrology, Dialysis & Transplantation, International Renal Research Institute, San Bortolo Hospital, Vicenza, Italy

Abstract

Ultrafiltration is the mechanical removal of fluid from the vasculature. Hydrostatic pressure is applied to blood across a semipermeable membrane to separate isotonic plasma water from blood. Because solutes in blood freely cross the semipermeable membrane, large amounts of fluid can be removed at the discretion of the treating physician without affecting any change in the serum concentration of electrolytes and other solutes. Ultrafiltration has been used to relieve congestion in patients with heart failure for almost four decades. In contrast to the adverse physiological consequences of loop diuretics, numerous studies have demonstrated favorable responses to ultrafiltration. Such studies have shown that removal of large amounts of isotonic fluid relieves symptoms of congestion, improves exercise capacity, improves cardiac filling pressures, restores diuretic responsiveness in patients with diuretic resistance, and has a favorable effect on pulmonary function, ventilatory efficiency, and neurohormonal activation. Ultrafiltration is the only fluid removal strategy shown to improve outcomes in randomized controlled trials of patients hospitalized with decompensated heart failure.

Ultrafiltration has been consistently shown to improve signs and symptoms of congestion, increase diuresis, lower diuretic requirements, and correct hyponatremia in patients with advanced heart failure [1]. It is not known if the clinical benefits of ultrafiltration translate into improved survival. Acute heart failure, which is most commonly due to decompensation of chronic heart failure, can also occur in the setting of circulatory collapse complicating myocardial infarction, hypertension, pericardial disease, cardiomyopathy, myocarditis, pulmonary embolus, or arrhythmias or after cardiac surgery. In these clinical settings ultrafiltration has been used predominantly after diuretics have failed or in the presence of acute renal failure. However, the results of some studies suggest that earlier utilization of ultrafiltration can expedite and maintain compensation of acute heart failure by simultaneously reducing volume overload without causing intravascular volume depletion and reestablishing acid-base and electrolyte balance [1].

The use of ultrafiltration in patients with advanced heart failure has also highlighted the risks associated with mechanical fluid removal. Overly aggressive ultrafiltration in patients with decompensated heart failure can convert non-oliguric renal dysfunction into oliguric failure by increasing neurohormonal activation and decreasing renal perfusion pressure, which may result in the requirement for permanent renal replacement therapy.

Intermittent isolated ultrafiltration has been described in more than 100 NYHA Class IV heart failure patients who have failed aggressive vasodilator, diuretic, and inotropic therapy [1]. Restoration of diuresis and natriuresis after intermittent ultrafiltration helps to identify patients with recoverable cardiac functional reserve. Intermittent isolated ultrafiltration is valuable even in partial responders because it improves quality of life and may be used as a bridge to heart transplantation.

A device that permits both withdrawal of fluid and blood return through peripheral veins has been available for more than 5 years (Aquadex System 100, CHF-Solutions, Minneapolis, Minn., USA). However, central venous access remains an option with this device. Fluid removal can range from 10 to 500 ml/h, blood flow can be set at 10–40 ml/min, and total extracorporeal blood volume is only 33 ml. The device consists of a console, an extracorporeal blood pump, and venous catheters. The device has recently been improved with the insertion of an on-line hematocrit sensor.

To date, four clinical trials of intermittent peripheral veno-venous ultrafiltration have been completed. In the first study in 21 fluid-overloaded patients, removal of an average of 2,611 ± 1,002 (range 325–3,725) ml over 6.43 ± 1.47 h reduced weight from 91.9 ± 17.5 to 89.3 ± 17.3 kg ($p < 0.0001$), and also reduced signs and symptoms of pulmonary and peripheral congestion without associated changes in heart rate, blood pressure, electrolytes, or hematocrit [2]. The aim of the second study was to determine if ultrafiltration with this same Aquadex System 100 before intravenous diuretics in patients with decompensated heart failure and diuretic resistance results in euvolemia and hospital discharge in 3 days, without hypotension or worsening renal function. Ultrafiltration was initiated within 4.7 ± 3.5 h of hospitalization and before intravenous diuretics in 20 heart failure patients with volume overload and diuretic resistance (age 74.5 ± 8.2 years; 75% ischemic disease; ejection fraction 3 ± 15%), and continued until euvolemia. Patients were evaluated at each hospital day, and at 30 and 90 days. An average of 8,654 ± 4,205 ml was removed with 2.6 ± 1.2 eight-hour ultrafiltration courses. Twelve patients (60%) were discharged in ≤3 days. One patient was readmitted in 30 and 2 patients in 90 days. Weight ($p = 0.006$), Minnesota Living with Heart Failure scores ($p = 0.003$), and Global Assessment (p = 0.00003) were improved after ultrafiltration, at 30 and 90 days. B-type natriuretic peptide levels were decreased after ultrafiltration (from 1,236 ± 747 to 988 ± 847 pg/ml) and at 30 days (816 ± 494 pg/ml; p = 0.03. Blood pressure, renal function, and medications were unchanged. The results of this study suggest that in heart failure patients with volume overload and diuretic resistance, early ultrafiltration before intravenous diuretics effectively and safely decreases length of stay and readmissions. Clinical benefits persisted at 3 months after treatment [3].

The aim of the third study was to compare the safety and efficacy of ultrafiltration with the Aquadex System 100 device versus those of intravenous diuretics in patients with decompensated heart failure. Compared to the 20 patients randomly assigned to intravenous diuretics, the 20 patients randomized to a single 8-hour ultrafiltration session had greater median fluid removal (2,838 vs. 4,650 ml; p = 0.001) and median weight loss (1.86 vs. 2.5 kg; p = 0.24). Ultrafiltration was well tolerated and not associated with adverse hemodynamic renal effects. The results of this study show that an initial treatment decision to administer ultrafiltration in patients with decompensated congestive heart failure results in greater fluid removal and improvement of signs and symptoms of congestion than those achieved with traditional diuretic therapies [4].

The Ultrafiltration versus Intravenous Diuretics for Patients Hospitalized for Acute Decompensated Heart Failure (UNLOAD) trial was designed to compare the safety and efficacy of veno-venous ultrafiltration and standard intravenous diuretic therapy for hospitalized heart failure patients with ≥2 signs of hypervolemia [5]. Two hundred patients (63 ± 15 years; 69% men; 71% ejection fraction, ≤40%) were randomized to ultrafiltration or intravenous diuretics. At 48 h, weight (5.0 ± 3.1 vs. 3.1 ± 3.5 kg; p = 0.001) and net fluid loss (4.6 vs. 3.3 liters; p = 0.001) were greater in the ultrafiltration group. Dyspnea scores were similar. At 90 days, the ultrafiltration group had fewer patients re-hospitalized for heart failure (16 of 89 [18%] vs. 28 of 87 [32%]; p = 0.037), heart failure re-hospitalizations (0.22 ± 0.54 vs. 0.46 ± 0.76; p = 0.022), re-hospitalization days (1.4 ± 4.2 vs. 3.8 ± 8.5; p = .022) per patient, and unscheduled visits (14 of 65 [21%] vs. 29 of 66 [44%]; p = 0.009; fig. 1). Changes in serum creatinine were similar in the 2 groups throughout the study. No clinically significant changes in serum blood urea nitrogen, sodium, chloride, and bicarbonate occurred in either group. Serum potassium <3.5 mEq/l occurred in 1 of 77 (1%) patients in the ultrafiltration and in 9 of 75 (12%) patients in the diuretics group (p = 0.018). Episodes of hypotension during 48 h after randomization were similar (4 of 100 [4%] vs. 3 of 100 [3%]).

Thus, the UNLOAD trial demonstrated that in decompensated heart failure, ultrafiltration safely produces greater weight and fluid loss than intravenous diuretics, reduces 90-day resource utilization for heart failure, and is an effective alternative therapy. It is also important to recognize the limitations of the UNLOAD trial. The treatment targets for both diuretics and ultrafiltration were

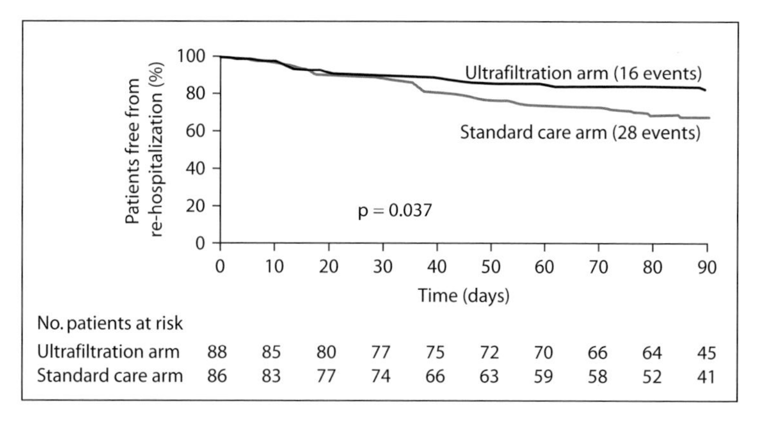

Fig. 1. Freedom from heart failure re-hospitalization at 90 day in the UNLOAD clinical trial. Kaplan-Meier estimate of freedom from re-hospitalization for heart failure within 90 days after discharge in the ultrafiltration and standard care groups. From Costanzo et al. [5], with permission.

not specified beforehand. Although treatment was not blinded, it is unlikely that a placebo effect influenced either weight loss or the improved 90-day outcomes associated with ultrafiltration.

The possibility that standard care patients were inadequately treated is diminished by the observation that 43% of patients in the standard care group lost at least 4.5 kg during hospitalization, a weight loss greater than that observed in 75% of patients enrolled in the Acute Decompensated Heart Failure National Registry (ADHERE).

In a subsequent analysis from the UNLOAD trial the outcomes of 100 patients randomized to ultrafiltration were compared to those of patients randomized to standard intravenous diuretic therapy with continuous infusion (n = 32) or bolus injections (n = 68). Choice of diuretic therapy was by the treating physician [6].

At 48 h weight loss was 5.0 ± 3.1 kg in the ultrafiltration group, 3.6 ± 3.5 kg in patients treated with continuous intravenous diuretic infusion, and 2.9 ± 3.5 kg in those given bolus diuretics (p = 0.001 ultrafiltration versus bolus diuretic; p > 0.05 for the other comparisons). Net fluid loss was 4.6 ± 2.6 liters in the ultrafiltration group, 3.9 ± 2.7 liters in patients treated with continuous intravenous diuretic infusion, and 3.1 ± 2.6 liters in those given bolus diuretics (p < 0.001 ultrafiltration versus bolus diuretic; p > 0.05 for the other comparisons). At 90 days, re-hospitalizations plus unscheduled visits for heart failure per patient (re-hospitalization equivalents) were fewer in the ultrafiltration group (0.65 ± 1.36) than in continuous infusion (2.29 ± 3.23; p = 0.016 versus ultrafiltration) and bolus diuretics (1.31 ± 1.87; p = 0.050 vs. ultrafiltration) groups. No serum creatinine differences occurred between groups up to 90 days. The results of this analysis from the UNLOAD clinical trial show that despite the lack of statistical difference in weight and fluid loss by ultrafiltration and intravenous diuretics administered by continuous infusion, more patients treated with ultrafiltration had a sustained clinical benefit, as indicated by fewer re-hospitalizations and unscheduled heart failure office or emergency department visits. These findings support the hypothesis that removal of isotonic fluid by ultrafiltration, rather than hypotonic urine by intravenous diuretics, may contribute to the sustained clinical benefit associated with this treatment strategy. Among 15 acutely decompensated heart failure patients treated first with a furosemide intravenous bolus and then with ultrafiltration because congestion persisted despite therapy with the intravenous loop diuretics, sodium concentration in the ultrafiltrate sampled after 8 h of therapy was significantly higher than the urinary sodium concentration after the intravenous furosemide bolus (134 ± 8.0 vs. 60 ± 47 mmol/l; $p = 0.000025$) [7]. Importantly, while the sodium concentration of the ultrafiltrate was similar in all 15 patients, urinary sodium concentration was highly variable, ranging from <20 to 100 mmol/l in 13 (87%) of the subjects. Only 2 patients had urinary sodium concentrations comparable to those of the ultrafiltrate [7]. Volume overload in heart failure patients is inevitably related to an increase and abnormal distribution of total body sodium [1, 7]. Thus a treatment strategy which is simultaneously more effective in reducing total body sodium and excess fluid may improve outcomes better than therapies which remove either hypotonic fluid or free water (fig. 2). Another key finding of this analysis is that hypokalemia, defined as serum potassium <3.5 mEq/l, occurred less frequently in the ultrafiltration-treated patients than in those treated with intravenous continuous diuretic infusion (1 vs. 22%; $p = 0.003$). The observation that potentially deleterious excessive potassium losses associated with diuretic therapy may not occur with ultrafiltration raises the possibility that avoidance of electrolyte abnormalities may also account for the improved outcomes associated with ultrafiltration.

The absence of a relationship in all 3 groups between the volume of fluid removed and outcomes supports the hypothesis that the *composition* of the fluid removed may have a greater effect than its *quantity* in improving outcomes of congested heart failure patients.

Edema in fluid-overloaded heart failure patients is isotonic, and therefore eunatremic patients with edema have appreciable total body sodium excess. Thus, loop diuretic-induced diuresis of hypotonic fluid will reduce excess total body water while failing to eliminate excess total body sodium [1, 7].

The economic impact of ultrafiltration as an initial strategy for decompensated heart failure was not addressed in the UNLOAD trial.

Additional prospective, randomized studies are needed to confirm or refute the hypothesis that removal of isotonic rather than hypotonic fluid is a key factor in producing sustained clinical benefit in congested heart failure patients, and to determine if clinical features predictive of natriuretic failure in response

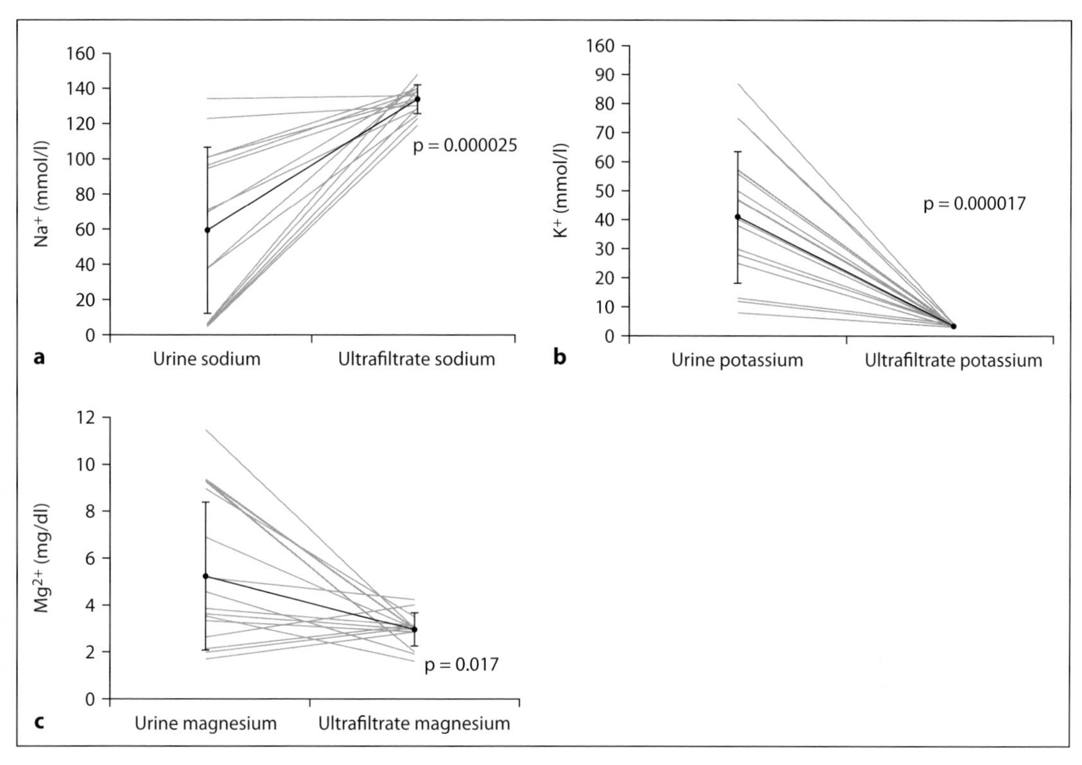

Fig. 2. Differential effects of ultrafiltration and furosemide on electrolytes. Sodium (**a**), potassium (**b**) and magnesium (**c**) concentrations in urine prior to ultrafiltration and in the ultrafiltrate 8 h after initiation of ultrafiltration. From Ali et al. [7], with permission.

to diuretics can be identified and if diuretic strategies can be developed to effectively reduce total body sodium excess.

Conclusions

Of the ultrafiltration approaches described, the most practical are veno-venous ultrafiltration techniques in which isotonic plasma is propelled through the filter by an extracorporeal pump. These approaches avoid arterial puncture, remove a predictable amount of fluid and are not associated with significant hemodynamic instability. Ultrafiltration has been used in patients with decompensated heart failure and volume overload refractory to diuretics. These patients generally have preexisting renal insufficiency and, despite daily oral diuretic doses, develop signs of pulmonary and peripheral congestion. Ultrafiltration and diuretic holiday may restore diuresis and natriuresis. Some patients with volume overload refractory to all available intravenous vasoactive therapies have had significant improvements in symptoms, hemodynamics, and renal function following ultrafiltration. A strategy of *early* ultrafiltration and diuretic holiday can result in more effective weight reduction and can shorten hospitalization. Patients should not be considered for ultrafiltration if the following conditions exist: venous access cannot be obtained; there is a hypercoagulable state; systolic blood pressure is <85 mm Hg or there are signs or symptoms of cardiogenic shock; intravenous pressors are required to maintain adequate blood pressure, or there is end-stage renal disease, as documented by a requirement for dialysis approaches.

Ultrafiltration can be carried out in patients with hematocrit levels of >40% only if it can be proven that hypovolemia is absent.

Many questions regarding the use of ultrafiltration in heart failure patients remain unanswered and must be addressed in future studies. These include: optimal fluid removal rates in individual patients; effects of ultrafiltration on cardiac remodeling; influence of a low oncotic pressure occurring in patients with cardiac cachexia on plasma refill rates, and the economic impact of ultrafiltration to determine whether the expense of disposable filters is offset by the cost savings caused by reduced re-hospitalization rates.

The number of patients with symptomatic congestive heart failure continues to increase in North America and Europe. As cardiac output falls, the natural compensatory response to arterial underfilling is an increased neurohormonal activation, which paradoxically can lead to further reduction in cardiac output, compromising renal and gut blood flow [8]. This may result in deteriorating renal function and diuretic resistance. In these patients the re-hospitalization rates for acutely decompensated heart failure is high and the resulting costs for society are becoming enormous. One future approach could be to create a truly wearable device that would allow patients to ambulate, while being treated. To allow mobility, patients will require a dual lumen central venous catheter, coupled to a miniaturized blood pump, with an accurate battery-powered minipump to regulate the ultrafiltration flow, and a heparin infusion [9]. We have performed the first human trial on such a prototype and published encouraging results [10]. Six volume overloaded patients were treated for 6 h with a wearable ultrafiltration system. Blood flow through the device was around 116 ml/min with an ultrafiltration rate ranging from 120 to 288 ml/h, leading to an average of 151 mmol of sodium removed during the treatment. More importantly during the study, all patients maintained cardiovascular stability. This device, designed to operate continuously, can remove fluid at a slower hourly rate, compared to standard intermittent hemodialysis, so that refilling of plasma volume from the extravascular spaces can be maintained and intravascular depletion avoided. Further developments are underway creating new devices designed to allow fluid overloaded patients with symptomatic congestive heart failure to be managed in either the outpatient setting or in observation units.

References

1 Costanzo MR: Ultrafitration therapies for congestive heart failure; in Hosenpud JD, Greenberg BH (eds): Congestive Heart Failure, ed 3. Philadelphia, Lippincott, William and Wilkins, 2007, chapt 44.

2 Jaski BE, Ha J, Denys BG, Lamba S, Trupp RJ, Abraham WT: Peripherally inserted veno-venous ultrafiltration for rapid treatment of volume overloaded patients. J Card Fail 2003; 9:227–231.

3 Costanzo MR, Saltzberg M, O'Sullivan J, Sobotka P: Early ultrafiltration in patients with decompensated heart failure and diuretic resistance. J Am Coll Cardiol 2005; 46:2047–2051.

4 Bart BA, Boyle A, Bank AJ, et al: Randomized controlled trial of ultrafiltration versus usual care for hospitalized patients with heart failure: relief for acutely fluid overloaded patients with decompensated congestive heart failure. J Am Coll Cardiol 2005;46:2043–2046.

5 Costanzo MR, Guglin ME, Saltzberg MT, et al: Ultrafiltration versus intravenous diuretics for patients hospitalized for acute decompensated heart failure. J Am Coll Cardiol 2007;49:675–683.

6 Costanzo MR, Saltzberg MT, Jessup M, et al: Ultrafiltration is associated with fewer re-hospitalizations than continuous diuretic infusion in patients with decompensated heart failure: results from UNLOAD. J Card Fail 2010, in press.

7 Ali SS, Olinger CC, Sobotka PA, et al: Loop diuretics can cause clinical natriuretic failure: a prescription for volume expansion. Congest Heart Fail 2009;15:1–4.

8 Clark WR, Paganini E, Weinstein D, Bartlett R, Sheinfeld G, Ronco C: Extracorporeal ultrafiltration for acute exacerbation of chronic heart failure: report from the acute dialysis quality initiative. Int J Artif Organs 2005;28:466–476.

9 Ronco C, Davenport A, Gura V: A wearable artificial kidney: dream or reality? Nat Clin Pract Nephrol 2008;4:604–605.

10 Gura V, Ronco C, Nalesso F, Brendolan A, Beizai M, Ezon C, Davenport A, Rambod E: A wearable hemofilter for continuous ambulatory ultrafiltration. Kidney Int 2008;73: 497–502.

Maria Rosa Costanzo, MD, FACC, FAHA
Edward Heart Hospital, 4th Floor
801 South Washington Street, PO Box 3226
Naperville, IL 60566 (USA)
Tel. +1 630 527 2730, Fax +1 630 527 2754, E-Mail mcostanzo@midwestheart.com

Ronco C, Bellomo R, McCullough PA (eds): Cardiorenal Syndromes in Critical Care.
Contrib Nephrol. Basel, Karger, 2010, vol 165, pp 244–250

What Is 'BEST' RRT Practice?

Shigehiko Uchino

Intensive Care Unit, Department of Anesthesiology, Jikei University School of Medicine, Tokyo, Japan

Abstract

Acute kidney injury (AKI) is a common complication among critical illnesses. In severe cases, renal replacement therapy (RRT) is required. It has been reported that hospital mortality of the patients who require RRT is more than 60%. Because of the high mortality, it is quite important to conduct RRT appropriately to improve outcome of patients with severe AKI. However, RRT is not a single homogeneous therapy but rather there are diverse modes of therapy and various ways of providing RRT that might affect its efficacy and safety. The Acute Dialysis Quality Initiative (ADQI) reviewed the available evidence and recommended that more research should be conducted on such technical aspects of RRT in order to determine if certain techniques are preferred for certain indications. As a response to the recommendation by ADQI, the BEST Kidney (Beginning and Ending Supportive Therapy for the Kidney) study was conducted. This study is a multicenter, multinational, prospective, epidemiological study with the aim of understanding multiple aspects of AKI at an international level. This study was conducted at 54 centers in 23 countries from September 2000 to December 2001. The study included more than 1,700 patients including 1,260 who were treated with RRT. Using the large database, several aspects related to RRT have been analyzed, including comparison of IRRT and continuous RRT (CRRT), timing of RRT initiation and discontinuation, and practice variations for CRRT around the world. This study found that RRT practice was quite varied around the world. RRT practice is not aligned with the best evidence and variations in practice may be responsible for significant morbidity. The BEST Kidney Study has generated several hypotheses related to RRT practice in the intensive care unit. Such hypotheses will need to be tested in future clinical trials and hopefully help reduce practice variations for patients with AKI requiring RRT.

Acute kidney injury (AKI) is a common complication among critical illnesses. In severe cases, renal replacement therapy (RRT) is required. It has been reported that the incidence of RRT requirement is approximately 4% and hospital mortality of the patients who require RRT is more than 60% [1]. Because of the high mortality, it is quite important to conduct RRT appropriately to improve outcome of patients with severe AKI.

However, RRT is not a single homogeneous therapy but rather there are diverse modes of therapy and various ways of providing RRT that might affect its efficacy and safety. There are several forms of RRT, including peritoneal dialysis (PD), intermittent RRT (IRRT) and continuous RRT (CRRT). There are also variations for target patient population, timing of treatment, convection vs. diffusion, treatment dose (estimated small solute clearance), choice of membrane and mode of anticoagulation. However, a limited number of studies have analyzed some of these technical aspects in randomized controlled trials. The Acute Dialysis Quality Initiative (ADQI) reviewed the available evidence and recommended that more research should be conducted on such technical aspects of RRT in order to determine if certain techniques are preferred for certain indications [2]. As a response to the recommendation by ADQI, the BEST Kidney (Beginning and Ending Supportive Therapy for the Kidney) study was conducted [3].

What Is the BEST Kidney Study?

The BEST Kidney Study is a multicenter, multinational, prospective, epidemiological study with the aim of understanding multiple aspects of AKI at an international level [3]. This study was conducted at 54 centers in 23 countries from September 2000 to December 2001 (table 1). All patients who were older than 12 years and who were admitted to one of the participating intensive care units (ICUs) during the observational period were considered for study inclusion. From this population, only patients who were treated with RRT other than for drug poisoning or who had at least one of the predefined criteria for AKI were included in the study. The criteria for AKI were oliguria defined as urine output <200 ml in 12 h and/or marked azotemia defined as a blood urea nitrogen level >84 mg/dl (urea 30 mmol/l). Patients with any dialysis treatment before admission to the ICU or patients with end-stage renal failure and receiving dialysis were excluded.

Information was prospectively obtained at study inclusion and was recorded on a standardized case report form developed for this study. Data were collected by means of an electronically prepared Excel-based data collection tool, which was made available to participating centers with instructions. All centers were asked to complete data entry and e-mail the data to the central office, where the data were screened in detail by a dedicated intensive care specialist for any missing information, logical errors, insufficient detail, or addition of queries. Any queries generated an immediate e-mail inquiry and were to be resolved within 48 h.

The study recruited more than 1,700 patients including 1,260 who were treated with RRT. Using the large database, several aspects related to RRT have been analyzed, including comparison of IRRT and CRRT [4], timing of RRT initiation and discontinuation [5, 6], and practice variations for CRRT around the world [7].

Table 1. List of participating centers to the BEST Kidney Study

Region	Country	Centers, n
Asia	Japan	4
	China	2
	Singapore	2
	Indonesia	1
Australasia	Australia	6
North America	USA	6
	Canada	2
South America	Brazil	4
	Uruguay	1
Europe	Italy	6
	Belgium	2
	Germany	2
	Netherlands	2
	Norway	2
	Portugal	2
	Spain	2
	Czech Republic	1
	Greece	1
	Israel	1
	Russia	1
	Sweden	1
	Switzerland	1
	UK	1

Choice of RRT Modality

Mortality and Renal Recovery

CRRT is often the preferred choice over IRRT in the ICU, usually because of improved hemodynamic stability and better solute control. Despite its physiological benefits, randomized controlled trials have not shown benefit of CRRT in mortality over IRRT. However, lack of survival benefit does not imply that IRRT and CRRT are equal. Renal recovery is another important outcome for patients with AKI and might be affected differently by IRRT and CRRT. Some patients with AKI who require RRT in the ICU become dialysis-dependent even after recovery from their sickness and hospital discharge.

In the BEST Kidney Study, among the 1,260 patients treated with RRT for AKI, 1,006 were initially treated with CRRT and 212 with IRRT (40 with other forms of RRT and 2 did not have information for RRT choice). Compared to CRRT, the IRRT group had a higher incidence of pre-admission chronic renal dysfunction and worse renal function at ICU admission. On the other hand, patients in the CRRT group were more likely to be hypotensive and to require vasopressor/inotropic drugs, had worse pulmonary gas exchange and required mechanical ventilation more frequently. Although CRRT had a lower hospital survival than IRRT (36 vs. 52%), it did not remain a significant independent predictor of ICU or hospital survival or dialysis-free hospital survival on multivariable logistic regression analysis. However, CRRT was an independent predictor of recovery from dialysis dependence at hospital discharge (odds ratio [OR] 3.333, $p < 0.0001$) among survivors.

Intra-treatment hypotension was reported in 19% of CRRT patients and 28% of IRRT patients ($p = 0.0044$). Therefore, hypotension during IRRT could be a mechanism of renal injury and delayed recovery. Although randomized controlled trials conducted so far do not support CRRT over IRRT for benefit of renal recovery, all of those studies have significant limitations including the sample size, study design and randomization. On the other hand, our finding is based on an observational but large database. Another large observational study also found the same result [8]. Appropriately planned trials will be required to solve this issue.

Cost

Some people argue that IRRT should be chosen because CRRT does not have a favorable impact on mortality and is costly. However, although several authors have attempted to estimate costs of each therapy, they involve multiple factors that may not be generalizable to other sites and only single-center cost studies have been performed. Therefore, we estimated the cost difference between CRRT and IRRT using the BEST kidney database [unpubl. data]. We estimated costs based on staffing, as well as fluid, dialyzer and anticoagulant use and found that the theoretical range of costs were from USD 5,700 more with CRRT to USD 230 more with IRRT. The median difference in cost between CRRT and IRRT was USD 910 per day (greater with CRRT). Reducing replacement fluid volumes in CRRT to not more than 25 ml/min (approx. 25 ml/kg/h) would result in USD 340 mean savings and could largely eliminate cost differences between therapies.

Timing of RRT Initiation and Discontinuation

Timing of RRT Initiation

It is unknown that earlier initiation of RRT (e.g. before developing severe azotemia) is related to better outcome in patients with AKI or not. If RRT is started too early, one might treat some patients with RRT who would actually not require RRT, which is not a treatment without complications, e.g. bleeding tendency due to anticoagulation, immobilization, blood loss in clotted filters, thrombocytopenia, activation of white blood cells, etc. On the other hand, if initiation of RRT is delayed, it also might cause problems, e.g. fluid overload, electrolyte and acid-base abnormalities, immunosuppression due to uremia, etc. There have been several studies looking at this issue and one meta-analysis concluded that early initiation of RRT in patients with AKI might be associated with improved survival [9]. However, the decision to start RRT can depend on numerous factors and is, therefore, a complex process. Accordingly, timing of RRT initiation has been difficult to study.

Using the BEST kidney database, we sought to study the relationship between the timing of RRT initiation and clinically important outcomes. The timing of RRT initiation was assessed using several approaches. First, timing was categorized by serum biomarker values into early or late RRT, according to the median values of serum urea and serum creatinine at the time of initiation of RRT. Second, we assessed timing based on acute changes to kidney function and patients were stratified into early or late RRT according to the median change in urea and creatinine from ICU admission to start of RRT. Finally, we assessed the temporal relationship between the start of RRT relative to the date of ICU admission.

RRT timing by urea showed no significant difference in crude or covariate-adjusted mortality (OR 0.92 and $p = 0.48$, OR 1.25 and $p = 0.16$, respectively). When stratified by creatinine, late RRT was associated with lower crude and covariate-adjusted mortality (OR 0.46 and $p < 0.001$, OR 0.51 and $p < 0.001$, respectively). However, for timing relative to ICU admission, late RRT was associated with greater crude and covariate-adjusted mortality (OR 1.87 and $p < 0.001$, OR 2.20 and $p < 0.001$, respectively). Overall, late RRT was associated with a longer duration of RRT and stay in hospital and greater dialysis dependence. Therefore, although timing of RRT initiation might exert an important influence on patient survival, this largely depends on its definition.

Timing of CRRT Discontinuation

Although both CRRT and mechanical ventilation are common organ support therapies in the ICU, there is a gap between the two therapies in terms of amount and quality of evidence. In particular, there are negligible data about the process of discontinuation of CRRT. This lack of evidence is different from the field of mechanical ventilation, where many studies dealing with the process of 'weaning' from mechanical ventilation have been conducted. Therefore, we also sought to identify which factors present at the time of discontinuation may assist physicians in predicting successful cessation of CRRT.

Among 529 patients who survived initial CRRT, 313 patients were removed successfully from CRRT and did not require any RRT for at least 7 days and were classified as the 'success' group. The rest (216 patients) were classified as the 'repeat-RRT' group. Patients in the 'success' group had lower hospital mortality (29 vs. 43%, $p < 0.0001$) compared with patients in the 'repeat-RRT' group. They also had lower creatinine and urea concentrations and a higher urine output at the time of CRRT discontinuation. Multivariate logistic regression analysis for successful discontinuation of CRRT identified urine output (during the 24 h before CRRT discontinuation) and creatinine as significant predictors of successful discontinuation. The area under the receiver operating characteristic curve (AUROC) to predict successful discontinuation of CRRT was 0.81 for urine output and 0.64 for creatinine. The predictive ability of urine output was negatively affected by the use of diuretics (AUROC 0.67 with diuretics and 0.85 without diuretics). Therefore, urine output at the time of initial discontinuation of CRRT was the most important predictor of successful discontinuation, especially if occurring without the administration of diuretics.

Practice Variations for CRRT Around the World

We also focused on the use of CRRT and its technical aspects and associated outcomes, which was the first international study describing the practice of CRRT and cataloguing its variations. Among 1,006 patients treated with CRRT, all patients except 1 were treated with a venovenous technique. The most common mode was continuous venovenous hemofiltration (53%). More than half of all patients were treated with predilution techniques (59%). Polyacrylonitrile was the most commonly used membrane material (47%). Polysulfone, polyamide, and cellulose triacetate were also common (21, 17, and 9%, respectively). Bioincompatible membranes (cuprophane) were not used for any patient. Approximately one-third of patients did not receive any drugs for circuit anticoagulation (33%). Among patients who received anticoagulation, unfractionated heparin was the most common choice (43%). The remaining patients were treated with various kinds of anticoagulant (e.g. sodium citrate, nafamostat mesilate, and low-molecular-weight heparin). The median unadjusted CRRT dose was 2,000 ml/h and the corrected dose was 20.4 ml/kg/h.

Hypotension related to CRRT occurred in 19% of patients. Bleeding complications that physicians considered were related to CRRT occurred in 3% of patients, with the indwelling vascular catheter site being the most common site of bleeding. Arrhythmias were observed in 4% of patients. Critical arrhythmias (cardiac arrest, ventricular tachycardia or fibrillation) that were considered to be related to RRT occurred in 8 patients, and were fatal in 3. Approximately one-third of patients (33%) died on CRRT. A similar number of patients (30%) recovered renal function after a median of 4 days of treatment and were discharged from ICU without need for RRT. After initial cessation of CRRT, 23% either switched to IRRT or needed to go back to CRRT. In total, hospital mortality was 64%. Renal recovery occurred in 86% of survivors at hospital discharge.

None of the technical features of CRRT (mode, dilution site for replacement fluid, filter material, anticoagulation, and corrected dose) remained significant as an independent variable for hospital mortality in multivariate logistic regression analysis.

Conclusion

We have conducted a multinational, multicenter, prospective, epidemiological study of critically ill patients with severe AKI, most of whom required RRT. We found that RRT practice was quite varied around the world. RRT practice is not aligned with the best evidence and variations in practice may be responsible for significant morbidity. The BEST Kidney Study has generated several hypotheses related to RRT practice in the ICU. Such hypotheses will need to be tested in future clinical trials and hopefully help reduce practice variations for patients with AKI requiring RRT.

References

1 Uchino S: The epidemiology of acute renal failure in the world. Curr Opin Crit Care 2006;12:538–543.

2 Kellum JA, Mehta RL, Angus DC, Palevsky P, Ronco C, ADQI Workgroup: The first international consensus conference on continuous renal replacement therapy. Kidney Int 2002; 62:1855–1863.

3 Uchino S, Kellum JA, Bellomo R, Doig GS, Morimatsu H, Morgera S, Schetz M, Tan I, Bouman C, Macedo E, Gibney N, Tolwani A, Ronco C; Beginning and Ending Supportive Therapy for the Kidney (BEST Kidney) Investigators: Acute renal failure in critically ill patients. A multinational, multicenter study. JAMA 2005;294:813–818.

4 Uchino S, Bellomo R, Kellum JA, Morimatsu H, Morgera S, Schetz MR, Tan I, Bouman C, Macedo E, Gibney N, Tolwani A, Oudemans-Van Straaten HM, Ronco C; Beginning and Ending Supportive Therapy for the Kidney (BEST Kidney) Investigators Writing Committee: Patient and kidney survival by dialysis modality in critically ill patients with acute kidney injury. Int J Artif Organs 2007; 30:281–292.

5 Bagshaw SM, Uchino S, Bellomo R, Morimatsu H, Morgera S, Schetz M, Tan I, Bouman C, Macedo E, Gibney N, Tolwani A, Oudemans-van Straaten HM, Ronco C, Kellum JA; Beginning and Ending Supportive Therapy for the Kidney (BEST Kidney) Investigators: Timing of renal replacement therapy and clinical outcomes in critically ill patients with severe acute kidney injury. J Crit Care 2009;24:129–140.

6 Uchino S, Bellomo R, Morimatsu H, Morgera S, Schetz M, Tan I, Bouman C, Macedo E, Gibney N, Tolwani A, Straaten HO, Ronco C, Kellum JA: Discontinuation of continuous renal replacement therapy: a post-hoc analysis of a prospective multicenter observational study. Crit Care Med 2009;37:2576–2582.

7 Uchino S, Bellomo R, Morimatsu H, Morgera S, Schetz M, Tan I, Bouman C, Macedo E, Gibney N, Tolwani A, Oudemans-van Straaten H, Ronco C, Kellum JA: Continuous renal replacement therapy: a worldwide practice survey. The Beginning and Ending Supportive Therapy for The kidney (BEST Kidney) Investigators. Intensive Care Med 2007;33:1563–1570.

8 Bell M, SWING, Granath F, Schön S, Ekboml A, Martling CR: Continuous renal replacement therapy is associated with less chronic renal failure than intermittent haemodialysis after acute renal failure. Intensive Care Med 2007;33:773–780.

9 Seabra VF, Balk EM, Liangos O, Sosa MA, Cendoroglo M, Jaber BL: Timing of renal replacement therapy initiation in acute renal failure: a meta-analysis. Am J Kidney Dis 2008;52:272–284.

Shigehiko Uchino, Assoc. Prof.
Department of Anesthesiology, Jikei University School of Medicine
3-19-18, Nishi-Shinbashi, Minato-ku, Tokyo 105-8471 (Japan)
Tel. +81 3 3433 1111, Fax +81 3 5401 0454
E-Mail s.uchino@jikei.ac.jp

Ronco C, Bellomo R, McCullough PA (eds): Cardiorenal Syndromes in Critical Care.
Contrib Nephrol. Basel, Karger, 2010, vol 165, pp 251–262

Anticoagulation for Renal Replacement Therapy: Different Methods to Improve Safety

Heleen M. Oudemans-van Straaten[a] · Enrico Fiaccadori[b] · Ian Baldwin[c]

[a]Department of Intensive Care Medicine, Onze Lieve Vrouwe Gasthuis, Amsterdam, The Netherlands; [b]Department of Intensive Care, Austin Hospital, Melbourne, Vic., Australia, and [c]Intensive Care Unit, Internal Medicine & Nephrology Department, Parma University Medical School, Parma, Italy

Abstract

The different methods for anticoagulation of the extracorporeal circuit for renal replacement therapy (RRT) vary in safety profile, bleeding being the most important side effect. Avoiding severe bleeding is a key aim of RRT prescription. The present paper is a clinical review comparing regional anticoagulation with citrate to heparin for continuous RRT, and different anticoagulant strategies for prolonged intermittent therapies. Regional anticoagulation with citrate or the use of prostacyclins provide high safety because they do not increase the patient's risk of bleeding. The use of citrate may additionally increase biocompatibility. However, both confer other risks when used without understanding and therefore require carefully designed protocols for bedside use with guidelines for early detection and management of complications. If so, the use of citrate may improve patient and kidney survival. Further studies are needed to confirm and explain this benefit.

Acute kidney injury (AKI) in the setting of multiple organ failure confers high mortality [1]. To support the failing kidney, renal replacement therapy (RRT) is required, with preference for continuous (CRRT) or prolonged intermittent modes if the circulation is vasopressor-dependent. Anticoagulation of the extracorporeal circuit is generally needed. Heparins confer anticoagulation of the patient as well, increasing the risk of bleeding. Bleeding events are reported in 10–50% of the patients [2–4]. A serious bleeding event is one associated with anemia, shock and need for blood transfusion. Avoiding severe bleeding is a key aim of CRRT prescription.

CRRT can be performed without anticoagulation. The method is safest with regard to bleeding, but early clotting of the circuit is the consequence. Another option is to antagonize heparin anticoagulation after the filter by administering protamine, thus providing anticoagulation in the circuit, 'regional' anticoagulation [5]. The continuous formation of heparin-protamine complexes may however cause harm [6, 7]. Alternatives are to use prostaglandins as an anticoagulant, apply regional anticoagulation with citrate before and calcium replacement after the filter, or give RRT (and anticoagulation) intermittently, achieving treatment in a shorter time with less anticoagulation. This paper provides a clinical review comparing citrate to heparin use in CRRT and different anticoagulation strategies when prolonged intermittent, 'hybrid' RRT modalities are used.

Regional Anticoagulation with Citrate

Sodium citrate is the basic salt of citric acid ($Na_3C_6H_5O_7$). When infused in the circuit, citrate chelates calcium inducing anticoagulation by reducing ionized calcium (iCa^{2+}) level. As iCa^{2+} is essential for the progression of the coagulation cascade to form a stable clot, the anticoagulant effect is achieved once calcium is chelated [8]. With the use of citrate, the circuit is anticoagulated, but the patient is not as the patient blood calcium level is kept normal with a calcium infusion. For this reason, citrate is considered safer than heparin.

For anticoagulation, the dose of citrate is commonly set to achieve a reduction of iCa^{2+} level in the circuit blood to <0.3 mmol/l [8–10]. Part of the calcium bound to citrate is lost in the waste fluid and requires replacement by a separate calcium infusion to maintain serum iCa^{2+} level to normal [10].

Citrate: Removal, Metabolism, Accumulation and Toxicity

Both free and calcium-bound citrate is eliminated by convection or diffusion across the membrane. The (relative) amount depends on filtrate fraction and dialysis dose. The sieving coefficient is between 0.87 and 1.0, and is not different between hemofiltration and hemodialysis [11, 12]. Citrate not eliminated enters the patient's circulation and is metabolized in the mitochondrial citric acid cycle of liver, skeletal muscle and renal cortex.

The use of citrate anticoagulation is limited by the patient's capacity to metabolize citrate, which is decreased if liver function or tissue perfusion fail [13, 14]. If metabolism falls short, citrate accumulation is evidenced by metabolic acidosis. Due to chelation with citrate, iCa^{2+} decreases whereas the bound fraction increases. Total calcium increases further by calcium infusion to correct low iCa^{2+}. Total to iCa^{2+} ratio is the most sensitive marker of citrate accumulation [15–17]. Toxicity of citrate is mediated by hypocalcemia which causes myocardial depression if severe. Monitoring of total to iCa^{2+}

ratio and adjustment of citrate dose in case of a rise >2.25 is a safe way to avoid toxicity.

Citrate accumulation can also cause metabolic alkalosis. This occurs if a high citrate dose is infused (e.g. protocol violation) or removal declines due to loss of filter function. To prevent this, the coupling of citrate and blood pumps in modern CRRT devices is a good safety design feature.

Citrate as a Buffer

Citrate also provides buffer base. One mole of sodium citrate consumes 3 mol of hydrogen, providing 3 mol of sodium bicarbonate if trisodium citrate is used [10], or less if part of the cations in the citrate solution are composed of hydrogen: one third in the acid citrate dextrose solution-A (ACD-A) [12, 18, 19] and one tenth in a solution used in the Netherlands [20]. Because citrate provides sodium and buffer, replacement and dialysis fluids contain less sodium and bicarbonate. Metabolic control is usually very stable, however this depends on the protocol used and on unintended violations. Metabolic acidosis [21] and alkalosis [22] are both described. In the largest randomized controlled trial comparing citrate to low molecular weight heparin, metabolic control was better with citrate than with heparin due to a strict algorithm defining the use of a buffer-free and/or a bicarbonate-buffered replacement fluid in the protocol [20].

Citrate: Different Protocols for CRRT

After the first report of the Mehta group [23], various citrate protocols are reported. A common theme to the reports is the complex preparation process of formula or custom-made solutions by the hospital pharmacy. The US Food and Drug Administration (FDA) mandates fluids required for fluid replacement associated with CRRT be classified as a drug and in this context the fluids must be made by the hospital's pharmacy upon physician-written prescription prior to use [24]. Nowadays, several fluids are commercially available. Choice of the protocol depends on local legislation, expertise, availability of fluids and preference for simplicity or flexibility, and for CRRT mode (convective and/or diffusive) and dose. Training, understanding and compliance to the protocol are crucial.

Continuous Venovenous Hemofiltration (CVVH) Post- or Predilution

Citrate is either administered by continuous prefilter infusion of a more or less concentrated sodium citrate solution (133–1,000 mmol/l), together with a low sodium and low-buffer or buffer-free replacement fluid in postdilution [25, 26]. Citrate can also be included in an isotonic calcium-free predilution replacement fluid in a concentration corresponding to the desired bicarbonate equivalent (11–23 mmol/l) [9, 27]. This fluid is now commercially prepared [28]. Calcium is replaced separately (fig. 1). The latter method is simple and

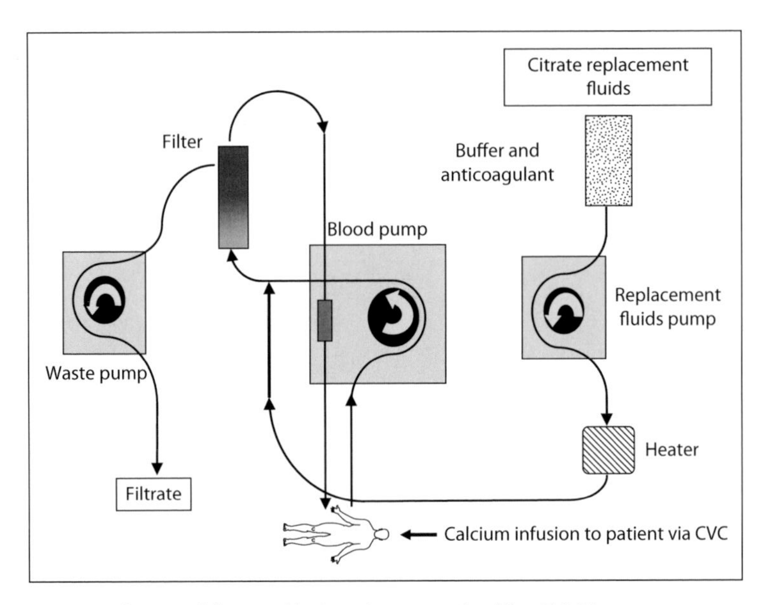

Fig. 1. Scheme of the predilution citrate method for CVVH.

safe, but less flexible to correct acid base balance due to the fixed buffer content [21, 29].

Continuous Venovenous Hemodialysis (CVVHD) or Hemodiafiltration (CVVHDF)

For CVVHD(F), some authors use an isotonic citrate replacement fluid pre-filter [22, 30, 31], others the ACD-A solution [32–34] generally together with a calcium-free dialysate [8, 35]. Calcium is infused separately. However, the successful use of a calcium-containing dialysate is reported as well [29]. Replacement and dialysis fluids vary widely regarding electrolyte and buffer content [25]. Acidosis and alkalosis are controlled by adjusting citrate, dialysate or replacement rates and relies on strict adherence to the specific protocol [22, 36].

Citrate Compared to Heparin

Bleeding Complications and Circuit Survival

Several studies compare circuit life and bleeding complications with citrate to non-randomized controls on heparin [8, 25, 37]. Groups are generally not comparable, because the citrate patients generally had a higher bleeding risk. Nevertheless, bleeding complications were less in the citrate groups [8, 37]. Circuit survival with citrate was mostly longer, sometimes comparable,

Table 1. Randomized clinical studies comparing citrate to heparin anticoagulation for RRT

Reference	Design/ patients	Circuit life, h		Bleeding		Transfusion (RBCs/day)		3-Month survival	
		citrate	heparin	citrate	heparin	citrate	heparin	citrate	heparin
Monchi 2004 [10]	RCOT n = 20	70 (44–140) p = 0.0007	40 (17–48)	n = 0	n = 1	0.2 (0–0.4) p < 0.001	1.0 (0–2.0)		
Kutsogiannis 2005 [32]	RCT n = 30	125 (95–157) p < 0.001	38 (25–62)	RR 0.17 (0.03–1.04) p = 0.06		0.53 (0.24–1.20) p = 0.13			
Betjes 2007 [39]	RCT n = 48			0% p < 0.01	33%	0.43 p = 0.01	0.88		
Fealy[a] 2007 [5]	RCOT n = 10	17 12–20	13 9–28						
Oudemans 2009 [20]	RCT n = 200	27 (13–47)	26 (15–43)	6% p = 0.08	16%	0.27 (0–0.63) p = 0.31	0.36 (0–0.83)	45%[b] p = 0.02	62%[b]

RCOT = Randomized cross-over trial; RCT = randomized controlled trial, circuit life in median (IQR); RR = relative risk; RBCs/day = number of red cell units per day of CVVH.
[a] The study compared citrate to regional heparinization (heparin-protamine).
[b] Per-protocol analysis.

and in some studies shorter than with heparin [26]. Differences in circuit life between studies can be explained by the wide variety of citrate dose (2–6 mmol/l of blood flow) [25], citrate dose titrated on post-filter iCa^{2+} [10, 32] or a fixed citrate/blood flow infusion rate [20], by the use of dialysis or filtration, pre- or postdilution, differences in CRRT dose and filtration fraction, or if a reduction in citrate flow is used for control of metabolic alkalosis or acidosis [9, 38].

Four randomized controlled studies comparing anticoagulation with citrate to heparin have been published [10, 20, 32, 39] (table 1) – two of these report a significantly longer circuit survival with citrate, three report less bleeding, and two less red blood cell transfusion using citrate.

Patient and Kidney Survival

Unexpectedly, the largest trial found a 15% absolute 3-month survival benefit for citrate on intention to treat and a 17% survival benefit in the per-protocol treated patients (table 1). Multivariate analysis found that in addition to high age, severe organ failure and high transfusion rate, anticoagulation with heparin contributed to mortality, indicating that the benefit of citrate could not be fully explained by less bleeding. Post-hoc complementary subgroup analysis showed a particular benefit of citrate in patients after surgery, with sepsis, a high degree of organ failure or younger age. In none of the subgroups survival with citrate was worse. Among the higher proportion of surviving patients, more citrate patients were independent from chronic dialysis. The difference was significant for all patients ($p = 0.02$) and tended to significance for those who survived (97 vs. 86%, $p = 0.08$). A higher survival rate with citrate was also reported in the control arm of a clinical trial on a cytopheretic inhibitory device (SCD) in adjunction to CVVH [40]. Notably, citrate may additionally improve biocompatibility [41] either due to the protective effect of hypocalcemia at the membrane [42–44] or to the avoidance of heparin which stimulates the release of inflammatory neutrophil and platelet products in a dose-dependent way [45, 46]. Citrate may also confer mitochondrial protection [47, 48].

Prolonged Intermittent Renal Replacement Therapy

Available options for RRT in patients with AKI in the ICU have recently been expanded with prolonged intermittent RRT modalities [49–51]. These techniques are known under different acronyms, such as 'sustained low efficiency dialysis' (SLED) or 'extended daily dialysis' (EDD). They use lower low blood (100–200 ml/min) and dialysate (100–300 ml/min) flow rates compared with standard intermittent hemodialysis/hemofiltration. Treatment times are longer in duration (commonly 8–12 h), and either standard dialysis, or purpose-built machines with CRRT membranes are used [52, 53].

Key features are represented by procedural simplicity, flexible scheduling, high dialysis-dose delivery, excellent solute control, good hemodynamic tolerance, and lower cost: essentially, these 'hybrid' RRT modalities combine the advantages of both the continuous and the conventional intermittent forms of RRT, avoiding many of their problems [54–56], particularly frequent clotting and continuous anticoagulation.

However, despite shorter treatment time, anticoagulation of the extracorporeal circulation is often required. The optimal strategy is still a matter of debate with no consensus about this issue [26, 57].

In the case of SLED/EDD, several strategies for maintenance of extracorporeal circuit viability have been utilized in the past, such as saline flushes without any antihemostatic agent [49, 50, 58], unfractioned standard heparin [49, 50, 54, 55] and prostacyclin [59] (table 2). The reported incidence of circuit clotting in SLED is 26–46% with saline flushes and 17–26% with standard heparin at doses of 500–1,000 IU/h [49, 50, 58]; the best results obtained so far

Table 2. Anticoagulation strategies reporting clinical endpoints for prolonged intermittent renal replacement therapies

Reference	Modality/patients (treatments) duration	Antihemostatic strategy and extra-corporeal clotting rates		Hemorrhagic complications	
Kumar 2000 [49]	SLED/25 (367) median 7.5 (IQR 6–8) h	none 117/367 clotting 41/117 (35%)	heparin 250/367 clotting 43/250 (17.2%)	NA	
Marshall 2001 [50]	SLED/37 (145) mean 10.4 (SD 2.7) h	none 41/145 clotting NA[a]	Heparin 104/145 clotting NA[a]	no heparin 1	heparin 1
Marshall 2004 [58]	SLED-f/24 (56) 8 h	none 13/56 clotting 7/13 (54%)	heparin 40/56 clotting 9/40 (22.5%)	none 0	heparin 0
Berbece 2006 [54]	SLED/23 (165) mean 7.5 SD 1.1 h	saline flushes 105/165 clotting 31/105 (29%)	heparin 60/165 clotting 11/60 (18%)	saline 0	heparin 0
Fiaccadori 2007 [59]	SLED/64 (282) median 8 h	none 97/282 clotting 39/97 (40.2%)	prostacyclin[b] 185/282 clotting 19/185 (10%)	none 3/29 (10.3%)	prostacyclin 2/35 (5.7%)
Clark 2008 [60]	SLED/30 (117)[c] mean 7.32, SD 1.34 h	citrate 117 clotting 0/117 (0%)		0	
Fiaccadori 2009 [61]	SLED/37 (273)[d] median 8 h	citrate 273 clotting 21/273 (7.7%)		2/37 (5.4%)	

SLED = Sustained low-efficiency dialysis, SLED-f = sustained low-efficiency diafiltration, NA = not available; IQR = interquartile range; SD = standard deviation.

[a] Cumulative clotting rate (heparin or no heparin) 38/145 (26%); separate data not available.

[b] Prostacyclin 6 ng/kg/h.

[c] Not all ICU patients.

[d] APACHE-II 24.7 ± 4.6, mechanical ventilation 33/37.

(10% incidence of circuit interruption, half of them for circuit coagulation), are those with prostacyclin (PGI_2), a physiological product of human endothelium with antiaggregant effects at low doses and vasodilatory effects at higher doses [59]. PGI_2 is available for clinical use as its synthetic analogue epoprostenol, and inhibits platelet activation and aggregation induced by contact of platelets with the surfaces of the extracorporeal circuit during RRT; compared with other antihemostatic drugs for RRT, PGI_2 is easy to use by continuous infusion, and exerts an action which, due to the drug's short half-life, fades away very rapidly after withdrawal [59]. Doses of epoprostenol utilized for SLED are 6 ng/kg/h; with this infusion rate, vasodilatory effects are modest, and bleeding complications low [59].

With the introduction and the widespread use of citrate as a regional anticoagulation method in CRRT (see above), increasing interest has also been paid to the application of this strategy for SLED/EDD. Anticoagulation protocols for regional anticoagulation with citrate for SLED/EDD are based on different citrate solutions, and on the utilization of either CRRT or standard dialysis machines (table 2) [60–62].

Some preliminary results of a simplified citrate protocol for SLED have been recently presented using blood and dialysis fluid rates of 200 and 300 ml/min respectively, dialysis fluid with iCa^{2+} 1.25 mmol/l, and sodium citrate/citric acid solution (ACD-A®, citrate 113 mmol/l; Fresenius Kabi, Italy) as anticoagulant (table 2) [61]. Treatment was monitored by iCa^{2+} and ACT: targets for iCa^{2+} were >0.85 mmol/l in the patient and 0.25–0.45 mmol/l in the blood entering the filter; ACT targets were normal values in the patient, and at least 2–3 times the normal values before the filter. Calcium gluconate was infused if iCa^{2+} in the patient was <0.85 mmol/l. Most of the treatments (252/273, 92.3%) were concluded in the prescribed time (median duration 8 h); 4/273 (1.4%) were prematurely interrupted for ongoing circuit coagulation, 11 (4%) for elevated TMP values, 1/273 (0.3%) for central venous catheter malfunction, 1 (0.3%) for urgent diagnostics, and 3/283 (1%) for refractory hypotension. Systemic calcium infusion was needed in 1 case only; two major bleeding complications (2/37, 5.4%) were observed. These preliminary results suggest that simplified protocols with commercially available citrate solutions allow safe and effective regional anticoagulation for SLED, with limited laboratory monitoring and without need for systemic calcium infusion in most patients.

Conclusion

Different methods for anticoagulation of the RRT circuit vary in safety profile. Regional anticoagulation with citrate or the use of prostacyclins provide high safety because they do not increase the patient's risk of bleeding. However, both confer other risks when used without understanding and therefore require carefully designed protocols for bedside use with guidelines for early detection and management of complications. If so, the use of citrate may improve patient survival and recovery of kidney function.

References

1 Metnitz PG, Krenn CG, Steltzer H, Lang T, Ploder J, Lenz K, Le Gall JR, Druml W: Effect of acute renal failure requiring renal replacement therapy on outcome in critically ill patients. Crit Care Med 2002;30:2051–2058.
2 Abramson S, Niles JL: Anticoagulation in continuous renal replacement therapy. Curr Opin Nephrol Hypertens 1999;8:701–707.
3 Martin PY, Chevrolet JC, Suter P, Favre H: Anticoagulation in patients treated by continuous venovenous hemofiltration: a retrospective study. Am J Kidney Dis 1994;24:806–812.
4 Van de Wetering J, Westendorp RG, van der Hoeven JG, Stolk B, Feuth JD, Chang PC: Heparin use in continuous renal replacement procedures: the struggle between filter coagulation and patient hemorrhage. J Am Soc Nephrol 1996;7:145–150.
5 Fealy N, Baldwin I, Johnstone M, Egi M, Bellomo R: A pilot randomized controlled crossover study comparing regional heparinization to regional citrate anticoagulation for continuous venovenous hemofiltration. Int J Artif Organs 2007;30:301–307.
6 Carr JA, Silverman N: The heparin-protamine interaction. A review. J Cardiovasc Surg (Torino) 1999;40:659–666.
7 Rossmann P, Matousovic K, Horacek V: Protamine-heparin aggregates. Their fine structure, histochemistry, and renal deposition. Virchows Arch B Cell Pathol Incl Mol Pathol 1982;40:81–98.
8 Ward DM, Mehta RL: Extracorporeal management of acute renal failure patients at high risk of bleeding. Kidney Int Suppl 1993;41:S237–S244.
9 Palsson R, Niles JL: Regional citrate anticoagulation in continuous venovenous hemofiltration in critically ill patients with a high risk of bleeding. Kidney Int 1999;55:1991–1997.
10 Monchi M, Berghmans D, Ledoux D, Canivet JL, Dubois B, Damas P: Citrate vs. heparin for anticoagulation in continuous venovenous hemofiltration: a prospective randomized study. Intensive Care Med 2004;30:260–265.
11 Chadha V, Garg U, Warady BA, Alon US: Citrate clearance in children receiving continuous venovenous renal replacement therapy. Pediatr Nephrol 2002;17:819–824.
12 Swartz R, Pasko D, O'Toole J, Starmann B: Improving the delivery of continuous renal replacement therapy using regional citrate anticoagulation. Clin Nephrol 2004;61:134–143.
13 Apsner R, Schwarzenhofer M, Derfler K, Zauner C, Ratheiser K, Kranz A: Impairment of citrate metabolism in acute hepatic failure. Wien Klin Wochenschr 1997;109:123–127.
14 Kramer L, Bauer E, Joukhadar C, Strobl W, Gendo A, Madl C, Gangl A: Citrate pharmacokinetics and metabolism in cirrhotic and noncirrhotic critically ill patients. Crit Care Med 2003;31:2450–2455.
15 Meier-Kriesche HU, Gitomer J, Finkel K, DuBose T: Increased total to ionized calcium ratio during continuous venovenous hemodialysis with regional citrate anticoagulation. Crit Care Med 2001;29:748–752.
16 Bakker AJ, Boerma EC, Keidel H, Kingma P, van der Voort PH: Detection of citrate overdose in critically ill patients on citrate-anticoagulated venovenous haemofiltration: use of ionised and total/ionised calcium. Clin Chem Lab Med 2006;44:962–966.
17 Hetzel GR, Taskaya G, Sucker C, Hennersdorf M, Grabensee B, Schmitz M: Citrate plasma levels in patients under regional anticoagulation in continuous venovenous hemofiltration. Am J Kidney Dis 2006;48:806–811.
18 Mitchell A, Daul AE, Beiderlinden M, Schafers RF, Heemann U, Kribben A, Peters J, Philipp T, Wenzel RR: A new system for regional citrate anticoagulation in continuous venovenous hemodialysis. Clin Nephrol 2003;59:106–114.
19 Tobe SW, Aujla P, Walele AA, Oliver MJ, Naimark DM, Perkins NJ, Beardsall M: A novel regional citrate anticoagulation protocol for CRRT using only commercially available solutions. J Crit Care 2003;18:121–129.
20 Oudemans-Van Straaten HM, Bosman RJ, Koopmans M, van der Voort PH, Wester JP, van der Spoel JI, Dijksman LM, Zandstra DF: Citrate anticoagulation for continuous venovenous hemofiltration. Crit Care Med 2009;37:545–552.

21 Bihorac A, Ross EA: Continuous venovenous hemofiltration with citrate-based replacement fluid: efficacy, safety, and impact on nutrition. Am J Kidney Dis 2005;46:908–918.
22 Tolwani AJ, Prendergast MB, Speer RR, Stofan BS, Wille KM: A practical citrate anticoagulation continuous venovenous hemodiafiltration protocol for metabolic control and high solute clearance. Clin J Am Soc Nephrol 2006;1:79–87.
23 Mehta RL, McDonald BR, Aguilar MM, Ward DM: Regional citrate anticoagulation for continuous arteriovenous hemodialysis in critically ill patients. Kidney Int 1990;38:976–981.
24 Abramson S, Niles JL: Anticoagulation in continuous renal replacement therapy. Curr Opin Nephrol Hypertens 1999;8:701–707.
25 Oudemans-Van Straaten HM, Wester JP, de Pont AC, Schetz MR: Anticoagulation strategies in continuous renal replacement therapy: can the choice be evidence based? Intensive Care Med 2006;32:188–202.
26 Joannidis M, Oudemans-Van Straaten HM: Clinical review: patency of the circuit in continuous renal replacement therapy. Crit Care 2007;11:218.
27 Egi M, Naka T, Bellomo R, Cole L, French C, Trethewy C, Wan L, Langenberg CC, Fealy N, Baldwin I: A comparison of two citrate anticoagulation regimens for continuous veno-venous hemofiltration. Int J Artif Organs 2005;28:1211–1218.
28 Naka T, Egi M, Bellomo R, Cole L, French C, Botha J, Wan L, Fealy N, Baldwin I: Commercial low-citrate anticoagulation haemofiltration in high risk patients with frequent filter clotting. Anaesth Intensive Care 2005;33:601–608.
29 Gupta M, Wadhwa NK, Bukovsky R: Regional citrate anticoagulation for continuous venovenous hemodiafiltration using calcium-containing dialysate. Am J Kidney Dis 2004;43:67–73.
30 Gabutti L, Marone C, Colucci G, Duchini F, Schonholzer C: Citrate anticoagulation in continuous venovenous hemodiafiltration: a metabolic challenge. Intensive Care Med 2002;28:1419–1425.
31 Dorval M, Madore F, Courteau S, Leblanc M: A novel citrate anticoagulation regimen for continuous venovenous hemodiafiltration. Intensive Care Med 2003;29:1186–1189.
32 Kutsogiannis DJ, Gibney RT, Stollery D, Gao J: Regional citrate versus systemic heparin anticoagulation for continuous renal replacement in critically ill patients. Kidney Int 2005;67:2361–2367.
33 Bagshaw SM, Laupland KB, Boiteau PJ, Godinez-Luna T: Is regional citrate superior to systemic heparin anticoagulation for continuous renal replacement therapy? A prospective observational study in an adult regional critical care system. J Crit Care 2005;20:155–161.
34 Cointault O, Kamar N, Bories P, Lavayssiere L, Angles O, Rostaing L, Genestal M, Durand D: Regional citrate anticoagulation in continuous venovenous haemodiafiltration using commercial solutions. Nephrol Dial Transplant 2004;19:171–178.
35 Tobe SW, Aujla P, Walele AA, Oliver MJ, Naimark DM, Perkins NJ, Beardsall M: A novel regional citrate anticoagulation protocol for CRRT using only commercially available solutions. J Crit Care 2003;18:121–129.
36 Morgera S, Schneider M, Slowinski T, Vargas-Hein O, Zuckermann-Becker H, Peters H, Kindgen-Milles D, Neumayer HH: A safe citrate anticoagulation protocol with variable treatment efficacy and excellent control of the acid-base status. Crit Care Med 2009;37:2018–2024.
37 Van der Voort PH, Postma SR, Kingma WP, Boerma EC, Van Roon EN: Safety of citrate based hemofiltration in critically ill patients at high risk for bleeding: a comparison with nadroparin. Int J Artif Organs 2006;29:559–563.
38 Gabutti L, Marone C, Colucci G, Duchini F, Schonholzer C: Citrate anticoagulation in continuous venovenous hemodiafiltration: a metabolic challenge. Intensive Care Med 2002;28:1419–1425.
39 Betjes MG, van Oosterom D, van Agteren M, van de WJ: Regional citrate versus heparin anticoagulation during venovenous hemofiltration in patients at low risk for bleeding: similar hemofilter survival but significantly less bleeding. J Nephrol 2007;20:602–608.
40 Humes HD, Sobota JT, Ding F, Song JH: A selective cytopheretic inhibitory device to treat the immunological dysregulation of acute and chronic renal failure. Blood Purif 2010;29:183–190.

41 Oudemans-Van Straaten HM: Citrate anticoagulation for continuous renal replacement therapy in the critically ill. Blood Purif 2010;29:208–213.
42 Bohler J, Schollmeyer P, Dressel B, Dobos G, Horl WH: Reduction of granulocyte activation during hemodialysis with regional citrate anticoagulation: dissociation of complement activation and neutropenia from neutrophil degranulation. J Am Soc Nephrol 1996;7:234–241.
43 Bohler J, Donauer J, Birmelin M, Schollmeyer PJ, Horl WH: Mediators of complement-independent granulocyte activation during haemodialysis: role of calcium, prostaglandins and leukotrienes. Nephrol Dial Transplant 1993;8:1359–1365.
44 Gritters M, Grooteman MP, Schoorl M, Schoorl M, Bartels PC, Scheffer PG, Teerlink T, Schalkwijk CG, Spreeuwenberg M, Nube MJ: Citrate anticoagulation abolishes degranulation of polymorphonuclear cells and platelets and reduces oxidative stress during haemodialysis. Nephrol Dial Transplant 2006;21:153–159.
45 Leitienne P, Fouque D, Rigal D, Adeleine P, Trzeciak MC, Laville M: Heparins and blood polymorphonuclear stimulation in haemodialysis: an expansion of the biocompatibility concept. Nephrol Dial Transplant 2000;15:1631–1637.
46 Gritters M, Borgdorff P, Grooteman MP, Schoorl M, Schoorl M, Bartels PC, Tangelder GJ, Nube MJ: Platelet activation in clinical haemodialysis: LMWH as a major contributor to bioincompatibility? Nephrol Dial Transplant 2008;23:2911–2917.
47 Weinberg JM, Venkatachalam MA, Roeser NF, Nissim I: Mitochondrial dysfunction during hypoxia/reoxygenation and its correction by anaerobic metabolism of citric acid cycle intermediates. Proc Natl Acad Sci USA 2000;97:2826–2831.
48 Feldkamp T, Weinberg JM, Horbelt M, Von Kropff C, Witzke O, Nurnberger J, Kribben A: Evidence for involvement of nonesterified fatty acid-induced protonophoric uncoupling during mitochondrial dysfunction caused by hypoxia and reoxygenation. Nephrol Dial Transplant 2009;24:43–51.
49 Kumar VA, Craig M, Depner TA, Yeun JY: Extended daily dialysis: a new approach to renal replacement for acute renal failure in the intensive care unit. Am J Kidney Dis 2000;36:294–300.
50 Marshall MR, Golper TA, Shaver MJ, Alam MG, Chatoth DK: Sustained low-efficiency dialysis for critically ill patients requiring renal replacement therapy. Kidney Int 2001;60:777–785.
51 Faulhaber-Walter R, Hafer C, Jahr N, Vahlbruch J, Hoy L, Haller H, Fliser D, Kielstein JT: The Hannover Dialysis Outcome Study: comparison of standard versus intensified extended dialysis for treatment of patients with acute kidney injury in the intensive care unit. Nephrol Dial Transplant 2009;24:2179–2186.
52 Fliser D, Kielstein JT: Technology Insight: treatment of renal failure in the intensive care unit with extended dialysis. Nat Clin Pract Nephrol 2006;2:32–39.
53 Vanholder R, Van Biesen W, Lameire N: What is the renal replacement method of first choice for intensive care patients? J Am Soc Nephrol 2001(suppl 17):S40–S43.
54 Berbece AN, Richardson RM: Sustained low-efficiency dialysis in the ICU: cost, anticoagulation, and solute removal. Kidney Int 2006;70:963–968.
55 Kielstein JT, Kretschmer U, Ernst T, Hafer C, Bahr MJ, Haller H, Fliser D: Efficacy and cardiovascular tolerability of extended dialysis in critically ill patients: a randomized controlled study. Am J Kidney Dis 2004;43:342–349.
56 Marshall MR, Golper TA, Shaver MJ, Alam MG, Chatoth DK: Urea kinetics during sustained low-efficiency dialysis in critically ill patients requiring renal replacement therapy. Am J Kidney Dis 2002;39:556–570.
57 Oudemans-Van Straaten HM: Review and guidelines for regional anticoagulation with citrate in continuous hemofiltration. Neth J Crit Care 2004;8:146–156.
58 Marshall MR, Ma T, Galler D, Rankin AP, Williams AB: Sustained low-efficiency daily diafiltration for critically ill patients requiring renal replacement therapy: towards an adequate therapy. Nephrol Dial Transplant 2004;19:877–884.

59 Fiaccadori E, Maggiore U, Parenti E, Giacosa R, Picetti E, Rotelli C, Tagliavini D, Cabassi A: Sustained low-efficiency dialysis with prostacyclin in critically ill patients with acute renal failure. Nephrol Dial Transplant 2007;22:529–537.
60 Clark JA, Schulman G, Golper TA: Safety and efficacy of regional citrate anticoagulation during 8-hour sustained low-efficiency dialysis. Clin J Am Soc Nephrol 2008;3:736–742.
61 Fiaccadori E, Parenti E, Grego P: Simplified regional citrate anticoagulation for sustained low-efficiency dialysis in acute kidney injury. J Am Soc Nephrol 2009;20:597A.
62 Szamosfalvi B, Frinak S, Yee J: Automated regional citrate anticoagulation: technological barriers and possible solutions. Blood Purif 2010;29:204–209.

Heleen M. Oudemans-van Straaten, MD
Department of Intensive Care Medicine, Onze Lieve Vrouwe Gasthuis
PO Box 95500, NL–1090 HM Amsterdam (The Netherlands)
Tel. +31 20599 3007, Fax +31 20599 2128
E-Mail h.m.oudemans-vanstraaten@olvg.nl

Ronco C, Bellomo R, McCullough PA (eds): Cardiorenal Syndromes in Critical Care.
Contrib Nephrol. Basel, Karger, 2010, vol 165, pp 263–273

Renal Replacement Therapy in Adult Critically Ill Patients: When to Begin and When to Stop

Dinna N. Cruz[a] · Zaccaria Ricci[b] · Sean M. Bagshaw[c] · Pasquale Piccinni[d] · Noel Gibney[c] · Claudio Ronco[a]

[a]Department of Nephrology, Dialysis & Transplantation, International Renal Research Institute, and [d]Department of Anesthesiology and Intensive Care Medicine, San Bortolo Hospital, Vicenza, Vicenza, and [b]Department of Pediatric Cardiosurgery, Bambino Gesù Hospital, Rome, Italy, and [c]Division of Critical Care Medicine, Faculty of Medicine and Dentistry, University of Alberta, Edmonton, Alta., Canada

Abstract

Renal replacement therapy (RRT) is an important therapeutic and supportive measure for acute kidney injury (AKI) in the critical care setting. While RRT is extensively used in clinical practice, there remains uncertainty about the ideal circumstances of when to initiate RRT and for what indications. Many factors, including logistics, resource availability, physician experience and patient-related factors are involved in the decision of when to start and stop RRT for those with AKI. Among the patient-related factors, examples include 'dynamic' trends in AKI and/or non-kidney organ dysfunction, additional measures of acute physiology, such as fluid accumulation and relative oliguria. There currently exists a large variation in clinical practice regarding starting and stopping RRT, due in part to the lack of consensus on this issue. In this article, we briefly review a new opinion-based algorithm to aid in the decision on when to initiate RRT in adult critically ill patients. This algorithm was developed using available clinical evidence, recognizing the inherent limitations of observational studies. It aims to provide a starting point for clinicians and future prospective studies. We also review the available literature on discontinuation of RRT and propose a few simple recommendations on how to 'wean' patients from RRT.

Acute kidney injury (AKI) is an important complication of critical illness with an important impact on morbidity, mortality and health resource utilization. Between 5 and 15% of patients admitted to the intensive care unit (ICU) will need renal replacement therapy (RRT) [1]. However, there is wide variation among physicians and institutions regarding timing and indications for initiation of RRT in the ICU [2–4]. Many factors, including those that are patient-specific, clinician-specific and those related to organizational/logistical issues, are involved in the decision to initiate RRT.

Many studies have evaluated the timing of RRT initiation, in which 'early' or 'late' RRT initiation was variably defined: based on biochemical criteria, urine output criteria, or by 'door-to-dialysis' time [2, 5–9]. A recent meta-analysis [8] summarized these studies, including 5 randomized and quasi-randomized controlled trials. Their pooled analysis failed to show a significant mortality reduction with early RRT (RR 0.64; 95% CI 0.40, 1.05), however this was underpowered. Among the cohort studies, early RRT appeared to reduce mortality (RR 0.72; 95% CI 0.64, 0.82). Overall, there was a strong trend towards a beneficial effect of early RRT [8]. Nevertheless the definition of 'early' and 'late' remains highly subjective. This limits our ability to come up with more concrete recommendations.

In 2004, a consensus-driven classification scheme for AKI (RIFLE: acronym for Risk-Injury-Failure-Loss-End-stage renal disease) classification was published [10]. It was intended to define the presence or absence of the clinical syndrome of AKI in a given patient, and to describe the severity of the AKI. RIFLE uses two criteria: (a) change in blood creatinine from a baseline value, and (b) urine flow rates per body weight over a specified time period. Risk is the mildest category of AKI, followed by Injury, and Failure is the most severe category. In 2007, this was modified by the Acute Kidney Injury Network [11]. Risk, Injury, and Failure were replaced with Stages 1, 2 and 3, respectively, and an absolute increase in creatinine of at least 0.3 mg/dl was added to Stage 1. The RIFLE/AKIN criteria have been validated and proven robust for clinically relevant outcomes in patients with AKI across numerous studies [12–15].

The RIFLE/AKIN criteria provide the possibility of a more 'quantitative' characterization of timing. We recognize that these criteria have not been formally evaluated as a tool for guiding clinicians on when to initiate RRT, but nevertheless they may serve as an attractive starting point for a more systematic approach to initiation of RRT. The RIFLE/AKIN criteria also have the important advantage for being able to classify and follow the trend in AKI severity over time [16]. These are both vital to evaluate in the context of whether and when RRT initiation should be considered.

A proposed algorithm for RRT initiation in adult critically ill patients utilizing RIFLE/AKIN was recently published [17]. This opinion-based algorithm (fig. 1) incorporates several patient-specific factors, based on clinical evidence when available, that may influence when to initiate RRT. The objectives of this algorithm are to provide a starting point not only for guiding clinicians on when to consider use of RRT in such patients, but also for further prospective evaluation and study to understand the ideal timepoint/circumstances for when to initiate RRT.

When to Begin RRT in Critically Ill Patients

A summary of the algorithm is shown (fig. 1), and is based on the assumption that there are no contraindications to RRT. When a patient has 'absolute' or traditional indications for RRT, the decision is fairly straightforward [17]. It is important, however, to recognize that RRT initiation for these indications may be viewed as 'rescue therapy' where delays may have deleterious consequences for the patient.

In the absence of 'absolute' indications for RRT, the patient is then assessed for the presence and severity of AKI (as defined by RIFLE/AKIN criteria). 'Dynamic' trends in AKI and/or non-kidney organ dysfunction, as well as additional measures of acute physiology, e.g. fluid accumulation, relative oliguria (i.e. urine output >200 ml/12 h, but insufficient to prevent fluid accumulation) all factor into the decision of when to initiate RRT for those with AKI.

Severe AKI

In the presence of severe AKI (i.e. RIFLE category F or AKIN category III) and/or rapidly deteriorating kidney function, we would consider RRT initiation, particularly if there was failure to respond to initial therapy [18]. Data to support consideration of early RRT in these patients is largely provided by observational data [14, 15, 19, 20]. Bell et al. [19] performed a 7-year retrospective analysis of 207 patients with AKI receiving RRT. When stratified by RIFLE class at the time RRT was initiated, those with RIFLE class F had considerably higher 30-day mortality when compared to those initiating RRT at either RIFLE class R or I (adjusted HR 3.4, 95% CI 1.2–9.3; crude 30-day mortality: 57.9% for F vs. 23.5% for R vs. 22.0% for I). It should be recognized that the RIFLE class should not be used in isolation to decide on RRT initiation – but rather the RIFLE class plus interaction with the overall goals of therapy and other important clinical variables should be carefully weighed. We recognize that additional prospective evaluation on this issue is needed to guide clinical practice, however, in many circumstances, the risks of not providing RRT may exceed those of initiation of RRT.

Mild to Moderate AKI

The decision of if, and when, to initiate RRT in critically ill patients with mild-moderate AKI (i.e. RIFLE category R/I or AKIN category I/II) is often the most challenging. It is important to recognize that the decision to initiate RRT in these patients is most likely to be multifactorial and there may not be any single indication. The presence of one or more mitigating factors, such as rapidly worsening AKI and/or overall severity of illness, severe sepsis, and reduced

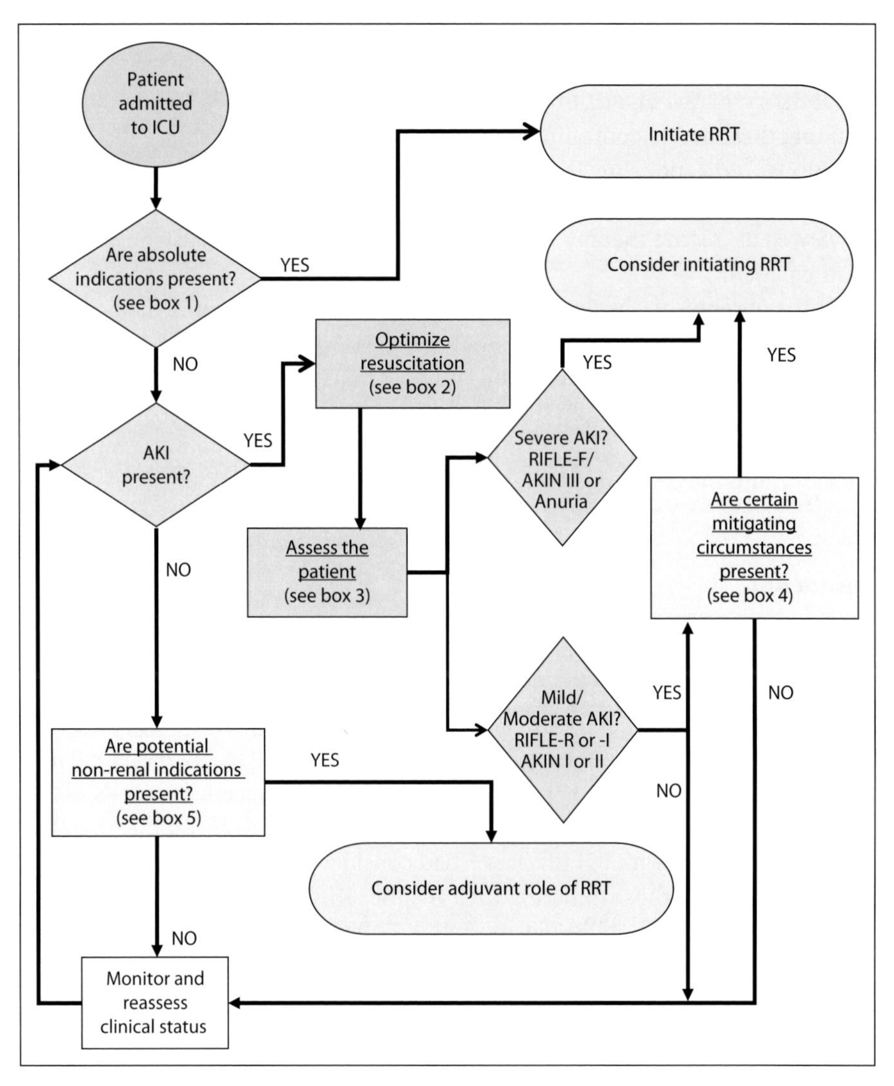

Fig. 1. Algorithm for RRT initiation in the adult critically ill patient.

Box 1

Are absolute indications for RRT present?

- Uremic complications, e.g. encephalopathy, pericarditis, bleeding
- Serum urea ≥36 mmol/l (100 mg/dl)
- K^+ ≥6 mmol/l and/or ECG abnormalities
- Mg ≥4 mmol/l and/or anuria/absent deep tendon reflexes
- Serum pH ≤7.15
- Urine output <200 ml/12 h or anuria
- Diuretic-resistant organ edema (i.e. pulmonary edema) in the presence of AKI

Box 2
Optimize resuscitation:
- Intravascular volume
- Cardiac output
- Mean arterial pressure
- Intra-abdominal pressure

Box 3
Assess the patient:
- AKI severity and trend
- Illness severity and trajectory
- Initial response to initial resuscitation therapy

Box 4
Are certain mitigating circumstances present?
- Rapidly worsening AKI
- Rapidly worsening illness severity
- Hypercatabolic state
- Refractory fluid overload and/or accumulation
- Severe sepsis
- Permissive hypercapnia
- Reduced renal reserve
- Low probability for rapid renal recovery

Box 5
Are potential non renal indications present?
- Refractory fluid overload
- Refractory septic shock
- Acute liver failure
- Severe tumor lysis syndrome
- Severe electrolyte disturbances
- Dysthermia
- Selected toxins (e.g. lithium, aminoglycosides, rhabdomyolysis)

renal reserve (fig. 1, Box 4) would push us to consider RRT in earlier stages of AKI. Primary diagnoses associated with high catabolic rates (e.g. septic shock, major trauma, burn injury) or those likely to place considerable demand on kidney function (i.e. gastrointestinal bleeding, rhabdomyolysis) should be identified in the context of potential need for early initiation of RRT. Critically ill patients with acute lung injury/acute respiratory distress syndrome receiving lung-protective ventilation may intentionally develop respiratory acidosis due to permissive hypercapnia [21]. Coexistent and/or evolving AKI in these patients will significantly impair capacity for kidney bicarbonate regeneration to buffer systemic acidemia. Earlier RRT may prove beneficial in these patients prior to the development of severe acidemia, worsening acute respiratory distress syndrome and/or volume overload.

A positive fluid balance and overt clinical fluid overload, when refractory to medical therapy (i.e. diuretics), is also an important circumstance where RRT initiation may prove beneficial. Other than fluid 'overload', in which the patient has obvious clinical signs of excess fluid (e.g. peripheral or pulmonary edema, weight gain) usually in association with oliguria, we also recognize the concept of fluid 'accumulation'. With the latter case, these clinical signs may not yet be readily apparent and the patient may have a urine output above the traditional definition of 'oliguria', but this urine output is inadequate for keeping up with the patient's daily intake, resulting in an increasingly positive fluid balance. In critically ill patients, fluid overload may be under-recognized as an important contributor to morbidity and mortality [22–26]. Longer duration of mechanical ventilation, weaning failure, delayed tissue healing, and cardiopulmonary complications have all been associated with fluid overload [25–27]. Likewise, a positive fluid balance has been shown to be associated with higher mortality in critically ill adults and children [22, 23, 28]. RRT initiation should therefore be considered an important therapeutic measure for the prevention (and not only the treatment) of refractory fluid overload.

Extracorporeal Blood Purification for 'Non-Renal' Indications

In a minority of patients who do not fulfill RIFLE/AKIN criteria for AKI, but who have certain mitigating medical issues, the adjuvant role of RRT, although controversial, may be considered. Key examples of such conditions are briefly discussed below. In this context, the term extracorporeal blood purification (EBP), rather than RRT, may be more appropriate. For example, high-volume hemofiltration has been advocated as a potential adjuvant immunomodulatory therapy in refractory septic shock, by some consensus groups [29, 30], but not others [31]. Several small clinical trials have shown promising results for improvements in hemodynamics, metabolic parameters and survival [32–34]. In their consensus statement the Acute Dialysis Quality Initiative (ADQI) Working Group concluded that the use of EBP in sepsis has a biological rationale that merits further investigation. While confirmatory data from multicenter randomized trials are needed to inform clinical practice on this issue, the authors believe that patients with refractory septic shock may benefit from high-volume EBP [29].

Acute liver failure is another potential 'non-kidney' indication for RRT. Acute liver failure is characterized by profound metabolic and regulatory hepatic dysfunction contributing to life-threatening complications including encephalopathy, cardiovascular collapse, AKI, bleeding and susceptibility to infection that may culminate into multiorgan failure. EBP and albumin dialysis have been used to bridge these patients to native liver recovery or provide opportunity for liver transplantation [35]. Moreover, many acute liver failure patients develop AKI as their illness evolves, may have underlying or unrecognized hepatorenal syndrome, and thus, should be considered candidates for EBP that may or may not include albumin dialysis.

Although likewise controversial, EBP for selected toxins (e.g. lithium, methanol) is sometimes performed [36]. EBP is however more likely to be performed when the intoxication is complicated by kidney dysfunction which may further reduce clearance of the toxin and/or its metabolites. In addition to the exogenous toxins, EBP can also be performed to aid in clearance of selected endogenous toxins, particularly in the context of concomitant kidney dysfunction (i.e. rhabdomyolysis, tumor lysis syndrome). Electrolyte disturbances and metabolic acidosis induced by certain toxins also can be readily corrected with EBP.

Initiation of RRT: Risks vs. Benefits

We also accept that RRT initiation is not without risk for adverse consequences including hypotension (and exacerbation of kidney injury), bleeding, dialysis catheter-related complications, etc. 'Early' RRT initiation has the potential to expose a large number of patients to this therapy, who may have otherwise spontaneously recovered kidney function and/or survived without having received it. Unfortunately, there is a paucity of data on this issue, and recognizing these gaps in knowledge are important along with the careful weighing of the potential risks of RRT initiation against the goals of therapy and proposed benefits.

When to Stop RRT

About 10 years ago, a passionate debate on ventilation weaning strategies (pressure support ventilation vs. T-piece spontaneous ventilation vs. continuous pressure airway pressure vs. synchronized intermittent mandatory ventilation) was ongoing: the scientific community finally concluded that it was impossible to show the superiority of a method over the other: the *manner* in which the mode of ventilation weaning is applied (timing, length, clinical and technical knowhow) may have a greater effect on the likelihood of weaning than the *mode* itself [37]. In other words, just like stereotyped approaches to ventilation are anachronistic and inappropriately try to fit the patient into a fixed therapy rather than a tailoring the therapy to the patient, so should RRT be adjusted to fulfill the needs of the patients. If this fundamental concept may be applied to all therapeutic efforts in the care of critical illness, we find it is useful to explain how, in the specific setting of 'weaning from RRT', no good evidence exists at present and it is unlikely to be produced in the near future. Nevertheless, some insights may be gleaned from recent literature. An interesting report from the BEST (Beginning and Ending Supportive Therapy) Kidney study group described current practice for the discontinuation of CRRT in a multinational setting in order to identify variables associated with successful discontinuation and if the approach to discontinue continuous RRT may affect patient outcomes [38]. 313 patients were weaned from CRRT for at least 7 days and were classified as the 'success' group; 216 patients were classified as the 'repeat RRT' group. Multivariate logistic regression analysis for successful discontinuation of CRRT identified urine output (during the 24 h before stopping continuous RRT: OR 1.078 per 100 ml/day increase) and creatinine (OR 0.996 per mmol/l increase) as significant predictors of successful cessation. The predictive ability of urine output was negatively affected by the use of diuretics. As it always happens with observational studies, it is not possible to establish if the 'repeat RRT' population would have achieved a different outcome by waiting for a better creatinine or urine output before stopping RRT.

Risk factors for re-dialysis were also analyzed by the National Taiwan University Surgical ICU Acute Renal Failure Study Group [39]. In this study, RRT weaning of 94 postoperative patients was considered successful when prolonged for at least 30 days (21%) and was correlated with SOFA score, age, dialysis duration and, again, urine output. Interestingly, of the patients who remained 'RRT-free' for 5 days after RRT discontinuation, more than two-thirds remained RRT-free up to the 30th day.

As a general recommendation, before weaning from RRT, physicians should wait for adequate urine output (without diuretic therapy) and optimized creatinine values (the additional effect of patient glomerular filtration rate and treatment clearance should lead to normal or subnormal creatinine values while on RRT). Once renal function appears close to the baseline or 'pre-AKI' level, it seems reasonable to interrupt the treatment without any specific weaning protocol. It is possible, on the other hand, that patients with signs of only partial renal recovery may benefit from more specific and prolonged weaning algorithms. Examples would be a decrease of ultrafiltration rate, or prescription of intermittent treatments where the therapy was previously continuous. Future trials are needed to design, if possible, such protocols. Furthermore, evaluation of new renal biomarkers as prognostic factors is intriguing in order to explore if they can predict when patients have recovered sufficient renal function to allow them to remain RRT-free once RRT is stopped.

Conclusions

RRT is an important therapeutic and supportive measure for AKI in the critical care setting. Currently, there exists large variation in clinical practice regarding starting and stopping RRT, due in part to the lack of consensus on this issue. An opinion-based algorithm to aid in the decision on when to initiate RRT in critically ill patients has recently been proposed based on available clinical evidence. This algorithm aims to provide a starting point for clinicians, as well as for further prospective evaluation and development of broad consensus on this important issue. In terms of 'weaning' from RRT, it may be prudent to wait until urine output is adequate (without diuretic therapy) and normal creatinine values are present. If renal recovery is only partial, 'weaning' protocols may be considered, such as a decrease of ultrafiltration rate, or prescription of intermittent treatments where the therapy was previously continuous.

References

1 Uchino S, Kellum JA, Bellomo R, Doig GS, Morimatsu H, Morgera S, Schetz M, Tan I, Bouman C, Macedo E, Gibney N, Tolwani A, Ronco C: Acute renal failure in critically ill patients: a multinational, multicenter study. JAMA 2005;294:813–818.

2 Bagshaw SM, Bellomo R: Early diagnosis of acute kidney injury. Curr Opin Crit Care 2007;13:638–644.

3 Ricci Z, Ronco C, D'Amico G, De Felice R, Rossi S, Bolgan I, Bonello M, Zamperetti N, Petras D, Salvatori G, Dan M, Piccinni P: Practice patterns in the management of acute renal failure in the critically ill patient: an international survey. Nephrol Dial Transplant 2006;21:690–696.

4 Overberger P, Pesacreta M, Palevsky PM: Management of renal replacement therapy in acute kidney injury: a survey of practitioner prescribing practices. Clin J Am Soc Nephrol 2007;2:623–630.

5 Gettings LG, Reynolds HN, Scalea T: Outcome in post-traumatic acute renal failure when continuous renal replacement therapy is applied early vs. late. Intensive Care Med 1999;25:805–813.

6 Kresse S, Schlee H, Deuber HJ, Koall W, Osten B: Influence of renal replacement therapy on outcome of patients with acute renal failure. Kidney Int Suppl 1999;S75–78.

7 Liu KD, Himmelfarb J, Paganini EP, Ikizler A, Soroko S, Mehta RL, Chertow GM: Timing of initiation of dialysis in critically ill patients with acute kidney injury. Clin J Am Soc Nephrol 2006;1:915–919.

8 Seabra VF, Balk EM, Liangos O, Sosa MA, Cendoroglo M, Jaber BL: Timing of renal replacement therapy initiation in acute renal failure: a meta-analysis. Am J Kidney Dis 2008;52:272–284.

9 Splendiani G, Mazzarella V, Cipriani S, Zazzaro D, Casciani CU: Continuous renal replacement therapy: our experience in intensive care unit. Ren Fail 2001;23:259–264.

10 Bellomo R, Ronco C, Kellum JA, Mehta RL, Palevsky P: Acute renal failure – definition, outcome measures, animal models, fluid therapy and information technology needs. Second International Consensus Conference of the Acute Dialysis Quality Initiative (ADQI) Group. Crit Care 2004;8:R204–R212.

11 Mehta RL, Kellum JA, Shah SV, Molitoris BA, Ronco C, Warnock DG, Levin A: Acute Kidney Injury Network: report of an initiative to improve outcomes in acute kidney injury. Crit Care 2007;11:R31.

12 Ricci Z, Cruz D, Ronco C: The RIFLE criteria and mortality in acute kidney injury: a systematic review. Kidney Int 2008;73:538–546.

13 Joannidis M, Metnitz B, Bauer P, Schusterschitz N, Moreno R, Druml W, Metnitz PG: Acute kidney injury in critically ill patients classified by AKIN versus RIFLE using the SAPS-3 database. Intensive Care Med 2009;35:1692–1702.

14 Cruz DN, Bolgan I, Perazella MA, Bonello M, de Cal M, Corradi V, Polanco N, Ocampo C, Nalesso F, Piccinni P, Ronco C: North East Italian Prospective Hospital Renal Outcome Survey on Acute Kidney Injury (NEiPHROS-AKI): targeting the problem with the RIFLE Criteria. Clin J Am Soc Nephrol 2007;2:418–425.

15 Thakar CV, Christianson A, Freyberg R, Almenoff P, Render ML: Incidence and outcomes of acute kidney injury in intensive care units: a Veterans Administration study. Crit Care Med 2007;37:2552–2558.

16 Cruz DN, Ricci Z, Ronco C: Clinical review: RIFLE and AKIN – time for reappraisal. Crit Care 2009;13:211.

17 Bagshaw SM, Cruz DN, Gibney RN, Ronco C: A proposed algorithm for initiation of renal replacement therapy in adult critically ill patients. Crit Care 2009;13:317.

18 Himmelfarb J, Joannidis M, Molitoris B, Schietz M, Okusa MD, Warnock D, Laghi F, Goldstein SL, Prielipp R, Parikh CR, Pannu N, Lobo SM, Shah S, D'Intini V, Kellum JA: Evaluation and initial management of acute kidney injury. Clin J Am Soc Nephrol 2008; 3:962–967.

19 Bell M, Liljestam E, Granath F, Fryckstedt J, Ekbom A, Martling CR: Optimal follow-up time after continuous renal replacement therapy in actual renal failure patients stratified with the RIFLE criteria. Nephrol Dial Transplant 2005;20:354–360.

20 Hoste EA, Clermont G, Kersten A, Venkataraman R, Angus DC, De Bacquer D, Kellum JA: RIFLE criteria for acute kidney injury are associated with hospital mortality in critically ill patients: a cohort analysis. Crit Care 2006;10:R73.

21 ARDS NET Investigators: Ventilation with lower tidal volumes as compared with traditional tidal volumes for acute lung injury and the acute respiratory distress syndrome. The Acute Respiratory Distress Syndrome Network. N Engl J Med 2000;342:1301–1308.

22 Alsous F, Khamiees M, DeGirolamo A, Amoateng-Adjepong Y, Manthous CA: Negative fluid balance predicts survival in patients with septic shock: a retrospective pilot study. Chest 2000;117:1749–1754.

23 Payen D, de Pont AC, Sakr Y, Spies C, Reinhart K, Vincent JL: A positive fluid balance is associated with a worse outcome in patients with acute renal failure. Crit Care 2008;12:R74.

24 Sakr Y, Vincent JL, Reinhart K, Groeneveld J, Michalopoulos A, Sprung CL, Artigas A, Ranieri VM: High tidal volume and positive fluid balance are associated with worse outcome in acute lung injury. Chest 2005; 128:3098–3108.

25 Upadya A, Tilluckdharry L, Muralidharan V, Amoateng-Adjepong Y, Manthous CA: Fluid balance and weaning outcomes. Intensive Care Med 2005;31:1643–1647.

26 Wiedemann HP, Wheeler AP, Bernard GR, Thompson BT, Hayden D, deBoisblanc B, Connors AF Jr, Hite RD, Harabin AL: Comparison of two fluid-management strategies in acute lung injury. N Engl J Med 2006;354:2564–2575.

27 Brandstrup B, Tonnesen H, Beier-Holgersen R, Hjortso E, Ording H, Lindorff-Larsen K, Rasmussen MS, Lanng C, Wallin L, Iversen LH, Gramkow CS, Okholm M, Blemmer T, Svendsen PE, Rottensten HH, Thage B, Riis J, Jeppesen IS, Teilum D, Christensen AM, Graungaard B, Pott F: Effects of intravenous fluid restriction on postoperative complications: comparison of two perioperative fluid regimens: a randomized assessor-blinded multicenter trial. Ann Surg 2003;238:641–648.

28 Goldstein SL, Currier H, Graf C, Cosio CC, Brewer ED, Sachdeva R: Outcome in children receiving continuous venovenous hemofiltration. Pediatrics 2001;107:1309–1312.

29 Bellomo R, Honore PM, Matson J, Ronco C, Winchester J: Extracorporeal blood treatment methods in SIRS/Sepsis. Int J Artif Organs 2005;28:450–458.

30 Honore PM, Joannes-Boyau O: High-volume hemofiltration in sepsis: a comprehensive review of rationale, clinical applicability, potential indications and recommendations for future research. Int J Artif Organs 2004; 27:1077–1082.

31 Dellinger RP, Levy MM, Carlet JM, Bion J, Parker MM, Jaeschke R, Reinhart K, Angus DC, Brun-Buisson C, Beale R, Calandra T, Dhainaut JF, Gerlach H, Harvey M, Marini JJ, Marshall J, Ranieri M, Ramsay G, Sevransky J, Thompson BT, Townsend S, Vender JS, Zimmerman JL, Vincent JL: Surviving Sepsis Campaign: international guidelines for management of severe sepsis and septic shock: 2008. Crit Care Med 2008;36:296–327.
32 Cornejo R, Downey P, Castro R, Romero C, Regueira T, Vega J, Castillo L, Andresen M, Dougnac A, Bugedo G, Hernandez G: High-volume hemofiltration as salvage therapy in severe hyperdynamic septic shock. Intensive Care Med 2006;32:713–722.
33 Honore PM, Jamez J, Wauthier M, Lee PA, Dugernier T, Pirenne B, Hanique G, Matson JR: Prospective evaluation of short-term, high-volume isovolemic hemofiltration on the hemodynamic course and outcome in patients with intractable circulatory failure resulting from septic shock. Crit Care Med 2000;28:3581–3587.
34 Joannes-Boyau O, Rapaport S, Bazin R, Fleureau C, Janvier G: Impact of high volume hemofiltration on hemodynamic disturbance and outcome during septic shock. ASAIO J 2004;50:102–109.
35 Sen S, Williams R, Jalan R: Emerging indications for albumin dialysis. Am J Gastroenterol 2005;100:468–475.
36 Holubek WJ, Hoffman RS, Goldfarb DS, Nelson LS: Use of hemodialysis and hemoperfusion in poisoned patients. Kidney Int 2008;74:1327–1334.
37 Butler R, Keenan SP, Inman KJ, Sibbald WJ, Block G: Is there a preferred technique for weaning the difficult-to-wean patient? A systematic review of the literature. Crit Care Med 1999;27:2331–2336.
38 Uchino S, Bellomo R, Morimatsu H, Morgera S, Schetz M, Tan I, Bouman C, Macedo E, Gibney N, Tolwani A, Straaten HO, Ronco C, Kellum JA: Discontinuation of continuous renal replacement therapy: a post-hoc analysis of a prospective multicenter observational study. Crit Care Med 2009;37:2576–2582.
39 Wu VC, Ko WJ, Chang HW, Chen YW, Lin YF, Shiao CC, Chen YM, Chen YS, Tsai PR, Hu FC, Wang JY, Lin YH, Wu KD: Risk factors of early redialysis after weaning from postoperative acute renal replacement therapy. Intensive Care Med 2008;34:101–108.

Dinna N. Cruz, MD, MPH
Department of Nephrology, Dialysis & Transplantation, International Renal Research Institute
San Bortolo Hospital
Viale Rodolfi 37, IT–36100 Vicenza (Italy)
Tel. +39 0 444753650, Fax +39 0 444753973, E-Mail dinnacruzmd@yahoo.com

Ronco C, Bellomo R, McCullough PA (eds): Cardiorenal Syndromes in Critical Care.
Contrib Nephrol. Basel, Karger, 2010, vol 165, pp 274–283

Urine Abnormalities in Acute Kidney Injury and Sepsis

Sean M. Bagshaw[a] · Rinaldo Bellomo[b]

[a]Division of Critical Care Medicine, University of Alberta Hospital, Edmonton, Alta., Canada, and
[b]Department of Intensive Care Medicine, Austin Hospital, Heidelberg, Vic., Australia

Abstract

Acute kidney injury (AKI) is a common complication of critical illness. While the etiology of AKI in critically ill patients is likely often multifactorial, sepsis has consistently been found an important contributing factor and has been associated with high attributable morbidity and mortality. Accordingly, the timely identification of septic AKI in critically ill patients is clearly a clinical priority. The diagnosis of AKI has traditionally depended upon biochemical measurements such as serum creatinine, urea, and urine output. In addition, several urinary biochemical tests, derived indices and microscopy have also been widely cited as valuable in the diagnosis and classification of AKI. However, the value of these urinary tests in the diagnosis, classification, prognosis and clinical management in septic AKI remains unclear, due in part to a lack of kidney morphologic changes and histopathology in human studies of septic AKI. This review will summarize the urinary biochemistry and microscopy in septic AKI.

Acute kidney injury (AKI) is a common complication of critical illness [1]. While the etiology of AKI in critically ill patients is likely often multifactorial, sepsis has consistently been found an important contributing factor. In fact, sepsis represents the single most common predisposing factor for its development, accounting for close to 50% of cases [2]. Moreover, the morbidity and mortality attributable to septic AKI remains high [3]. Therefore, the timely identification of septic AKI in critically ill patients is clearly a clinical priority.

The diagnosis of AKI depends upon biochemical measurements as outlined by the consensus statement of the Acute Dialysis Quality Initiative (http://www.ADQI.net) [4]. In addition to the above biochemical diagnostic criteria, many textbooks of internal medicine, nephrology and critical care advocate the use of a number of widely cited urinary biochemical and microscopy-based tests as valuable in the diagnosis and classification of AKI [5–7]. However, the value of these urinary tests in the diagnosis, classification, prognosis and clinical management in septic AKI remains unclear, due in part to a paucity of data on kidney morphologic changes and histopathology in human studies of septic AKI.

Epidemiology of Septic AKI

Sepsis is a common condition that precipitates critical illness and often requires management in an intensive care unit (ICU) setting [8, 9]. In the multicenter European SOAP study, Vincent et al. [9] found that 37.4% of adult patients admitted to an ICU had a primary septic diagnosis. Moreover, the incidence of sepsis continues to increase. In a 25-year surveillance of sepsis diagnoses in the USA, Martin et al. [8] described an 8.7% annual increase in incidence of primary sepsis diagnoses.

Several observational studies of septic critically ill patients have described concomitant AKI occurring in 11–64% [2, 3, 9–11]. Likewise, observational studies focused on AKI in critical illness have described the rates of sepsis as a primary contribute factor in 26–58% [1, 12, 13]. Overall, an abundance of data now show septic AKI is common and increasingly encountered in critically illness.

Importantly, critically ill patients with septic AKI would appear to have important differences in baseline characteristics, acute physiology, laboratory parameters and treatment intensity compared to those with non-septic AKI. Septic AKI patients are generally older, have a higher burden of co-morbid illness, are characterized by higher acuity of illness, greater organ dysfunction and higher treatment intensity. These patients are more likely admitted to hospital/ICU for medical indications, however, if surgical, they were more likely to have received an emergency procedure [2, 3, 14]. Septic AKI patients have higher rates of oliguria, receive more fluid and/or diuretic therapy and are more likely to accumulate fluid early in their clinical course when compared with non-septic AKI or those with sepsis only [2, 14–16].

These distinguishing features of septic AKI translate into differences in clinical outcomes, including higher mortality, longer durations of hospitalization, and trends for higher rates of renal recovery to dialysis independence. The BEST Kidney Study found septic, as compared with non-septic AKI, was independently associated with higher risk of hospital mortality in multivariable analysis (OR 1.5, 95% CI 1.2–1.9) [2]. Hoste et al. [10] found duration of ICU stay was approximately doubled in septic patients developing AKI (20 days for AKI vs. 10 days for non-AKI, $p = 0.001$). Moreover, lengths of stay increased significantly with increasing severity of AKI when defined by the RIFLE criteria [3]. These data imply septic AKI is associated with a more complex and prolonged course and has important resource implications and higher attributable health-care costs.

Urinary Biochemistry in Septic AKI

There are numerous urinary biochemical tests and indices that have been described as surrogates for renal tubular function in AKI. The most common urinary tests and indices used in clinical practice include urinary sodium (UNa), fractional excretion of sodium (FeNa), and fractional excretion of urea (FeU). Like urinary microscopy, these indices have traditionally been used to distinguish between pre-renal azotemia (PRA) from acute tubular necrosis (ATN). However, these tests lack sensitivity and specificity for discriminating the underlying etiology and/or severity of AKI. This is especially true in critically ill septic patients, in whom numerous factors other than renal tubular dysfunction can impact urinary solute excretion [17, 18]. Furthermore, it is important to recognize that the tradition of classifying AKI into PRA and ATN is, to some extent, completely arbitrary. These two entities, in many clinical contexts, almost certainly exist on a continuum of injury, and their division into separate diagnostic terms offers limited clinical and prognostic value [19].

Urine Sodium (UNa). The UNa is a commonly cited measure for the classification of AKI with values <10–20 mmol/l suggestive of PRA and values >40 mmol/l consistent with ATN. However, the diagnostic value of UNa in septic AKI has rarely been described [17, 18]. In a systematic review of 27 clinical studies describing urine biochemistry in septic AKI, UNa was found to be highly variable and little data was available on the time-dependent changes [17]. For example, in one study, only 33% of patients had a UNa <20 mmol/l, and 23% of those classified as ATN had a UNa <20 mmol/l. The UNa has also commonly been found to be <20 mmol/l in patients with hyperdynamic and resuscitated septic shock. In a small cohort of critically ill patients with hyperdynamic septic shock, the UNa decreased from 48 to 33 mmol/l over 24 h in those with AKI, despite 72% having received diuretic therapy [15]. Moreover, additional data have suggested that the UNa poorly predicts the need for renal replacement therapy (RRT) or recovery of renal function in AKI.

Fractional Excretion of Sodium (FeNa). The use of FeNa is based on the physiological principle that filtered sodium remains avidly reabsorbed by the renal tubules in PRA (resulting in a FeNa <1%), but not in ATN (resulting in a FeNa >1%). While theoretically attractive, the true diagnostic value of FeNa in septic AKI in clinical practice is unclear and questionable. The FeNa is frequently >1% in patients having received diuretics, and is not uncommonly <1% in conditions ranging rhabdomyolysis to radiocontrast exposure, in addition to sepsis. Data from a systematic review found the FeNa was highly variable in patients with AKI, with estimates of FeNa <1% in the range of 33–100% across studies, translating into poor operative characteristics. In selected series of patients with hyperdynamic septic shock and severe AKI requiring RRT, the FeNa was <1% in all patients [17]. On the other hand, in a small cohort study, Brosius and Lau [20] demonstrated similar proportions of septic patients had a FeNa value <1% as did a value >1% (30 vs. 24%, p = NS). In this study, the FeNa was partially dependent on the timing of measures from onset of ARF with many converting from FeNa <1 to >1% on repeat testing. Overall, the studies in this review, individual patient data were inconsistently available and the proportion with FeNa <1% could only be assessed in 92 patients. Carvounis et al. [21] described a FeNa of 0.4, 2.1 and 8.9% for patients classified as PRA, PRA with diuretic exposure and ATN, respectively. This pattern was also reported in another small study where the average FeNa was 0.7 and 2.5% in presumed PRA and ATN, respectively [22]. Both these studies, however, are potentially limited due to the variable timing from onset of AKI and measurement of urinary biochemistry. Overall, the FeNa has been shown a poor predictor for need for RRT, renal recovery or mortality.

Fractional Excretion of Urea (FeU). A FeU <35% is cited as being consistent with PRA and a FeU >35% with ATN. The FeU has only been described in two studies that included septic patients [15, 21]. In one study, a FeU <35% was evident in 90, 89 and 4% for PRA, PRA with diuretics and ATN, respectively [21]. A FeU of ≤35% had greater sensitivity and specificity than a FeNa ≤1% in classifying AKI. However, there were notably limitations to this study. In the other study, a cohort of resuscitated critically ill patients with early non-oliguric septic AKI, Van Biesen et al. [15] described an average FeU <35% over 24 h despite the majority having received diuretics.

Other. Several additional urinary biochemical indices exist, including urinary sodium, urine/plasma creatinine ratio, serum urea/creatinine ratio, urine/serum urea ratio, urine uric acid/creatinine ratio, fractional excretion of uric acid, fractional excretion of chloride, and the renal failure index. Unfortunately, these tests are all of questionable value in the early diagnosis and classification of AKI. In this setting, they provide neither a quantitative measure of acute injury severity, nor any useful prognostic information regarding the likelihood of renal recovery, need for RRT, or overall mortality. A recent review of urinary biochemical indices in septic AKI similarly concluded that no single biochemical index was reliable in guiding the diagnosis, classification, or prognosis of AKI [17].

Unfortunately, no particular urinary biochemical test or index is reliable for the diagnosis, classification or prediction of the clinical course of septic AKI or identify patients with ATN. The major limitations to the current human studies are related to variable timing of insults, variable severity of insult, the lack of control groups, the presence of confounders and the lack of histopathological correlation [23].

Urine Microscopy in Septic AKI

Examination of the urinary microscopy is a long-held convention in AKI. Urine microscopy has inherent advantages in that it is inexpensive, readily available, and non-invasive. Clearly, urine microscopy has unquestionable diagnostic value for certain clinical conditions. For example, the observation of fragmented red blood cells or red blood cell casts in the urine is diagnostic of acute glomerulonephritis or vasculitis and can direct specific interventions. These conditions, however, are infrequently encountered in critically ill patients.

A common use of urinary microscopy is to attempt to categorize patients with AKI into one of two diagnostic categories: PRA and ATN. The classic urinary profile described in ATN contains renal tubular epithelial cells with coarse granular, muddy brown or mixed cellular casts, whereas the sediment in PRA is typically described as bland and sometimes revealing occasional hyaline or fine granular casts. The rationale for discriminating between PRA and ATN has largely centered on the idea that specific therapies and/or prognosis differ substantially between these syndromes. However, evidence establishing the diagnostic value of urine microscopy has largely been lacking.

Recently, Perazella et al. [7] performed a prospective observational study of urine microscopy in 267 consecutive hospitalized patients with a diagnosis of AKI. For each patient, a consultant nephrologist evaluated the probable diagnostic category of AKI (PRA or ATN) at two time points: first, at the time of initial clinical assessment (prior to urinary sediment examination) and, second, following patient discharge, kidney biopsy or patient death (final diagnosis). Each patient's urine sediment was analyzed and scored using a novel scale that ranged from 1 to 3 for the presence and number of red and white blood cells, renal tubular epithelial cells, and granular and hyaline casts. The final diagnosis was ATN in 125 patients (47%), PRA in 106 patients (40%) and 36 patients (13.5%) were excluded as having other causes of AKI. The urine sediment scoring system, based on the presence and quantity of granular casts and renal tubular epithelial cells, was found to be highly predictive for a final diagnosis of ATN. For example, a urine microscopy score of ≥2 (vs. score of 0) was associated with a 74-fold increase in the odds of a final diagnosis of ATN. In addition, the findings on urine microscopy led to a change of a pre-urine microscopy diagnosis of PRA to a final diagnosis of ATN in 27 patients (23%), and a change from ATN to PRA in 15 patients (14%). There were limitations to this study. For example, data were not provided on the primary contributing factors for AKI (i.e. sepsis, toxin, ischemia), about therapies received by patients (i.e. the specifics of fluid therapy, hemodynamic manipulation, vasopressor therapy and monitoring) or the impact of the final diagnosis – either PRA or ATN – on clinical outcomes such as peak serum creatinine level, duration of AKI, proportion receiving RRT, and renal recovery. While this study had limited practical application, it represented evidence to suggest there are detectable differences in the urine sediment of hospitalized patients with AKI, and that these differences may be useful for predicting severity of kidney injury and declines in function [24].

In a follow-up study using the same urine microscopy scoring system, Perazella et al. [25] evaluated the association between urine microscopy at the time of nephrology consultation in 249 hospitalized patients with AKI and worsening AKI, defined as progressing to a higher Acute Kidney Injury Network (AKIN) stage, need for RRT or death. Of these, 197 (79.1%) were categorized as PRA or ATN, with 40% classified as AKIN stage 1, 27% as stage 2, and 33% as stage 3. From the time of consultation, worsening AKI occurred in 40% of patients. A urinary microscopy score ≥3 was associated with a significantly higher adjusted relative risk of worsening AKI compared with a score of 0 (adjusted RR 7.3, 95% CI 4.5–9.7). These findings further suggest urinary microscopy may be further useful for predicting the clinical course of hospitalized patients with AKI and correlate with clinical outcomes.

Recently, the idea that PRA and ATN are distinct syndromes defined arbitrarily by changes occurring over 48 h has been challenged. In many circumstances, the distinction between PRA and ATN does not inform about the precipitating etiology of AKI. Rather, these syndromes in many clinical circumstances, likely exist on a continuum of acute injury and reflect varying severity (and duration) of AKI [26].

Numerous studies have shown urinary sediment for classifying AKI into PRA or ATN can be inconsistent, and often fails to correlate with traditional measures of urinary biochemistry or derived indices [7, 17]. For example, in the study by Perazella et al. [7], discordance between initial and final AKI diagnosis was not uncommon. For example, 23% of patients with a final diagnosis of PRA had evidence of varying numbers of renal tubular epithelial cells and granular casts (i.e. score of ≥2) in the urine sediment, whereas in those with a final diagnosis of ATN, 17% had no cells or casts (i.e. score of 1).

Regrettably, few studies have evaluated for differences in the urinary sediment between septic and non-septic AKI [17, 18]. Likewise, few experimental studies have evaluated the urinary microscopy findings in septic AKI [18]. In a systematic review of urinary findings in human septic AKI, urinary microscopy was described in only 7 studies enrolling 174 patients [17] (table 1). While muddy brown or epithelial cell casts, renal tubular cells, and variable trace hematuria and pyuria were commonly described (52–100%), normal microscopy was also reported, even 5 days after onset of septic AKI. Inferences from these studies are grossly limited largely due to urinary microscopy being performed at variable times during the course of AKI and not all studies provided individual data. However, in one small study, no significant differences in urinary microscopy were described when dichotomized by a FeNa value of <1 or >1% [20].

Urinary microscopy may prove useful for predicting the severity of kidney injury and clinical course in patients with AKI. However, there is a paucity of

Table 1. Summary of urinary microscopy and sediment from included articles reporting on urine in sepsis

Study (first author)	Proportion %	Description of urinary sediment
Chesney, 1981	52–74	Culture-negative pyuria (74%) with sheets of leukocytes; trace hematuria (52%)
Brosius, 1986	95	Renal tubular cells; cellular casts; pigmented casts; culture-negative pyuria; unexplained hematuria; normal microscopy in 1 patient
Diamond, 1982	100	Muddy brown casts; renal tubular cells; occasional RBC and WBC
Zager 1980	52	Brown granular casts; renal tubular cells; normal microscopy in 9 patients
Graber, 1991	57	Bubble cells described as large, bizarre single nucleated cells containing fluid filled vesicles; sediment also mixed with renal tubular cells; muddy brown casts, and oval fat bodies
Gay, 1997	88	Evidence of interstitial or tubular damage; normal microscopy in 1 patient
Marotto, 1997	81	Hematuria; no formal microscopy described

data on the diagnostic and predictive value of urinary microscopy in septic AKI.

Histopathological Correlation in Septic AKI

The pathophysiology of septic AKI remains incompletely understood. However, renal hypoperfusion, maladaptive neurohormonal activation and ischemia, followed by ATN, have long been promoted as central to its development [27]. Recent systematic reviews and experimental studies on septic AKI have challenged this paradigm [28, 29].

An experimental model of *Escherichia coli*-induced septic shock, characterized by a hyperdynamic circulatory state, described significantly increased renal vasodilatation and renal blood flow compared with baseline [29]. The lack of renal hypoperfusion, and thus global kidney ischemia observed in this study, would suggest the probably of ATN less likely; however, there is a dearth of

Table 2. Summary of human studies describing histopathology in septic AKI

Study (first author)	Population	Method	AKI (%)	ATN (%)
Hotchkiss, 1999	sepsis/septic shock	PM	12/20 (60)	1 (5)
Sato, 1978	sepsis	PM	6/6 (100)	1 (17)
Mustonen, 1984	sepsis/septic shock	biopsy	57/57 (100)	4 (17)
Rosenberg, 1971	sepsis	biopsy	1/1 (100)	0 (0)
Zappacosta, 1997	sepsis	biopsy	1/1 (100)	0 (0)
Diaz de Leon, 2006	severe sepsis	biopsy	107/332 (32)	20 (50)

knowledge about the histopathological correlation of AKI in sepsis. This was further demonstrated in a recent systematic review of studies describing the kidney histology in septic AKI, of which only 6 human and 14 experimental studies were found [30]. The human studies included only 77 patients, in whom only 7.8% shown histological evidence of ATN (table 2). While there were a variety of kidney morphologic changes described in septic AKI, the majority reported normal tubular histology rather than marked tubular necrosis consistent with ATN. These findings, coupled with the limited available data on urinary sediment in septic AKI, question the value of urinary microscopy for the diagnosis, classification and prognosis in septic AKI.

Conclusions

Septic AKI is increasingly encountered in critically ill patients, is associated with a more complex and prolonged course, has higher morbidity and mortality, and has higher attributable healthcare costs. Based on the available data, no specific urinary biochemical test or index is reliable for the diagnosis, classification or prediction of the clinical course of septic AKI. Urine microscopy may be readily advantageous and prove useful for predicting the severity of kidney injury and clinical course in patients with AKI, however there is currently a limited amount of data on its diagnostic and prognostic value in septic AKI. The majority of studies describing the urinary findings in septic AKI are plagued by limitations related to variable timing between insult and urinary evaluation, variable severity of insult, the lack of control groups, the presence of confounders and the lack of histopathological correlation. Additional research is clearly needed to better define the role and value of urinary biochemistry and urinalysis in septic AKI.

References

1 Uchino S, Kellum JA, Bellomo R, Doig GS, Morimatsu H, Morgera S, Schetz M, Tan I, Bouman C, Macedo E, et al: Acute renal failure in critically ill patients: a multinational, multicenter study. JAMA 2005;294:813–818.

2 Bagshaw SM, Uchino S, Bellomo R, Morimatsu H, Morgera S, Schetz M, Tan I, Bouman C, Macedo E, Gibney N, et al: Septic acute kidney injury in critically ill patients: clinical characteristics and outcomes. Clin J Am Soc Nephrol 2007;2:431–439.

3 Bagshaw SM, George C, Bellomo R: Early acute kidney injury and sepsis: a multicentre evaluation. Crit Care 2008;12:R47.

4 Bellomo R, Ronco C, Kellum JA, Mehta RL, Palevsky P: Acute renal failure – definition, outcome measures, animal models, fluid therapy and information technology needs: the Second International Consensus Conference of the Acute Dialysis Quality Initiative (ADQI) Group. Crit Care 2004;8:R204–R212.

5 Espinel CH: The FENa test. Use in the differential diagnosis of acute renal failure. JAMA 1976;236:579–581.

6 Miller TR, Anderson RJ, Linas SL, Henrich WL, Berns AS, Gabow PA, Schrier RW: Urinary diagnostic indices in acute renal failure: a prospective study. Ann Intern Med 1978;89:47–50.

7 Perazella MA, Coca SG, Kanbay M, Brewster UC, Parikh CR: Diagnostic value of urine microscopy for differential diagnosis of acute kidney injury in hospitalized patients. Clin J Am Soc Nephrol 2008;3:1615–1619.

8 Martin GS, Mannino DM, Eaton S, Moss M: The epidemiology of sepsis in the United States from 1979 through 2000. N Engl J Med 2003;348:1546–1554.

9 Vincent JL, Sakr Y, Sprung CL, Ranieri VM, Reinhart K, Gerlach H, Moreno R, Carlet J, Le Gall JR, Payen D: Sepsis in European intensive care units: results of the SOAP study. Crit Care Med 2006;34:344–353.

10 Hoste EA, Lameire NH, Vanholder RC, Benoit DD, Decruyenaere JM, Colardyn FA: Acute renal failure in patients with sepsis in a surgical ICU: predictive factors, incidence, comorbidity, and outcome. J Am Soc Nephrol 2003;14:1022–1030.

11 Oppert M, Engel C, Brunkhorst FM, Bogatsch H, Reinhart K, Frei U, Eckardt KU, Loeffler M, John S: Acute renal failure in patients with severe sepsis and septic shock – a significant independent risk factor for mortality: results from the German Prevalence Study. Nephrol Dial Transplant 2008;23:904–909.

12 Bagshaw SM, Laupland KB, Doig CJ, Mortis G, Fick GH, Mucenski M, Godinez-Luna T, Svenson LW, Rosenal T: Prognosis for long-term survival and renal recovery in critically ill patients with severe acute renal failure: a population-based study. Crit Care 2005;9:R700-R709.

13 Cruz D, Bolgan I, Perazella MA, Bonello M, de Cal M, Corradi V: North East Italian Prospective Hospital Renal Outcome Survey on Acute Kidney Injury (NEiPHROS-AKI): targeting the problem with the RIFLE criteria. Clin J Am Soc Nephrol 2007;2:418–425.

14 Bagshaw SM, Lapinsky S, Dial S, Arabi Y, Dodek P, Wood G, Ellis P: Acute kidney injury in septic shock: clinical outcomes and impact of duration of hypotension prior to initiation of antimicrobial therapy. Intensive Care Med 2009;35:871–881.

15 Van Biesen W, Yegenaga I, Vanholder R, Verbeke F, Hoste E, Colardyn F, Lameire N: Relationship between fluid status and its management on acute renal failure in intensive care unit patients with sepsis: a prospective analysis. J Nephrol 2005;18:54–60.

16 Payen D, de Pont AC, Sakr Y, Spies C, Reinhart K, Vincent JL: A positive fluid balance is associated with a worse outcome in patients with acute renal failure. Crit Care 2008;12:R74.

17 Bagshaw SM, Langenberg C, Bellomo R: Urinary biochemistry and microscopy in septic acute renal failure: a systematic review. Am J Kidney Dis 2006;48:695–705.

18 Bagshaw SM, Langenberg C, Wan L, May CN, Bellomo R: A systematic review of urinary findings in experimental septic acute renal failure. Crit Care Med 2007;35:1592–1598.

19 Uchino S, Bellomo R, Bagshaw SM, Goldsmith D: Transient azotaemia is associated with a high risk of death in hospitalized patients. Nephrol Dial Transplant 2010 (in press).
20 Brosius FC, Lau K: Low fractional excretion of sodium in acute renal failure: role of timing of the test and ischemia. Am J Nephrol 1986;6:450–457.
21 Carvounis CP, Nisar S, Guro-Razuman S: Significance of the fractional excretion of urea in the differential diagnosis of acute renal failure. Kidney Int 2002;62:2223–2229.
22 Du Cheyron D, Daubin C, Poggioli J, Ramakers M, Houillier P, Charbonneau P, Paillard M: Urinary measurement of Na^+/H^+ exchanger isoform 3 (NHE3) protein as new marker of tubule injury in critically ill patients with ARF. Am J Kidney Dis 2003;42:497–506.
23 Rosen S, Heyman SN: Difficulties in understanding human 'acute tubular necrosis': limited data and flawed animal models. Kidney Int 2001;60:1220–1224.
24 Kanbay M, Kasapoglu B, Perazella MA: Acute tubular necrosis and pre-renal acute kidney injury: utility of urine microscopy in their evaluation – a systematic review. Int Urol Nephrol 2009 (in press).
25 Perazella MA, Coca SG, Hall IE, Iyanam U, Koraishy M, Parikh CR: Urine microscopy is associated with severity and worsening of acute kidney injury in hospitalized patients. Clin J Am Soc Nephrol 2010 (in press).
26 Bagshaw SM, Gibney N: Clinical value of urine microscopy in acute kidney injury. Nat Rev Nephrol 2009;5:1–2.
27 Schrier RW, Wang W: Acute renal failure and sepsis. N Engl J Med 2004;351:159–169.
28 Langenberg C, Bellomo R, May C, Wan L, Egi M, Morgera S: Renal blood flow in sepsis. Crit Care 2005;9:R363–R374.
29 Langenberg C, Wan L, Egi M, May CN, Bellomo R: Renal blood flow in experimental septic acute renal failure. Kidney Int 2006;69:1996–2002.
30 Langenberg C, Bagshaw SM, May CN, Bellomo R: The histopathology of septic acute kidney injury: a systematic review. Crit Care 2008;12:R38.

Dr. Sean M. Bagshaw
Division of Critical Care Medicine, University of Alberta Hospital
3C1.12 Walter C. Mackenzie Centre
8440-122 Street, Edmonton, Alta T6G2B7 (Canada)
Tel. +1 780 407 6755, Fax +1 780 407 1228, E-Mail bagshaw@ualberta.ca

Ronco C, Bellomo R, McCullough PA (eds): Cardiorenal Syndromes in Critical Care.
Contrib Nephrol. Basel, Karger, 2010, vol 165, pp 284–291

Recovery from Acute Kidney Injury: Determinants and Predictors

Nattachai Srisawat[a] · Raghavan Murugan[a] · Xiaoyan Wen[a] · Kai Singbartl[a] · Gilles Clermont[a] · Somchai Eiam-Ong[b] · John A. Kellum[a]

[a]The CRISMA (Clinical Research, Investigation, and Systems Modeling of Acute Illness) Laboratory, Department of Critical Care Medicine, University of Pittsburgh School of Medicine, Pittsburgh, Pa., USA, and [b]Division of Nephrology, Department of Medicine, Faculty of Medicine, Chulalongkorn University, Bangkok, Thailand

Abstract

Predicting recovery of renal function following acute kidney injury (AKI) is one of the top ten questions in the field of AKI research. Accurate prediction would help physicians distinguish patients with poor renal prognosis in whom further therapy is likely to be futile from those who are likely to have good renal prognosis. Proper stratification of patients with AKI is also critical to design clinical trials to target patients with poor prognosis. Unfortunately, current general clinical severity scores (APACHE, SOFA, etc.) and AKI-specific severity scores (Mehta's score, Liano's score, Chertow's score, etc.) are not the good predictors of renal recovery. Recent progress on the pathophysiology of renal injury and recovery is encouraging. Repopulation of surviving renal tubular epithelial cell with the assistance of certain renal epithelial cell and specific growth factors such as neutrophil gelatinase-associated lipocalin (NGAL), hepatocyte growth factor (HGF), epidermal growth factor, and insulin-like growth factor-1, etc., play a major role in the recovery process. Such findings provide a great opportunity to test and validate these potential biomarkers as candidate markers of renal recovery. This review will describe the current understanding of the renal recovery process, and the role of clinical severity scores and novel biomarkers such as NGAL, HGF, and cystatin C in predicting renal recovery.

The prevalence of renal recovery varies by the definition and the study population (ICU vs. non-ICU). Results from the Beginning and Ending Supportive Therapy (BEST) Kidney Study group, a large multicenter, multinational, prospective observational study in critically ill patients, showed that 13.8% of surviving patients with AKI requiring renal replacement therapy (RRT) still required it at hospital discharge [1]. A recent, large multicenter clinical trial of intensive versus less intensive renal support for patients with acute kidney injury (AKI) found that only 16% of patients did not require RRT at discharge to home. Moreover, only 25% of the alive participants were dialysis-independent on day 60 and there was no difference between the two treatment strategies in terms of renal recovery [2]. Non-recovery of renal function can have tremendous negative effects on the quality of life and healthcare costs. Therefore, treatments that hasten renal recovery or shorten the duration of AKI are eagerly sought off.

Currently there are no effective treatment strategies to improve renal recovery and one of the most important barriers has been the inability to risk-stratify patients in terms of risk for non-recovery of renal function. During the past few years, several novel urine biomarkers such as neutrophil gelatinase-associated lipocalin (NGAL), cystatin C, and kidney injury molecule-1 (KIM-1) which represent tubular epithelial cell damage, have been proposed and studied for early diagnosis of AKI in various settings (post-cardiac surgery, post-transplant, contrast administration, and sepsis). Hopefully, the emergence of these novel biomarkers will be useful to predict renal recovery. For example, we have preliminarily shown that plasma NGAL can predict recovery after sepsis-induced AKI with an area under the ROC curve of 0.74 [3].

Why We Need to Predict Recovery of Renal Function?

Evidence from the epidemiological study has pointed out that AKI is an important risk factor for ESRD. The ability to predict renal recovery could help identify high-risk patients and possibly lead to the start of specific interventions such as angiotensin-converting enzyme inhibitors. Furthermore, this information will help to determine the duration of follow-up (longer time follow-up in patients who have high risk of chronic kidney disease after AKI). Identifying poor prognosis patients can also help to target the patients who will receive the most benefit from early RRT (fig. 1). We have summarized the benefits of understanding and predicting renal recovery in table 1.

Physiology of Renal Recovery

The pathophysiology of AKI appears to involve a complex interplay between tubular injury, renal hemodynamics, and inflammation. After the acute insult to tubular epithelial cells, there are three main types of cells which contribute to

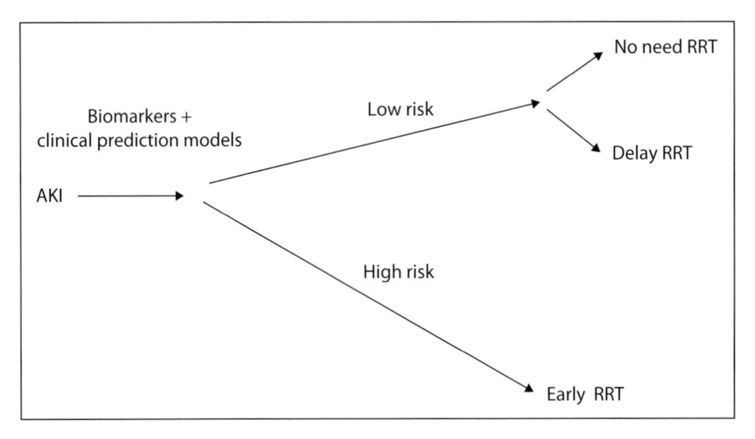

Fig. 1. Schematic diagram of using biomarkers and clinical prediction models to define AKI patients who require early RRT.

Table 1. Potential benefits of predicting renal recovery

– Identification of patients who will receive the most benefit from interventions
– Determine the optimal timing to initiate intervention (e.g. early vs. late initiation of RRT)
– Design type of intervention (e.g. use of CRRT vs. intermittent dialysis)
– Guide the duration of follow-up
– Initiation of intervention or medications to slow deterioration of renal function (e.g. angiotensin-converting enzyme inhibitor, angiotensin receptor blocker, statins, antioxidants)
– Determine the quality of life after hospital discharge and preparing for long-term RRT

renal repair: surviving renal tubular epithelial cells, renal specific stem cells, and mesenchymal stem cells. Most of the current evidence indicates that surviving renal proximal tubular epithelial cells (RPTCs) play a major role in the repair process and appear to follow a program which shares a number of similarities with the series of events observed during kidney development. RPTCs first appear to undergo dedifferentiation (i.e., loss of apical-basal polarity, lack of tight junctions, accompanied with a decrease in the expression of epithelial cell marker such as N-cadherin, E-cadherin, ZO-1 and an increase in the expression of mesenchymal cell or fibroblast markers such as vimentin, α-SMA, FSP1), and then proliferation. When the cell population has expanded sufficiently to physically replenish sloughed epithelium, cells undergo a redifferentiation process characterized by a decrease in mesenchymal cell markers and increase in epithelial cell markers to finally restore the physiologic function of RPTCs [4].

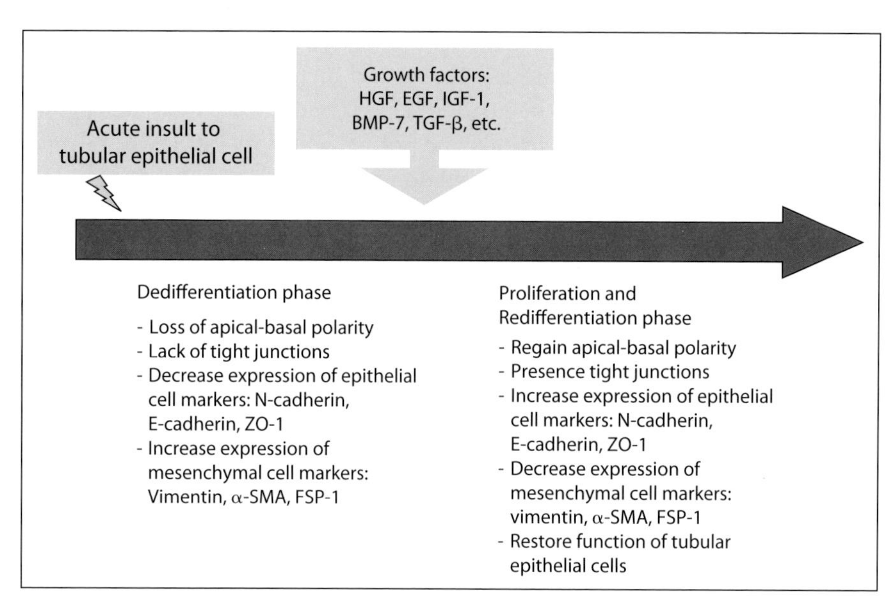

Fig. 2. Process of renal recovery after acute insult to tubular epithelial cell.

A number of growth factors such as the hepatocyte growth factor (HGF), epidermal growth factor (EGF), insulin-like growth factor-1 (IGF-1), bone morphogenic protein-7 (BMP-7), and the transforming growth factor-β (TGF-β) help renal repair in the process of dedifferentiation by interaction with the transmembrane receptors of the tubular epithelial cells. The consequence of this interaction leads to the induction of a number of signaling pathways including mitogen-activated protein kinase (MAPK), phosphoinositide-3-kinase (PI3-K), JAK/STAT and wnt/β-catenin pathway). The important effect of these signaling pathways and appropriate activation or inhibition causes migration of the dedifferentiated tubular epithelial cell to the injury area, cellular proliferation, promoting adhesion of tubular cells to the basement membrane (prevent cellular sloughing), anti-apoptosis, and anti-fibrotic effects in the late phase of recovery [5] (fig. 2).

Ultimately, the renal repair process requires redifferentiation of transformed renal epithelial cell function to restore tubular morphology. Evidence from in vitro studies showed HGF stimulated tubulogenesis by extending spindle-shaped fibroblasts and subsequently these cells regained apicobasolateral polarization and lumen formation by the expression of matrix metalloproteinase (MMP)-13. In addition, during the initial phase of renal injury, there is a loss of tubular epithelial cell polarity and the mislocalization of Na^+/K^+-ATPase from the basolateral surface to the apical surface.

Renal hemodynamic derangement also plays an important role in the pathophysiology of AKI. In a constitutive state, the endothelium regulates migration of inflammatory cells to the areas of injury via upregulating adhesion molecules such as L-selectin and E-selectin, and increasing vascular tone/permeabilitiy, and prevention of coagulation. Upon injury, the endothelium loses its ability to regulate these functions and subsequently its influence on renal function [6]. However, currently, we still do not know the specific mechanisms of endothelial repair and recovery.

Defining Renal Recovery

Before discussing the clinical predictors of renal recovery, we need to understand the definition of renal recovery. In 2004, the Acute Dialysis Quality Initiative (ADQI) proposed the following definition of renal recovery: 'complete renal recovery' was defined as return to pre-morbid renal function, and 'partial renal recovery' was defined as a change in RIFLE classification (R, I, or F) but not requiring RRT'. Non-recovery was therefore defined as persistent requirement of RRT, or no change in RIFLE score during hospitalization. Most of the studies evaluating AKI non-recovery have studied only patients receiving RRT. Therefore, AKI non-recovery was only defined as patients who are alive and dialysis-dependent. However, patients who die while still on dialysis should also be considered as AKI non-recovery.

The optimal duration of follow-up for patients with AKI needs to be defined. Some studies assessed the outcome 30, 60 or 90 days after hospital discharge. The consensus from the second ADQI conference suggests using from 60 to 90 days of follow-up for the evaluation of all-cause mortality. Finally, in the case of patients who survive AKI but still have impairment of renal function, which parameter should be assessed to define renal function? Serum creatinine might not be a good marker to use due to the decreasing of muscle mass after the severe illness and other confounding factors. Therefore, the measurement of glomerular filtration rate by radioisotope clearance might be an alternative.

In summary, renal recovery could be defined as complete or partial recovery based on returning of renal function, while non-recovery should include patients who do not survive or who require chronic dialysis. However, we still need further studies to find out the optimal time point to assess the AKI outcome.

Clinical Predictors of Renal Recovery

One of the current tools for predicting renal recovery is the clinical severity scores. We can classify severity scores into two types: general illness severity scores and AKI-specific severity scores. Acute Physiology and Chronic

Health Evaluation II (APACHE II), Simplified Acute Physiology Score (SAPS), Mortality Probability Model (MPM), Logistic Organ Dysfunction Score (LODS), Multiple Organ Dysfunction Score (MODS), and Sequential Organ Failure Assessment (SOFA) are the examples of current general illness clinical severity scores which are widely used, while AKI-specific scores include the SHARF-II score and scores proposed by Mehta, Liano, Chertow and Paganini. Recently, Uchino et al. [7] tested two general illness severity scores (APACHE II and SOFA), and four AKI-specific severity scores in 1,742 patients as part of the BEST Kidney Study. Unfortunately, none of these scoring systems tested had a high level of discrimination or calibration to predict outcome for AKI patients.

Thus, although not specifically tested, it seems unlikely that any of the current severity scores can provide good prediction for renal recovery. A more complete prediction model should include both mortality and non-mortality renal outcomes such as persistent AKI or dialysis requirement. Although factors such as advanced age, male gender, presence of sepsis/septic shock, oliguria, hypotension, respiratory failure, use of mechanical ventilation, presence of oliguria, septic shock, and high serum bilirubin, low serum creatinine, using vasoactive substances are risk factors for AKI, we do not have a study that is directly designed to identify the factors associated with renal recovery. Moreover, limitations of previous AKI-specific severity scores such as the fact that most of them were single-center designs and involved less than 700 patients makes them less generalizable. Therefore, study designs for the future should be multicentered and include a large cohort of patients.

Novel Biomarkers for Predicting Renal Recovery

Based on the physiology of renal recovery, a number of biomarkers have the potential to predict renal recovery.

Neutrophil Gelatinase-Associated Lipocalin (NGAL)

NGAL is one of the biomarkers which was extensively studied in the field of AKI. We still do not know the exact origin of urinary NGAL. Does it originate from systemic circulation, or is it locally synthesized by injured tubular epithelial cell, or is it from infiltrating inflammatory cells? In the late phase of AKI, NGAL however plays a role as a growth and differentiation factor for restoring tubular epithelial function with the assistance of siderophore-iron complexes. However, most of the previous studies have tested plasma/urine NGAL as a marker for early diagnosis of AKI, and only a few studies which have tried to test a plasma/urine NGAL as a prognosticate marker of clinical outcome. In the setting of post-cardiac surgery, Wagener et al. [8], Koyner et al. [9] and Bennett et al. [10] have reported that urinary NGAL at the cut-points 190, 150, and 480 ng/ml, respectively, have the sensitivity/specificity of about 81.3/42.9, 100/93.85, and 75/50%, respectively, for predicting in-hospital mortality. In a recent study, Cruz et al. [11] studied critically ill patients and found plasma NGAL at the cut-point of 150 ng/ml which had a sensitivity and specificity 59.6 and 67.5%. However, the number of patients who reached the end-point (mortality) from these studies was relatively small.

Hepatocyte Growth Factor
HGF has been demonstrated to have multipotent effects in the kidney including mitogenic, morphogenic, and differentiating effects. In animal models of toxic and ischemic ATN, HGF therapy markedly accelerated renal recovery. However, there is only a small single-center study which demonstrated upregulation of urinary HGF during the acute phase of AKI and a gradual decline with recovering from AKI [12].

Cystatin C
Cystatin C is a non-glycosylated 13-kDa basic protein which is a member of the cystatin superfamily of cysteine protease inhibitors, regularly produced by all nucleated cells, freely filtered at the glomerulus, 100% reabsorbed at proximal tubule via megalin-mediated endocytosis, and catabolized. Therefore, it is not normally found in urine. Ketamura et al. [13] found that overexpression of cystatin C can increase the MHC class II expression and T-cell response via suppression of cathepsin activity. In the serum, cystatin C is moderately useful for predicting the RRT requirement (AUC = 0.76) but it was minimally useful for predicting death (AUC = 0.62) [14]. In urine, Herget-Rosenthal et al. [15] showed cystatin C, α-microglobulin, retinal-binding protein, and *N*-acetyl-β-D-glucosaminidase are the good biomarkers for predicting the RRT requirement which had AUC = 0.92, 0.86, and 0.80, respectively. Moreover, *N*-acetyl-β-D-glucosaminidase was also a good marker for predicting composite outcome of death or for predicting the RRT requirement (AUC = 0.71). Unfortunately, the accuracy of these biomarkers to predict recovery in patients with established AKI was not tested.

Conclusions

Understanding renal recovery is one of the essential parts in improving the outcome of AKI. Unfortunately, the existing clinical risk prediction for renal recovery is limited, at best. Recent knowledge of renal recovery with the emergence of novel biomarkers could be one of the new prognostic tools to solve this issue, but further work is needed to establish reliable test characteristics.

References

1 Uchino S, Kellum JA, Bellomo R, et al: Acute renal failure in critically ill patients: a multinational, multicenter study. JAMA 2005;294:813–818.
2 Palevsky PM, Zhang JH, O'Connor TZ, Chertow GM, Crowley ST, Choudhury D, Finkel K, Kellum JA, Paganini E, Schein RM, et al: Intensity of renal support in critically ill patients with acute kidney injury. N Engl J Med 2008;359:7–20.
3 Srisawat N, Kong L, Carter M, Murugan R, Singbartl K, Angus DC, Kellum JA, for the GenIMS Investigators: Plasma neutrophil gelatinase-associated lipocalin predicts renal recovery from sepsis-induced acute kidney injury. J Am Soc Nephrol 2009;20:355A.
4 Ishibe S, Cantley LG: Epithelial-mesenchymal-epithelial cycling in kidney repair. Curr Opin Nephrol Hypertens 2008;17:379–385.
5 Liu KD, Brakeman PR: Renal repair and recovery. Crit Care Med 2008;36(suppl):S187–S192.
6 Sutton TA, Fisher CJ, Molitoris BA: Microvascular endothelial injury and dysfunction during ischemic acute renal failure. Kidney Int 2002;62:1539–1549.
7 Uchino S, Bellomo R, Morimatsu H, Morgera S, Schetz M, Tan I, Bouman C, Macedo E, Gibney N, Tolwani A, et al: External validation of severity scoring systems for acute renal failure using a multinational database. Crit Care Med 2005;33:1961–1967.
8 Wagener G, Gubitosa G, Wang S, Borregaard N, Kim M, Lee HT: Urinary neutrophil gelatinase-associated lipocalin and acute kidney injury after cardiac surgery. Am J Kidney Dis 2008;52:425–433.
9 Koyner JL, Bennett MR, Worcester EM, Ma Q, Raman J, Jeevanandam V, Kasza KE, O'Connor MF, Konczal DJ, Trevino S, et al: Urinary cystatin C as an early biomarker of acute kidney injury following adult cardiothoracic surgery. Kidney Int 2008;74:1059–1069.
10 Bennett M, Dent CL, Ma Q, Dastrala S, Grenier F, Workman R, Syed H, Ali S, Barasch J, Devarajan P: Urine NGAL predicts severity of acute kidney injury after cardiac surgery: a prospective study. Clin J Am Soc Nephrol 2008;3:665–673.
11 Cruz DN, de Cal M, Garzotto F, Perazella MA, Lentini P, Corradi V, Piccinni P, Ronco C: Plasma neutrophil gelatinase-associated lipocalin is an early biomarker for acute kidney injury in an adult ICU population. Intensive Care Med 2010;36:444–451.
12 Taman M, Liu Y, Tolbert E, Dworkin LD: Increase urinary hepatocyte growth factor excretion in human acute renal failure. Clin Nephrol 1997;48:241–245.
13 Kitamura H, Kamon H, Sawa S, Park SJ, Katunuma N, Ishihara K, Murakami M, Hirano T: IL-6-STAT3 controls intracellular MHC class II αβ dimer level through cathepsin S activity in dendritic cells. Immunity 2005;23:491–502.
14 Coca SG, Yalavarthy R, Concato J, Parikh CR: Biomarkers for the diagnosis and risk stratification of acute kidney injury: a systematic review. Kidney Int 2008;73:1008–1016.
15 Herget-Rosenthal S, Poppen D, Husing J, Marggraf G, Pietruck F, Jakob HG, Philipp T, Kribben A: Prognostic value of tubular proteinuria and enzymuria in non-oliguric acute tubular necrosis. Clin Chem 2004;50:552–558.

John A. Kellum, MD, FCCM
608, Scaife Hall, Department of Critical Care Medicine, University of Pittsburgh
3550 Terrace Street, Pittsburgh, PA 15261 (USA)
Tel. +1 412 647 7810, Fax +1 412 647 8060
E-Mail kellumja@ccm.upmc.edu

Ronco C, Bellomo R, McCullough PA (eds): Cardiorenal Syndromes in Critical Care.
Contrib Nephrol. Basel, Karger, 2010, vol 165, pp 292–298

Dysnatremias in the Intensive Care Unit

Mitchell H. Rosner[a] · Claudio Ronco[b]

[a]Division of Nephrology, University of Virginia Health System, Charlottesville, Va., USA, and
[b]Department of Nephrology, Dialysis & Transplantation, International Renal Research Institute, San Bortolo Hospital, Vicenza, Italy

Abstract

Dysnatremias (hypo- and hypernatremia) are common in patients admitted to the intensive care unit (ICU) with a prevalence approaching 20–30% in some studies. Recent data reveals that both hypo- and hypernatremia present on admission to or developing in the ICU are independent risk factors for poor prognosis. The origin of hypernatremia in the ICU is often iatrogenic and due to inadequate free water replacement of ongoing water losses. The pathogenesis of hyponatremia in the ICU is more complicated but often is related to the combination of dysregulated arginine vasopressin production and concomitant inappropriate hypotonic fluid administration. Both the dysnatremia itself and the treatment of the electrolyte disturbance can be associated with morbidity and mortality making careful monitoring for and treatment of sodium disorders an imperative in the critically ill patient. Formulae have been devised to guide the therapy of severe hypo- and hypernatremia, but these formulae regard the patient as a closed system and do not take into account ongoing fluid losses that can be highly variable. Thus, a cornerstone of proper therapy is serial measurements of serum and urine electrolytes. The appropriate use of hypertonic (3%) saline in the treatment of hyponatremic encephalopathy has also shown to be very effective and the use of this therapy is reviewed here. Vasopressin receptor antagonists have also been shown to be effective at increasing serum sodium levels in patients with either euvolemic or hypervolemic hyponatremia and represent another therapeutic option. Recent data demonstrates that proper correction of hyponatremia is associated with improved short- and long-term outcomes.

Sodium disturbances (dysnatremias) leading to hyponatremia and hypernatremia are a common problem in adult patients admitted to hospital and intensive care units (ICUs). In fact, the majority of these cases develop after the patient is admitted to the ICU. Because of their incapacitation, lack of access to free water, reliance on intravenous fluid and nutritional support, and the usually serious nature of their underlying disease which often leads to impaired renal water handling, patients in the ICU are at high risk of developing sodium disturbances. In many cases, the occurrence of the dysnatremia acquired in the ICU is largely preventable. Patients in the ICU are well monitored and blood samples are taken frequently. Furthermore, the maintenance of fluid and electrolyte balance is one of the focal points of critical care. However, the development of sodium disorders may be insidious and identification of these electrolyte disturbances may be difficult for clinicians preoccupied with more acute medical issues or other laboratory investigations. The imperative to develop strategies to detect, prevent or correct ICU-acquired sodium disorders is made clearer by a series of recent reports demonstrating a strong, independent correlation between the development of a dysnatremia and poor outcomes.

This review focuses on the epidemiology, impact, pathophysiology and treatment of sodium disorders acquired in the ICU.

Epidemiology

Several studies have looked at the incidence of abnormal serum sodium values on admission to the hospital or ICU. A recent study from 77 ICUs in Austria over a period of 10 years demonstrated that 75.4% of patients had a normal ($135 \leq Na \leq 145$ mmol/l) serum sodium on ICU admission [1]. The frequencies of borderline ($130 \leq Na < 135$ mmol/l), mild ($125 \leq Na < 130$ mmol/l) and severe ($Na < 125$ mmol/l) were 13.8, 2.7 and 1.2% respectively. The frequencies of borderline ($145 < Na \leq 150$ mmol//l), mild ($150 < Na \leq 155$ mmol/l) and severe hypernatremia ($Na > 155$ mmol/l) were 5.1, 1.2 and 0.6% respectively. Comparing this with the incidence of development of ICU-acquired hyponatremia ($Na < 133$ mmol/l) of 11% or hypernatremia ($Na > 145$ mmol/l) of 26% seen in a recent study, it becomes clear that the majority of sodium disturbances are acquired once the patient enters the hospital [2].

Abnormal serum sodium values have also been shown to be a powerful risk factor for both morbidity and mortality. Compared with patients with normal serum sodium values, hospital mortality was increased in patients with ICU-acquired hyponatremia (16 vs. 28%, $p < 0.001$) and ICU-acquired hypernatremia (16 vs. 34%, $p < 0.001$) [2]. In a recent study of 98,411 adults it was found that the risk for in-hospital, 1-year and 5-year mortality was highly related to the occurrence as well as the degree of hyponatremia [3]. Furthermore, this association between hyponatremia and mortality is particularly evident in certain subpopulations such as patients with pneumonia, congestive heart failure or cirrhosis. Those patients developing hyponatremia have longer length of hospital stays, increased risk of requiring an ICU admission and increased total cost per hospital admission [4].

These studies also identified patient characteristics that are associated with the development of sodium disturbances and could be used to help clinicians identify patients at increased risk. An elevated baseline creatinine is associated with a 50% increased risk of ICU-acquired hypernatremia and may be a marker for impaired renal sodium and water regulation or decreased intravascular volume [2]. Mechanical ventilation is also highly associated with ICU-acquired hypernatremia. Certainly, the need for mechanical ventilation is a marker for illness severity but it also inhibits patient-clinician communication and makes patients dependent upon others for their water needs. Other risk factors for both hypo- and hypernatremia include: length of ICU stay and increasing severity of illness (as measured by APACHE II scores).

Hyponatremia

For hyponatremia to develop, a relative excess of water in conjunction with an underlying condition that impairs the kidney's ability to excrete water is required. Stimuli for the release of arginine vasopressin (AVP) and hence the impairment of water excretion are so frequent in hospitalized patients, especially those in the ICU, that virtually all patients are at risk of hyponatremia. This is especially true in the postoperative period when non-osmotic stimuli such as nausea, pain, narcotics, stress, and volume depletion lead to higher AVP levels compared with preoperative values. Thus, the most important factor resulting in hospital-acquired hyponatremia is the administration of hypotonic fluids to a patient with impaired urine-diluting capacity. While a healthy adult male can excrete over 15 liters of free water a day and maintain sodium homeostasis, it has been found that in previously healthy women in the postoperative setting as little as 3–4 liters of hypotonic fluid per day can result in fatal hyponatremic encephalopathy due to excessive AVP levels and impairment in free water excretion [5]. Hyponatremia can even develop if excessive isotonic solutions are administered if urine tonicity is higher than the infused fluid. Thus, to prevent hyponatremia, serum electrolytes should be checked daily (or more frequently in some cases) in patients receiving intravenous fluids and isotonic saline should be used, unless clinically otherwise indicated. Intravenous fluids should be considered as pharmacological agents with specific indications and contraindications. Hyponatremia constitutes a contraindication to the administration of hypotonic fluids.

Hyponatremia can be asymptomatic, which is usually the case for chronic hyponatremia secondary to cirrhosis or heart failure. Symptoms of hyponatremia develop because of osmotic swelling of the intracellular space as extracellular tonicity decreases. Hyponatremic encephalopathy can present with headaches, nausea, and vomiting. However, in the ventilated and sedated ICU patient, these symptoms will not be apparent and worsening of brain edema may lead to decreased mental status, seizures, coma, brainstem herniation, respiratory arrest, and death. Hyponatremic encephalopathy accounts for 30%

of new-onset seizures encountered in the ICU setting [6]. The combination of hyponatremia with hypoxemia is particularly dangerous as the lack of oxygen further impairs the ability of the brain to adapt to the osmotic changes and lead to a vicious cycle of encephalopathy.

While symptomatic acute hyponatremia is a life-threatening medical emergency, the treatment of hyponatremia may be life-threatening as well, as it carries the risk of the osmotic demyelination syndrome (ODS). This complication occurs if correction takes place too fast as brain cells require time to adapt to changes in the osmotic environment. The pons is particularly vulnerable to this type of injury leading to symptoms of quadriparesis.

An assessment of the time frame in which hyponatremia developed can guide the time frame needed for its correction. Most reported cases of ODS have complicated treatment in patients who developed hyponatremia outside of the hospital and there are rare cases reported in patients treated for hyponatremia that developed during their hospitalization. Thus, rapid correction of the serum sodium seems to be the most beneficial and less dangerous in patients who develop acute hyponatremia in the hospital while the danger of ODS increases with the duration of hyponatremia (particularly >48 h in duration). Regardless of the numeric serum sodium levels, treatment with hypertonic (3%) saline should be reserved for patients with symptomatic hyponatremia, e.g. with hyponatremic encephalopathy. A rise in the serum sodium of about 3 mEq/l to a total of 4–6 mEq/l may terminate seizures; once the symptoms resolve, the remainder of the sodium deficit should be corrected slowly. An assumption that can be used to guide initial therapy is that an infusion of 1 ml/kg of 3% saline (514 mEq Na/l) will raise the serum sodium by approximately 1 mEq/l. Alternatively, a bolus of 100 ml of 3% saline can be given and repeated up to 2–3 times until seizure activity ceases. Any patient receiving 3% saline should have the serum sodium checked at least every 2 h to guide therapy.

There has been considerable controversy about what constitutes an optimal rate of correction of the serum sodium. A conservative approach is to correct by no more than 8 mEq/l per day or 15–20 mEq/l over 48 h [7]. Several formulae have been recommended to help determine a proper infusion rate of hypertonic or isotonic saline to treat hyponatremia [7]. It is critical to understand that all formulae consider patients as closed systems and do not take into account ongoing water and electrolyte losses that may vary considerably between patients. The following formula proposed by Adrogué and Madias [8], which estimates the effect of any infusate on serum sodium, is a useful starting point in therapy:

$$\text{Change in serum Na}^+ = \frac{\text{infusate Na}^+ - \text{serum Na}^+}{\text{total body water} + 1}$$

Despite the usefulness of this formula, there can occasionally be considerable differences between anticipated and achieved serum sodium values stressing the need for serial measurements of serum sodium in any patient receiving hypertonic saline. It should also be realized that the serum sodium may rise unexpectedly rapidly not only with hypertonic saline but also with normal saline. This may occur if there was unrecognized volume depletion with sudden removal of the AVP stimulus during treatment with resulting rapid water diuresis. Patients with psychogenic polydipsia who drink themselves into a hyponatremic coma are also at high risk for overcorrection of hyponatremia through massive water diuresis and in such a case administration of hypotonic fluids and intravenous desmopressin (dDAVP) may be indicated to slow the rise of the serum sodium. Desmopressin will increase urinary concentration and reduce free water losses.

Recently, selective antagonists of the arginine vasopressin receptor (V2 receptor) in the collecting tubule of the kidney have been approved for the treatment of euvolemic or hypervolemic hyponatremia. These drugs include agents such as tolvaptan, conivaptan, and lixivaptan ('vaptans'). Administration of these drugs leads to an increase in free water excretion with little or no sodium loss [9]. In studies of these agents in patients with euvolemic or hypervolemic hyponatremia, a significant rise in serum sodium levels is seen in a fairly predictable manner and much less than 8 mEq/day [9]. These drugs have not been specifically studied in ICU patients but they represent another therapeutic option for correction of hyponatremia in a controlled and safe fashion. It is important to realize that the underlying cause of the hyponatremia will also need to be corrected as cessation of these vaptan drugs will lead to continued free water retention in the presence of AVP. Caution for drug-drug interactions with conivaptan is required.

In some settings, severe hyponatremia complicates the management of acute kidney injury. When dialysis is needed, cautious use of relatively hypertonic (as compared to the patient's serum sodium) is indicated and appears to be safe.

Hypernatremia

Because thirst is a powerful protective mechanism, restricted access to water is nearly always necessary for the development of hypernatremia. Several factors can predispose patients in the ICU to hypernatremia: the administration of hypertonic sodium bicarbonate solutions; renal water loss through a concentrating defect from renal disease or the use of diuretics or solute diuresis from glucose or urea in patients on high protein feeds or in a hypercatabolic state; gastrointestinal fluid losses through nasogastric suction and lactulose administration, and water losses through fever, drainages, and open wounds. Thus, most etiologies of hypernatremia involve states of impaired water access in conjunction with excessive free water losses [10]. Thus, the prevention of hypernatremia should focus on recognition of at-risk patients and appropriate free water administration.

The majority of patients with hypernatremia are also hypovolemic, requiring the administration of isotonic saline prior to correction of the water deficit. The urine osmolality and electrolytes should be measured to assess urinary concentrating ability and to estimate the electrolyte free water losses in the urine. Caution needs to be exercised in the interpretation of the urine osmolality, as this is an area where error is common. The urine osmolality alone cannot be used to determine if there is free water loss in the urine. This is because water can be excreted with non-electrolyte osmoles (under physiologic conditions, this non-electrolyte osmole in the urine is typically urea). In cases of a high urea load, massive amounts of water can be lost in the urine despite maximal urinary concentration. Urinary free water loss occurs when the urine sodium plus potassium ($[Na^+]_u + [K^+]_u$) is less than the plasma sodium plus potassium ($[Na^+]_{pl} + [K^+]_{pl}$). Failure of a patient to concentrate the urine at a time when the patient is hypernatremic should raise suspicion of a urinary concentrating defect.

The current water deficit of the patient can be calculated using the following formula:

$$\text{Water Deficit} = \left(\frac{[Na^+]_{pl} - [Na^+]^{**}}{[Na^+]^{**}}\right) \times \text{Total}^* \text{ Body Water}$$

$$\begin{aligned} * &= 0.5 \times \text{weight in kg (female)} \\ & \quad 0.6 \times \text{weight in kg (male)} \\ ** &= \text{desired serum sodium} \end{aligned}$$

This value is a guide for therapy that represents the amount of water necessary to correct the serum sodium to a desired value. Once again, this formula assumes a closed system and thus, ongoing sodium and water losses need to be accounted for in the replacement solutions to achieve the goals of correction. In the absence of hypernatremic encephalopathy, the serum sodium should not be corrected more quickly than 1 mEq/h or 15 mEq/24 h. In severe cases (>170 mEq/l), sodium should not be corrected to below 150 mEq/l in the first 48–72 h. Once again, frequent (every 2 h) measurement of serum sodium values is critical to ensure proper correction.

Hypervolemic hypernatremia is unfortunately not uncommon in the ICU as patients often receive large amounts of saline in the course of their illness [11]. However, hypernatremic hypervolemia is an iatrogenic complication that can develop only if renal function is compromised as hypernatremia would normally lead to sodium diuresis. Renal replacement therapy (hemodialysis) is usually the only effective treatment. Prevention of hypernatremia would be a more advisable strategy, however, and a rising serum sodium should be considered a relative contraindication for further administration of saline and should prompt treatment with water either via a feeding tube or as a hypotonic intravenous solution.

Conclusion

Disorders of serum sodium are common in ICU patients and the majority of cases are acquired within the hospital. This makes close observation of serum chemistries and appreciation of intravenous fluids as pharmacological agents imperative. Furthermore, attention to the risk factors associated with the development of these dysnatremias has the potential to prevent their occurrence. The development of sodium disorders is associated with an increase in mortality, length of hospital stay and increased hospital costs. Appropriate therapy requires an understanding of urinary water handling, judicious use of correction formulae and most importantly close monitoring of the patient's serum sodium as changes in intravenous fluid therapy are made.

References

1 Funk G-C, Lindner G, Druml W, Metnitz B, Schwarz C, Bauer P, Metnitz PGH: Incidence and prognosis of dysnatremias present on ICU admission. Intensive Care Med 2010;36:304–311.
2 Stelfox HT, Ahmed SB, Khandwala F, Zygun D, Shahpori R, Laupland K: The epidemiology of intensive care unit-acquired hyponatremia and hypernatremia in medical-surgical intensive care units. Crit Care 2008;12:R162.
3 Waikar SS, Mount DB, Curhan GC: Mortality after hospitalization with mild, moderate and severe hyponatremia. Am J Med 2009;122:857–865.
4 Zilberberg MD, Exuzides A, Spalding J, Foreman A, Jones AG, Colby C, Shorr AF: Hyponatremia and hospital outcomes among patients with pneumonia: a retrospective cohort study. BMC Pulm Med 2008;8:16.
5 Ayus JC, Wheeler JM, Arieff AI: Postoperative hyponatremic encephalopathy in menstruating women. Ann Intern Med 1992;117:891–897.
6 Wijdicks EF, Sharbrough FW: New-onset seizures in critically ill patients. Neurology 1993;43:1042–1044.
7 Verbalis JG, Goldsmith SR, Greenberg A, Schrier RW, Sterns RH: Hyponatremia treatment guidelines 2007: expert panel recommendations. Am J Med 2007;20(suppl 1):S1–S21.
8 Adrogué HJ, Madias NE: Hyponatremia. N Engl J Med 2000;342:1581–1589.
9 Rosner MH: Hyponatremia in heart failure: the role of arginine vasopressin and diuretics. Cardiovasc Drugs Ther 2009;23:307–315.
10 Palevsky PM, Bhagrath R, Greenberg A: Hypernatremia in hospitalized patients. Ann Intern Med 1996;124:197–203.
11 Kahn T: Hypernatremia with edema. Arch Intern Med 1999;159:93–98.

Mitchell H. Rosner, MD
Division of Nephrology, Box 800133, University of Virginia Health System
Charlottesville, VA 22908 (USA)
Tel. +1 434 924 2187, Fax +1 434 924 5848
E-Mail mhr9r@virginia.edu

Ronco C, Bellomo R, McCullough PA (eds): Cardiorenal Syndromes in Critical Care.
Contrib Nephrol. Basel, Karger, 2010, vol 165, pp 299–309

Recent Trials in Critical Care Nephrology

Rinaldo Bellomo[a] · Paul M. Palevsky[b] · Sean M. Bagshaw[c] · Noel Gibney[c] · Finlay A. McAlister[d] · Patrick M. Honore[e] · Olivier Joannes-Boyau[f] · John Prowle[a] · Michael Haase[g] · Dinna N. Cruz[h] · Claudio Ronco[h]

[a]Department of Intensive Care, Austin Hospital, Melbourne, Vic., Australia; [b]VA Pittsburgh Healthcare System, University of Pittsburgh School of Medicine, Pittsburgh, Pa., USA; [c]Division of Critical Care Medicine and [d]Department of Medicine, Faculty of Medicine and Dentistry, University of Alberta, Edmonton, Alta., Canada; [e]Burns Center, Queen Astrid Military Hospital, Brussels, Belgium; [f]Haut Leveque University Hospital of Bordeaux, University of Bordeaux 2, Pessac, France; [g]Department of Intensive Care and Nephrology, Charité Hospital, Berlin, Germany, and [h]Department of Nephrology, Dialysis & Transplantation, International Renal Research Institute, San Bortolo Hospital, Vicenza, Italy

Abstract

Several large observational studies or randomized controlled trials in the field of critical care nephrology have been completed and reported, or recently completed or have recently begun recruitment. These studies provide important information to guide our appreciation of current practice and consider new potentially effective intervention for the prevention or attenuation of acute kidney injury or suggest new avenues for the use of renal replacement therapy (RRT) in the treatment of sepsis. In particular, two studies, the ATN study and the RENAL study (both multicenter randomized controlled trials of >1,000 patients) provide, for the first time, level I evidence to guide the practice of RRT in critically ill patients and to better define the optimal intensity of such RRT in this setting. Clinicians practicing in the field of critical care nephrology need to be aware of these trials, their details, their findings or design or current recruitment rate and likely time of completion to continue to offer their patients the highest level of evidence-based medical care.

Over the last few years, several important studies and/or randomized controlled trials have been completed or have started in the field of critical care nephrology. It is important for practitioners in this field to have knowledge of the design and findings of these key studies and to appreciate the features of ongoing investigations, which might help shape future large randomized controlled trials in the next decade. In this brief overview, we will present the salient features of these studies with the goal of updating clinicians on recently published level I evidence and providing insights on emerging evidence.

The ATN Study

The Department of Veterans Affairs/National Institutes of Health (VA/NIH) Acute Renal Failure Trial Network (ATN) study was a multicenter, prospective, randomized controlled trial comparing two strategies of intensity of renal replacement therapy (RRT) in critically ill patients with acute kidney injury (AKI) [1]. Unlike other published trials that have evaluated the impact of dose of RRT on outcomes in AKI, the design of the ATN study allowed patients to switch between modalities of RRT as their hemodynamic status varied over time. In both treatment arms, RRT was provided as intermittent hemodialysis (IHD) when a patient was hemodynamically stable and as either continuous venovenous hemodiafiltration (CVVHDF) or sustained low-efficiency dialysis (SLED) in the setting of hemodynamic instability. Patients switched back and forth between modalities of therapy as their hemodynamic status varied over time. Once the clinical decision to initiate RRT was made, patients were randomized to strategies providing either greater or lesser intensity of RRT. In the more-intensive strategy, IHD and SLED were provided daily except Sunday (6 days per week) and CVVHDF was prescribed to provide an effluent flow rate of 35 ml/kg/h; in the less-intensive strategy, IHD and SLED were provided every other day except Sunday (3 days per week) and CVVHDF was prescribed to provide an effluent flow rate of 20 ml/kg/h. In both treatment arms, IHD and SLED were prescribed to provide a target single-pool Kt/V_{urea} of between 1.2 and 1.4 per treatment. While small solute clearance was carefully protocolized, volume management was left to the discretion of the clinicians; when patients were receiving intermittent therapies isolated ultrafiltration could be provided on non-study days if additional volume removal was required.

Study therapy was provided for up to 28 days, or until kidney function recovered, the patient was discharged from the acute care hospital, life-sustaining therapy was withdrawn or the patient died. Recovery of kidney function was defined based on measured creatinine clearance using 6-hour timed urine collections when urine flow exceeded 30 ml/h or there was a spontaneous fall in serum creatinine concentration. RRT was continued if the creatinine clearance was <12 ml/min (0.2 ml/s) and was discontinued if the creatinine clearance was >20 ml/min (0.3 ml/s); decisions regarding discontinuation of therapy for intermediate values were left to the clinician.

The primary study end-point was 60-day all-cause mortality. Secondary end-points included in-hospital death, recovery of kidney function, duration of RRT, length of intensive care unit (ICU) and hospital stay, days free of non-renal organ failures, and a composite end-point of discharge to home off of dialysis by day 60.

Between November 2003 and July 2007, 1,124 patients were enrolled in the study at 27 Department of Veterans Affairs and university-affiliated medical centers in the USA: 563 were assigned to the more-intensive treatment regimen and 561 to the less-intensive treatment regimen.

Study therapy was provided for a mean (±SD) of 13.4 ± 9.6 days in the more-intensive strategy arm and for 12.8 ± 9.3 days in the less-intensive strategy arm. The median time per day receiving CVVHDF was 21 h in both treatment groups. The mean (±SD) delivered effluent flow was 35.8 ± 6.4 ml/kg/h in the more-intensive arm and 22.0 ± 6.1 ml/kg/h in the less-intensive arm. When patients were hemodynamically stable, IHD was provided 5.4 times per week with a median interval between treatments of 1.1 days in the more-intensive arm and 3.0 times per week with a median interval between treatment, so of 2.1 days in the less-intensive arm. The mean (±SD) delivered single-pool Kt/V_{urea} after the first hemodialysis session was 1.32 ± 0.36 with no difference between treatment arms. The mean (±SD) time-averaged urea nitrogen concentration was 33 ± 17 mg/dl during IHD and 33 ± 18 mg/dl during CVVHDF in the more-intensive arm and 48 ± 19 and 47 ± 23 mg/dl, respectively, in the loss-intensive arm.

Of the 563 patients (53.6%) in the more-intensive therapy arm, 302 died within 60 days of randomization as compared to 289 of the 561 patients (51.6%) in the less-intensive strategy arm (odds ratio (OR) 1.09; 95% confidence interval (CI) 0.86–1.40; p = 0.47), with no significant differences in all prespecified subgroups.

Rates of recovery of kidney function were not different across treatment arms: 74.6% of patients in the more-intensive arm who survived to day 60 were dialysis-independent as compared to 76.2% in the less-intensive arm (p = 0.67). Only 15.7 and 16.4% of patients, respectively, were alive, dialysis-independent and discharged home by day 60 (p = 0.75).

In summary, the study found no added benefit from a more-intensive (high-dose) treatment strategy as compared to a more-conventional, less-intensive treatment strategy for the management of RRT in critically ill patients with AKI. The more-intensive strategy did not decrease mortality, accelerate recovery of kidney function or alter the rate of non-renal organ failure. These results do not imply that the dose of RRT is not important, but suggest that there is no need to provide IHD more frequently than 3 times per week, as long as a target Kt/V_{urea} of 1.2–1.4 per treatment is achieved, or CVVHDF with an effluent flow rate of >20 ml/kg/h, as long as time on therapy in maximized.

The RENAL Study

The Randomized Evaluation of Normal vs. Augmented Level (RENAL) Replacement Therapy Study was a collaboration of the Australian and New Zealand Intensive Care Society Clinical Trials Group and the George Institute for International Health. It was a multicenter, prospective, randomized controlled

Table 1. Similarities and differences between RENAL and ATN trials

	RENAL trial n = 1,508	ATN trial n = 1,124
Age	64.5	59.7
Male, %	64.6	70.6
Weight, kg	80.6	84.1
Sepsis, %	49.5	63
Pre-randomization dialysis, %	0	64.3
ICU, days before randomization	2.1	6.7
Ventilation, %	73.9	80.6
Hospital length of stay, days	25.2	48
Mortality, %	44.7 (day 90)	52.5 (day 60)
Duration RRT, days	12.2	21.5
Dialysis-dependent, %	5.6 (day 90)	24.6 (day 60)

trial comparing two strategies of intensity of continuous RRT (CRRT) in critically ill patients with AKI. Unlike the ATN study, the design of the RENAL study required that patients should only receive CVVHDF until either death or ICU discharge or recovery or at least 3 days of treatment had been given and the clinician believed it safe to switch to intermittent therapy [2–4]. With this design only a minority (<10%) of patients received any IHD during the later phase of their AKI, typically after ICU discharge. Once the clinical decision to initiate RRT was made and in the presence of at least one preset criterion of CRRT initiation, patients were randomized to either greater or lesser intensity of CRRT: postdilution CVVHDF to provide an effluent flow rate of 40 ml/kg/h or an effluent flow rate of 20 ml/kg/h. Volume management was left to the discretion of the clinicians. Recovery of kidney function was defined as independence form RRT.

The primary study end-point was 90-day all-cause mortality. Secondary end-points included in-hospital death, recovery of kidney function, duration of RRT, and length of ICU and hospital stay.

Between December 2005 and November 2008, 1,508 patients were enrolled in the study at 35 ICUs in Australia and New Zealand: 747 were assigned to the more-intensive treatment regimen and 761 to the less-intensive treatment regimen (table 1). Study therapy was provided for a mean of 6.3 days in the more-intensive strategy arm and for 5.9 days in the less-intensive strategy arm. Because of filter clotting, down time and surgery or radiological investigations, the mean delivered effluent flow was 33.4 ml/kg/h in the more-intensive arm and 22 ml/kg/h in the less-intensive arm (fig. 1). The mean daily urea nitrogen concentration was 35.6 mg/dl in the more-intensive arm and 44.5 mg/dl in the loss-intensive arm.

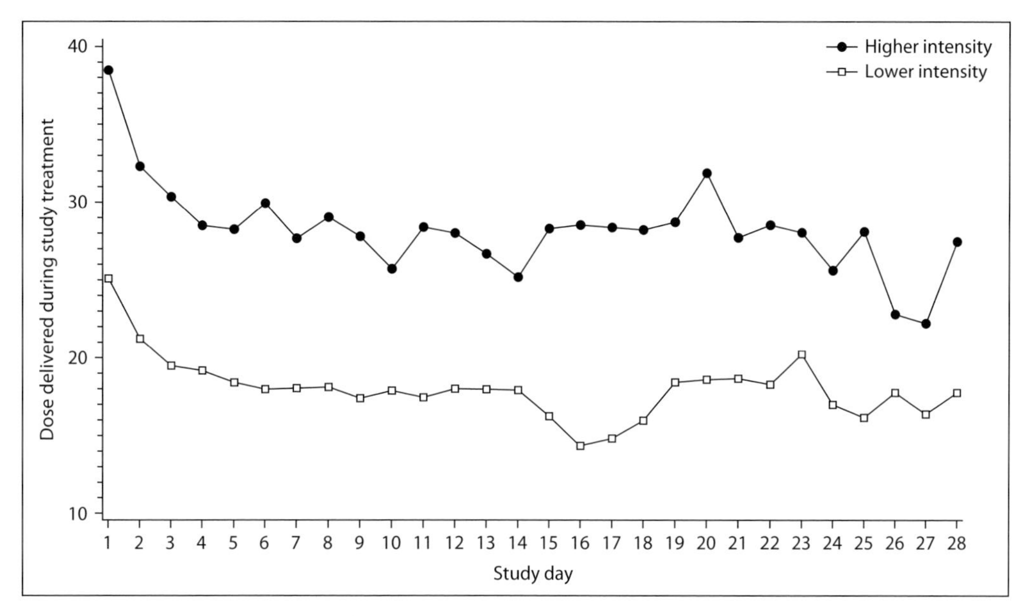

Fig. 1. Graphic representation of daily dose delivery expressed as ml/kg/h of effluent generation during the RENAL trial.

322 (44.7%) patients in the more-intensive therapy arm died within 90 days of randomization compared to 332 (44.7%) in the less-intensive strategy arm (OR 1.00; 95% CI 0.81–1.23; p = 0.99), with no significant differences in all prespecified subgroups. Rates of recovery of kidney function were not different across treatment arms: 93.2% of patients in the more-intensive arm who survived to day 90 were dialysis-independent as compared to 95.6% in the less-intensive arm (p = 0.14). A total of 49.8 and 51.6% of patients, respectively, were alive and dialysis-independent by day 90.

In summary, the study found no added benefit from a more-intensive (high-dose) CVVHDF strategy as compared to a more-conventional, less-intensive CVVHDF for the management of CRRT in critically ill patients with acute AKI. The more-intensive strategy did not decrease mortality, accelerate recovery of kidney function or alter the rate of non-renal organ failure. These results do not imply that the dose of RRT is not important, but suggest that there is no need to provide CVVHDF with a delivered effluent flow rate of >22 ml/kg/h.

The SPARK Study

Survey data have found loop diuretics are commonly prescribed to >60% critically ill patients with AKI [5]. Loop diuretics may theoretically attenuate kidney injury by reducing renal tubular oxygen demand and apoptosis, along with allowing more optimal management of fluids. Numerous studies have attempted to evaluate the impact of loop diuretics in established AKI; however, results have been inconsistent and conflicting [5]. Moreover, the evidence from the vast majority of these clinical trials either suffer from serious methodologic flaws, systematic bias, or are not generalizable to the AKI routinely encountered in modern ICU practice. Yet, the evidence generated from these studies largely forms the basis for the prevailing view on whether furosemide may have any benefit (or harm) in AKI. Interestingly, survey data would suggest that loop diuretics, while routinely used in critically ill patients with AKI, are not administered in a manner consistent with the studies. This would suggest a disconnect between existing evidence and clinical practice and further imply there is genuine equipoise for additional trials evaluating loop diuretics in critically ill patients with early AKI.

The SPARK Study (a phase II randomized blinded controlled trial of the effect of furoSemide in critically ill Patients with eARly acute Kidney injury – ClinicalTrials.gov NCT00978354) will evaluate the impact of a continuous furosemide infusion versus placebo, titrated to urine output, on kidney injury progression in critically ill patients with early AKI. Key inclusion criteria are admission to ICU; early AKI (defined by the RIFLE category – RISK [6]); presence of ≥2 criteria for the systemic inflammatory response syndrome within 24 h, and achievement of immediate resuscitation goals. Randomized will be stratified by sepsis. The secondary outcomes of the SPARK study include a comparison between furosemide and placebo on kidney injury biomarkers (i.e. neutrophil gelatinase-associated lipocalin (NGAL)), duration of AKI, fluid balance, electrolyte homeostasis, need for RRT, renal recovery and survival.

The SPARK study will have significance for providing insight into a key question about an intervention (furosemide) that is given to the majority of critically ill patients with AKI, but, to date, has no quality evidence to suggest benefit. Recruitment has started and the study should be completed by 2012.

The IVOIRE Study

The IVOIRE study (hIgh VOlume in Intensive caRE) aims to compare treatment with continuous venovenous hemofiltration (CVVH) at 35 versus 70 ml/kg/h in septic AKI for a period of 96 h [7]. The 96-hour period was based on the findings of the PROWESS study. As reviews showed indirect evidence that early timing of initiation of therapy could be associated with further improvement in mortality, it was decided to start therapy in septic shock at the injury (I) level of the RIFLE classification [6]. The first patient was randomized at the end November 2005 but most centers joined only in mid-2006.

The overall primary objective of this study was to assess the effects of early high-volume CVVH (70 ml/kg/h) on the mortality rate at 28 days in patients with septic shock complicated by acute renal insufficiency. The secondary objectives were to assess the consequences of hemofiltration on hemodynamic parameters, doses and duration of use of catecholamines, organ failure, duration of artificial ventilation and extravascular lung water, morbidity rate, duration of ICU stay and early mortality (96 h), and mortality after 60 and 90 days.

IVOIRE is randomized multicenter clinical trial. In all cases, patients are treated in accordance with current recommendations in the literature. This study is a superiority trial, aiming to detect a reduction of at least 15% in mortality rate. A sample size with 240 patients in each arm was calculated to give >80% power to detect such a difference at an α of 0.5. The team at the center coordinating the IVOIRE trial has conducted two pilot studies to prepare for this trial [2, 8].

The choice of hemofiltration volume in the control group (35 ml/kg/h) was motivated by the latest recommendations in the literature currently applied to intensive care in the early years of this millennium. The choice for the high-volume group (70 ml/kg/h) was made by consensus, taking into account both intensive care practices using this latest technique and the opinion of investigators who were consulted at various congresses, and taking in to account at present that there is no reference in the literature that allows a specific treatment volume to be selected.

The duration of treatment (96 h) was motivated by the experience of the team and the pilot studies, which were based on the same periods. This duration corresponds to the critical period of septic shock during which treatment has the most impact on survival. It is also the duration stipulated for other sepsis treatments in large studies at present.

Block randomization was used by means of a dedicated website. Data regarding losses of trace elements, vitamins, amino acids and other nutrients, apoptosis, oxidative stress, renal repair capacity index, cytokines and mediators and antibiotic pharmacokinetic are being collected as part of this study.

By the end of December 2009, more than 140 had been randomized with more than 45 patients in the last year. The first interim analysis will take place in January 2010. The expected mortality of the first 100 patients according to three severity scoring (SOFA, SAPS II & LOD) was 68% whereas the observed mortality at 28 days was 39% and at day 90 only 48%. So, in fact the observed global mortality was much lower than expected. Other details regarding the study can be found at the NCT website [9].

The CREAT Study

The Cardiopulmonary bypass, REnal injury & Atorvastatin Trial is a phase IIa single-center double-blind randomized controlled trial which aims to test whether the administration of a statin (atorvastatin) just before surgery and for a period of 3 days thereafter can attenuate AKI when compared to placebo. After cardiopulmonary bypass, AKI, manifest as an acute rise in plasma creatinine, is estimated to occur in 15% of patients with an otherwise normal preoperative renal function. Most of these patients do not require RRT, however the occurrence of even mild perioperative AKI is independently associated with postoperative mortality [1].

Cardiopulmonary bypass has been shown to induce markers of oxidative stress, stimulate leukocytes, activate endothelium and increase expression of inflammatory mediators. Such inflammatory states have been associated with AKI in animal models. HMG-CoA reductase inhibitors (statins) have been shown to have antioxidant and anti-inflammatory effects independent of cholesterol-lowering effects and their use has been associated with improved outcomes after cardiac surgery in observational and small prospective studies. Accordingly, we hypothesized that atorvastatin (40 mg once daily for 4 days) would reduce the severity of AKI in a prospective, double-blinded randomized placebo-controlled trial of 100 patients undergoing cardiopulmonary bypass with a higher risk of perioperative AKI (uncomplicated patients, <70 years having surgery limited to coronary artery bypass grafting were predefined as low risk and not included). Randomization was stratified by presence or absence of pre-admission statin therapy and data was analyzed on an intention-to-treat basis. Primary outcome was postoperative AKI in the first 5 days defined by the RIFLE consensus classification of AKI (RIFLE-R or greater). Additionally, NGAL, a biomarker of renal tubular injury, was measured in serum and urine assessed as a secondary outcome for AKI [8]. A series of other putative biomarkers of early AKI was also measured to gauge their association with the subsequent development of AKI in a defined population of high-risk patients. The study has just been completed and results are expected by early to mid-2010.

The BIC-AKI Study

The BICarbonate for AKI in cardiac surgery study (BIC-AKI study) is a phase IIa two-center double-blind randomized controlled trial which aimed to test whether the administration of intravenous bicarbonate with bolus loading just before surgery and continuous infusion for a period of 24 h thereafter can attenuate AKI when compared to placebo.

Rationale for the Trial

Urinary acidity may enhance the generation and toxicity of reactive oxygen species induced by cardiopulmonary bypass. Urinary alkalinization may protect from renal injury induced by oxidant substances, iron-mediated free radical pathways, and tubular hemoglobin cast formation. Of note, increasing urinary pH has recently been reported to attenuate acute renal dysfunction in patients undergoing contrast-media infusion. More alkaline urine reduces the generation of injurious hydroxyl radicals and lipid peroxidation. Bicarbonate directly scavenges hydroxyl ions and – as a not well absorbable anion compared to chloride – causes more rapid volume excretion and thereby might reduce the contact time between injurious radicals and renal tubules.

Accordingly, we hypothesized that urinary alkalinization might protect kidney function and conducted BIC-AKI to investigate whether sodium bicarbonate infusion with preoperative intravenous loading to achieve urinary alkalinization could reduce the incidence of AKI associated with cardiopulmonary bypass in cardiac surgical patients at increased renal risk.

In a cohort of 100 cardiac surgical patients [9], sodium bicarbonate treatment successfully alkalinized urine. There were no significant differences between the groups in baseline characteristics including duration of cardiopulmonary bypass and in hemodynamic and fluid management during and after cardiac surgery, nor in plasma creatinine, plasma urea or urinary NGAL. The mean dose of sodium bicarbonate was 307 ± 57 mmol and mean dose of sodium chloride was 309 ± 68 mmol ($p = 0.89$).

Fewer patients in the sodium bicarbonate group (16/50) developed AKI compared to the control group (26/50) (OR 0.43; 95% CI 0.19–0.98; ($p = 0.043$). Also, sodium bicarbonate infusion was associated with a significant attenuation in the postoperative increase of plasma urea, urinary NGAL and urinary NGAL/urinary creatinine ratio.

Perspective

As the study was not powered to detect group differences in need of RRT or assess mortality rates, and because there is justification for larger trials, the BIC-AKI investigators who are currently conducting such an international study (BIC-AKI II) with >400 patients enrolled at four centers in Europe, Canada and Australia and a simultaneous three-center study of >400 patients in Australia and New Zealand, still have to confirm or refute the findings of the first study. The final results of these two studies are expected by 2012.

The EUPHAS Study

The EUPHAS (Early Use of Polymyxin B Hemoperfusion in Abdominal Sepsis) study is a multicenter randomized controlled study conducted in 10 Italian ICUs comparing treatment with a polymyxin B column (two sessions of polymyxin B hemoperfusion) versus conventional treatment [10]. The rationale for the study was based on evidence that polymyxin B avidly binds endotoxin and that preliminary studies, mostly conducted in Japan, had shown evidence of benefit from such treatment. Accordingly, the investigators randomized 64 patients with severe sepsis/septic shock who had undergone emergency surgery for intra-abdominal infection. The primary outcome of the trial was a change in mean arterial pressure (MAP) and vasopressor requirement. Secondary outcomes included changes in organ function and 28-day mortality. The investigators found that polymyxin B hemoperfusion was associated with a significant 8 mm Hg increase in MAP at 72 h compared with only a non-significant 3 mm Hg increase in MAP in the conventional treatment group. They also found an improvement in SOFA organ failure scores (–3.4 vs. –0.1; $p < 0.001$) and in 28-day mortality (32 vs. 53%; unadjusted heart rate 0.43; 95% CI 0.20–0.94). This phase IIa study is an important first step towards larger and more sophisticated investigations of polymyxin B hemoperfusion in targeted patients with intra-abdominal sepsis.

CRRT in Sepsis Study

The CRRT in sepsis trial is a multicenter randomized controlled trial which compared the treatment of severe sepsis with either conventional therapy or conventional therapy with the addition of a 96-hour period of hemofiltration at 25 ml/kg/h. The concept of the study derives from experimental and clinical evidence suggesting that CVVH may remove inflammatory mediators and offer an adjunctive benefit in the treatment of sepsis [11]. In this study [12], a total of 80 patients were enrolled within 24 h of the development of organ failure related to sepsis. Patients randomized to intervention received CVVH at a dose of 2 l/h of effluent flow for 96 h with planned filter change at 12, 24, and 48 h. The primary end-point of the trial was the number, severity and duration of organ failure during 14 days as evaluated by the SOFA score. The trial was stopped at 76 patients because of inadequate recruitment. At the time of assessment, the number and severity of organ failure were actually *higher* in the hemofiltration group.

Conclusions

Over the last year, several important randomized controlled trials in the field of critical care nephrology have been reported, have just been completed, or have been initiated. Awareness of their design, execution, findings and/or status will help clinicians in their choice of therapy and appreciation of the latest evidence to guide practice.

References

1 Palevsky PM, Zhang JH, O'Connor TZ, et al: Intensity of renal support in critically ill patients with acute kidney injury. N Engl J Med 2008;359:7–20.
2 Bellomo R, Cass A, Cole L, et al: Intensity of continuous renal replacement therapy in critically ill patients. N Engl J Med 2009;361:1627–1638.
3 Finfer S, Cass A, Gallagher M, et al: The RENAL study: statistical analysis plan. Crit Care Resusc 2009;11:58–66.
4 RENAL Study Investigators: Renal replacement therapy for acute kidney injury in Australia and New Zealand intensive care units: a practice survey. Crit Care Resusc 2009;10:225–230.

5 Bagshaw SM, Delaney A, Haase M, et al: Loop diuretics in the management of acute renal failure: a systematic review and meta-analysis. Crit Care Resusc 2007;9:60–68.
6 Bellomo R, Ronco C, Kellum J, Mehta R, Palevsky P, the ADQI Workgroup: Acute renal failure – definition, outcome measures, animal models, fluid therapy and information technology needs: the Second International Consensus Conference of the Acute Dialysis Quality Initiative (ADQI) Group. Crit Care 2004;8:R204–R212.
7 Honore PM, Joannes-Boyau O, Boer W, Collin V: High-volume hemofiltration in sepsis and systemic inflammatory response syndrome: current concepts and future prospects. Blood Purif 2009;28:1–11.
8 Haase M, Bellomo R, Devarajan P, Schlattmann P, Haase-Fielitz A, NGAL Meta-Analysis Investigator Group: Accuracy of neutrophil gelatinase-associated lipocalin in diagnosis and prognosis in acute kidney injury: a systematic review and meta-analysis. Am J Kidney Dis 2009;54:1012–1024.
9 Haase M, Haase-Fielitz A, Bellomo R, et al: Sodium bicarbonate to prevent increases in serum creatinine after cardiac surgery: a pilot double-blind randomized controlled trial. Crit Care Med 2009;37:39–47.
10 Cruz DN, Antonelli M, Fumagalli R et al: Early use of polymyxin B hemoperfusion in abdominal septic shock: the EUPHAS randomized controlled trial. JAMA 2009;301:2445–2452.
11 Bellomo R, Ronco C: Renal replacement therapy in the intensive care unit. Crit Care Resusc 1999;1:13–24.
12 Payen D, Mateo J, Cavaillon JM, et al: Impact of continuous veno-venous hemofiltration on organ failure during the early phase of severe sepsis: a randomized controlled trial. Crit Care Med 2009;37:803–810.

Prof. Rinaldo Bellomo
Australian and New Zealand Intensive Care Research Centre, School of Public Health and Preventive Medicine, Monash University
Melbourne, Vic. (Australia)
Tel. +61 400504164, Fax +61 399030071, E-Mail rinaldo.bellomo@med.monash.edu.au

Ronco C, Bellomo R, McCullough PA (eds): Cardiorenal Syndromes in Critical Care.
Contrib Nephrol. Basel, Karger, 2010, vol 165, pp 310–314

Acute Kidney Injury and 2009 H1N1 Influenza-Related Critical Illness

Rinaldo Bellomo[a] · Ville Pettilä[a] · Steven A.R. Webb[b] · Michael Bailey[a] · Belinda Howe[a] · Ian M. Seppelt[c]

[a]Australian and New Zealand Intensive Care Research Centre, School of Public Health and Preventive Medicine, Monash University, Melbourne, Vic.; [b]Department of Intensive Care Medicine, Royal Perth Hospital, Perth, WA, and School of Population Health and School of Medicine and Pharmacology, University of Western Australia, Crawley, WA, and [c]Department of Intensive Care Medicine, Nepean Hospital, and Discipline of Intensive Care Medicine, Sydney Medical School – Nepean, University of Sydney, Sydney, NSW, Australia

Abstract

The year 2009 was characterized by a pandemic with a new virus, the 2009 H1N1 influenza virus. This pandemic was responsible for thousands of deaths worldwide, many more hospital admissions, and thousands of admissions to intensive care units (ICUs). Among those admitted to ICUs, the pandemic was associated with a mortality of approximately 16%, a high incidence of acute lung injury and, in some cases, acute respiratory distress syndrome severe enough to require support with extracorporeal membrane oxygenation. As part of such a critical illness, a percentage of patients developed acute kidney injury (AKI) which complicated their clinical course and, in some patients, required support by renal replacement therapy. In a case series from Mexico, the incidence of severe AKI was reported in about 30% of the patients. Similarly, at the Austin Hospital, of 13 cases, 8 developed AKI with 3 being classified in the failure category of the RIFLE classification. Among the patients with AKI, hospital mortality was approximately 25%. Of the AKI patients, 3 (37.5%) received renal replacement therapy and, among these, 1 died. In a case of severe AKI and multi-organ failure from whom histological material was obtained, the renal histopathological findings were typical of acute tubular necrosis. One patient who suffered from hypoxic brain injury due to cardiac arrest at home secondary to H1N1 pneumonia became a kidney and liver donor. There was no evidence of viral infiltration on kidney biopsy and the recipient did not develop H1N1 infection.

An influenza A (H1N1) pandemic emerged during 2009 and caused a significant burden of critical illness [1–3]. The virus, like other influenza viruses, tended to affect the upper airway and lungs and was responsible for thousands of deaths worldwide and for thousands of intensive care unit (ICU) admissions. Such 2009 H1N1 influenza-related critical illness was typically characterized by the presence of acute lung injury and, in some cases, severe acute respiratory distress syndrome. Finally, in some patients, acute respiratory distress syndrome was severe enough to require support by extracorporeal membrane oxygenation. Among patients admitted to the ICU, mortality varied from approximately 40% in a case series from Mexico [4] to approximately 16% among patients admitted to ICU in Australia and New Zealand (ANZ) [2].

Despite various reports of multi-organ failure in these patients [1–4], there has been limited information on the incidence of acute kidney injury (AKI) among critically ill patients, its severity, its need for renal replacement therapy (RRT) and patient outcome among those patients who developed AKI. Part of such limited interest arises from the fact that the 2009 H1N1 influenza virus, like other influenza viruses, mostly affects the upper and lower airways and lungs and seems unlikely to be responsible for renal injury. This view is supported by what is known about AKI in association with other influenza viruses. For example, influenza A infections rarely cause AKI and as such AKI seems mostly secondary to rhabdomyolysis [5]. However, in patients with avian influenza A (H5N1), an influenza virus which has appeared to be particularly aggressive from a clinical point of view, AKI has been reported in 17% of cases [6]. In contrast to such reports, a small single-center case series of 2009 H1N1 patients suggested a high risk (32%) of AKI [7]. This is clearly much greater than previously reported with other influenza viruses, but due to the limited scope of the above report, there remained uncertainty about the true incidence and outcome of such AKI. However, another case series from Mexico [4] also reported that 6 of 18 patients with serious illness related to 2009 H1N1 influenza developed AKI. Of these patients 5 died. This additional observation further supports a relatively high incidence of AKI and a high level of associated mortality among critically ill patients. However, the relevance of such data to developed countries, where ICU resources are more widely available and where RRT can be implemented with greater ease, is uncertain. In response to these uncertainties, investigators from ANZ recently used a prospectively collected influenza A (H1N1) 2009 national database [2] to evaluate the incidence and hospital outcome of AKI among critically ill adult patients with 2009 H1N1 infection. As part of this project, the Austin Hospital collected data on thirteen 2009 H1N1 ICU admissions.

ANZ 2009 H1N1 Influenza Database

Investigators from ANZ had previously performed a multicenter cohort study in 187 ICUs in ANZ comprising all adult, pediatric and combined adult and

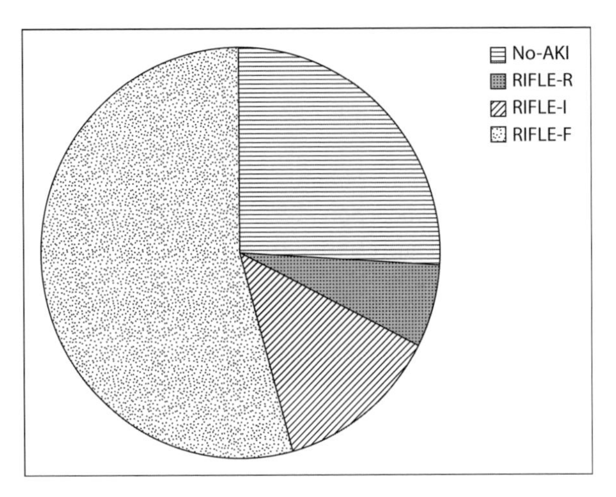

Fig. 1. Distribution of levels of renal dysfunction in patients with 2009 H1N1 admitted to the Austin Hospital ICU during the winter period in 2009 with classification according to the RIFLE system.

pediatric ICUs in the 2 countries [2]. As part of this project, between June 1 and August 31, 2009, all patients admitted to ICUs with confirmed influenza A were identified. Influenza A was confirmed by PCR, antigen detection or serology. From this prospectively collected database, the investigators identified all adult patients admitted with 2009 H1N1. For these patients, whenever available, admission and/or peak creatinine values were obtained to identify and classify patients as having 2009 H1N1 influenza-related AKI. To identify and classify these patients, the ANZ investigators used the creatinine criteria of the RIFLE classification [8]. In addition, they collected data on daily use of RRT and, because of concern about the possible role of rhabdomyolysis in the pathogenesis of AKI during 2009 H1N1 influenza infection, they also collected data on serum creatine kinase (CK) values at ICU admission and peak values during ICU stay. Among these patients, the Austin Hospital ICU contributed 13 cases which will be discussed below. A full report on the entire cohort is likely in 2010.

Incidence and Severity of AKI at Austin Hospital

Austin Hospital admitted 13 cases of 2009 H1N1 infection to the ICU. Among these adult patients, 8 had a degree of AKI according to the RIFLE classification. The distribution of AKI severity according to the RIFLE classification is shown in figure 1. Of these AKI patients, 4 already had AKI on ICU admission.

The median CK was 1,035 IU in patients with AKI compared with 668 in patients without AKI. Among AKI patients for whom data were available, 1 had an elevated serum CK level exceeding 3,000 IU/l. Three patients received RRT. Of these 1 patient died. Overall hospital mortality for AKI was 2 patients. The patient who died had evidence of acute tubular necrosis at postmortem.

One patient suffered a cardiac arrest at home because of H1N1-related hypoxemia. She became an organ donor. Her kidneys were biopsied and assessed for viral infiltration prior to transplantation. There was no evidence of viral infiltration. The recipient did not develop H1N1 infection.

Histopathology of 2009 H1N1 Influenza-Related AKI

In a report of postmortem findings in 5 patients who died of 2009 H1N1 influenza [9], 1 patient was found to have acute tubular necrosis (ATN). No information was reported on more subtle findings in the remaining patients. In another study of 21 patients who died of 2009 H1N1 influenza in Brazil [10], there were no signs of direct virus-induced injury in any organ other than the lungs. However, all patients had evidence of mild to moderate ATN. In 4 patients, there was evidence of myoglobin pigment in the tubules and thrombotic angiopathy was seen in 1 patient. Thus, it appears that the two most likely histopathological processes at work in these patients are, as suspected, the development of ATN and rhabdomyolysis-induced tubular injury.

Conclusions

A case series from the Austin Hospital and case series from elsewhere are essentially concordant in that a significant proportion of adult 2009 H1N1 patients admitted to the ICU developed AKI and that these patients had a high hospital mortality rate. RRT appeared necessary in approximately one third of the patients. 2009 H1N1 influenza-related AKI is also associated with higher CK levels. Moreover, the limited histopathological data available show that, although ATN is the most common postmortem finding, tubular deposition of myoglobin pigment occurs in some patients. This observation further suggests that, in selected patients and as is the case with other types of influenza infection, rhabdomyolysis may contribute to the development of AKI. Recent reports have also demonstrated elevated CK values in almost two thirds [11] and approximately three fourths [12] of critically ill 2009 H1N1 patients further supporting the notion that muscle cell injury may contribute to the development of AKI in some cases. In their aggregate, however, the above findings suggest that the severity of AKI in 2009 H1N1 patients is similar to that seen in other severely ill ICU patients with comparable hospital mortality rates and that it mostly reflects a process with strong similarities to AKI seen with other forms of severe sepsis and associated multi-organ failure.

References

1 Carr MJ, Gunson R, Maclean A, et al: Development of a real-time RT-PCR for the detection of swine-lineage influenza A (H1N1) virus infections. J Clin Virol 2009; 45:196–199.

2 ANZIC Influenza Investigators; Webb SA, Pettilä V, Seppelt I, et al: Critical care services and 2009 H1N1 influenza in Australia and New Zealand. N Engl J Med 2009;361:1925–1934.

3 Webb SA, Seppelt IM: Pandemic (H1N1) 2009 influenza ('swine flu') in Australian and New Zealand intensive care. Crit Care Resusc 2009;11:170–172.

4 Perez-Padilla R, de la Rosa-Zamboni D, Ponce de Leon S, et al: Pneumonia and respiratory failure from swine-origin influenza A (H1N1) in Mexico. N Engl J Med 2009; 361:680–689.

5 Shenouda A, Hatch FE: Influenza A viral infection associated with acute renal failure. Am J Med 1976;61:697–702.

6 Yu H, Gao Z, Feng Z, Shu Y, et al: Clinical characteristics of 26 human cases of highly pathogenic avian influenza A (H5N1) virus infection in China. PLoS One 2008;3:e2985.

7 Raffo L: Influenza A(H1N1) epidemic in Argentina. Experience in a national general hospital (Hospital Nacional Alejandro Posadas) (in Spanish). Medicina (B Aires) 2009;69:393–423.

8 Bellomo R, Ronco C, Kellum JA, Mehta RL, Palevsky P: Acute renal failure – definition, outcome measures, animal models, fluid therapy and information technology needs: the Second International Consensus Conference of the Acute Dialysis Quality Initiative (ADQI) Group. Crit Care 2004;8: R204–R212.

9 Soto-Abraham MV, Soriano-Rosas J, Diaz-Quinonez A, et al: Pathological changes associated with the 2009 H1N1 virus. N Engl J Med 2009;361:2001–2003.

10 Maud T, Abrahao Hajjar L, de Sanctis Callegari G, et al: Lung pathology in fatal novel human influenza A (H1N1) infection. AJRCCM e-published October 29, 2009

11 Ayala E, Kagawa FT, Wehner JH, et al: Rhabdomyolysis associated with 2009 influenza A(H1N1). JAMA 2009;302:1863–1864.

12 Kumar A, Zarychanski R, Pinto R, et al: Critically ill patients with 2009 influenza A(H1N1) infection in Canada. JAMA 2009; 302:1872–1879.

Prof. Rinaldo Bellomo
Australian and New Zealand Intensive Care Research Centre, School of Public Health and Preventive Medicine, Monash University
Melbourne, Vic. (Australia)
Tel. +61 400504164, Fax +61 399030071, E-Mail rinaldo.bellomo@med.monash.edu.au

Ronco C, Bellomo R, McCullough PA (eds): Cardiorenal Syndromes in Critical Care.
Contrib Nephrol. Basel, Karger, 2010, vol 165, pp 315–321

Acute Kidney Injury in the Elderly: A Review

Alexandra Chronopoulos[a,b] · Mitchell H. Rosner[c] · Dinna N. Cruz[a] · Claudio Ronco[a]

[a]Department of Nephrology, Dialysis & Transplantation, International Renal Research Institute, San Bortolo Hospital, Vicenza, Italy; [b]Department of Medicine, Hôpital Hôtel-Dieu de Lévis, Lévis, Qué., Canada; and [c]Division of Nephrology, University of Virginia Health System, Charlottesville, Va., USA

Abstract

The risk of developing acute kidney injury (AKI) is significantly increased in the elderly. It is the age-related renal and systemic changes as well as frequent comorbidities that render older individuals greatly susceptible to acute renal impairment. Although most often multifactorial, specific etiologies such as renal hypoperfusion due to cardiac failure, dehydration or hypotension of any cause, as well as sepsis, drug toxicity, surgery, or obstructive causes are often present. Contrast-induced nephropathy and atheroembolic disease are also frequently seen, especially in an acute care setting. Serum creatinine is most commonly used for diagnosis, despite it having several limitations, especially in the elderly. The mainstay of management is prevention of further deterioration, as the chances of renal recovery may be lower in older patients.

Advanced age has usually been characterized as a condition in which most physiological functions are shutting down, and life expectancy is short. Specific age limits have varied through time, and from one society to another. Today it is more a subjective term than a definition as it is increasingly being recognized that an individual's chronological age does not necessarily reflect physiological and health status.

The definitions for the elderly used in the medical literature vary greatly, the most common limits being 60, 65, 70, 75 and 80.

The definition of acute kidney injury (AKI) is also evolving. In recent years, through the work of groups such as the International Acute Dialysis Quality Initiative and the Acute Kidney Injury Network (AKIN), the RIFLE and AKIN classifications have emerged [1].

Epidemiology of AKI in the Elderly

AKI has long suffered from the lack of a collective definition. This has impacted greatly on the reported incidences and prevalences which vary significantly in the literature. These large variations also depend on the population studied. The incidences of AKI described in community-based studies vary from 140 to 620 per million population. In hospitalized patients, the reported incidences vary between 5 and 7% [2].

With the aging of the population, the demand for intensive care unit (ICU) admission for older patients will continue to rise, and this clinical entity will likely become increasingly common [1].

The incidence of AKI in the general ICU population is much higher than that in the community or in hospitalized patients and ranges from 20 to 40% in most recent large epidemiological studies. Although these studies were conducted in mixed ICU populations including patients of all age groups, about half of all studied patients were older than 65 years.

Factors Contributing to AKI Development in the Elderly

Age-Related Changes

Several of the changes reported with aging are in fact due to the associated comorbidities. There is compelling evidence, however, of structural as well as functional changes occurring in the aging kidney, independent of related illnesses [3].

Alterations in the renal vasculature include arteriosclerosis, intimal and medial hypertrophy, and fibrointimal hyperplasia; they lead to cortical glomerulosclerosis, tubular atrophy, interstitial fibrosis, and compensatory hypertrophy, and hyperfiltration of the medullary glomeruli which results in hyperfiltration injury, and eventually global glomerulosclerosis.

Age-related tubular changes compromise functions in both proximal and distal tubules. This results in a decreased capacity to concentrate and dilute urine, and compromised glomerulotubular feedback.

The changes in volume distribution (increase in adipose tissue and decrease in muscle mass), reduced drug excretion, imbalances between vasoconstricting and vasodilating substances within the kidney favoring renal vasoconstriction, and overall reduced drug metabolism contribute to a heightened susceptibility to drug toxicity.

Moreover, older individuals also seem to show a reduced capacity for renal regeneration when facing acute insults due to various cellular and molecular mechanisms that have recently been described in the elderly.

Sepsis
Alterations in immunological functions with aging can lead to an augmented inflammatory response to renal injury, as well as increased susceptibility to infections, and a tendency for occult infections. This has enormous implications as sepsis is one of the leading causes of AKI development.

Coexisting Illness
The most important risk factor for the development of AKI is the presence of comorbidities. Chronic kidney disease, in particular, dramatically increases the odds of developing renal injury after an acute insult. Diabetes, hypertension, atherosclerotic vascular disease, and cardiac insufficiency are some of the illnesses most frequently leading to chronic kidney disease in older patients. Urinary tract obstruction is also a major cause of AKI almost exclusively seen in the elderly [3, 4].

Cardiorenal Syndrome
Today the term cardiorenal syndrome represents a well-established disorder, thanks to the new definition/classification [5]. It refers to a pathophysiological disorder of the heart and kidneys whereby acute or chronic dysfunction of one organ may induce acute or chronic dysfunction of the other. As cardiac disease is highly prevalent in the elderly, cardiorenal syndrome is probably also more common in older patients. Although several studies have been published on this topic, few were in relation to the elderly in an acute setting.

Medication-Related Toxicity
Hospitalized patients are at a high risk of drug toxicity, as most studies show that older people are prescribed an average of 5 new medications at discharge for various problems. This predisposition is due to the above-mentioned age-related kidney and systemic changes [3]. Non-steroidal anti-inflammatory drugs, nephrotoxic antibiotics, angiotensin-converting enzyme inhibitors, and loop diuretics are the most common offenders [4].

Contrast-Induced Nephropathy
Numerous studies have analyzed the risk factors for developing contrast-induced nephropathy in various patient populations, the most popular being the use of radiocontrast for cardiac catheterization. Most of these studies have reported advanced age as an independent risk factor for the development of contrast-induced nephropathy, although other factors such as underlying renal dysfunction and female sex carry a higher risk [6].

Related Perioperative Factors
The high prevalence of comorbidities leads to an increased need for surgical procedures in the elderly. The incidence of AKI ranges from about 0.1% in

general surgery to up to 31% or more in cardiac surgery. Advanced age has been identified as a risk factor for AKI development in most studies; however, other elements such as emergent surgery and comorbidities are certainly more important. Pathophysiological and iatrogenic events occurring in the preoperative period, during the surgery, or after, can all have an impact on renal function [7].

Etiology of AKI in the Elderly

Renal hypoperfusion secondary to cardiac failure, dehydration, bleeding, or hypotension of any etiology is often the cause of AKI development in the elderly population, especially in the critically ill. Sepsis and medication-related toxicity are also frequent culprits, both in and outside the ICU. As in most critically ill patients, however, AKI in the elderly is often multifactorial [4]. As mentioned above, obstructive renal failure is especially common. Atheroembolic renal disease and contrast-induced nephropathy are also frequent causes, especially in hospitalized patients at risk of undergoing diverse radiologic examinations and technical or surgical procedures. Moreover, renal causes of AKI, especially rapidly progressive glomerulonephritis and acute interstitial nephritis, are not uncommon in the elderly patient and need to be considered.

Strategies for AKI Prevention in the Elderly

Prevention of acute kidney injury is a crucial consideration in the management of all elderly patients, as they are particularly prone to a variety of iatrogenic renal insults [4].

Whenever feasible, the key elements in any prevention strategy are avoidance of hypovolemia, nephrotoxic drugs, and contrast medium. Hypovolemia, if present, must be recognized promptly and treated aggressively. With advanced age and increasing comorbidity, patients are often subjected to polypharmacy. Efforts must be directed at avoiding the combination of various nephrotoxic drugs, as well as the non-judicious use of drugs that can potentially cause hypovolemia, such as diuretics or laxatives. Whenever possible, drug levels must be measured in search of toxicity. As noted, the elderly are especially susceptible to developing infections. Particular attention must also be paid to the appearance of signs of sepsis, which is often occult in elderly frail patients. Infections are often accompanied by a state of hypovolemia, and usually require early and adequate fluid resuscitation as well as timely and appropriate antibiotic therapy.

Contrast-induced nephropathy is a complication that has been the subject of numerous studies on prophylactic strategies. In addition to the general approach described above, several specific preventative therapies have proven effective, such as adequate hydration and use of the smallest amount possible of a low or iso-osmolar non-ionic contrast agent [6]. Specific therapies such as the use of bicarbonate solutions for hydration instead of 0.9% saline, and supplementation with N-acetylcysteine are probably effective, although conflicting data exist in the literature [6].

Although intra-abdominal hypertension is most often seen in the ICU in abdominal surgery patients and in trauma or sepsis patients with massive fluid resuscitation, it can sometimes be encountered in the wards [7]. Of particular relevance to the elderly, advanced age is a major risk factor for the development of Ogilvie's syndrome, and cases of severe acute colonic pseudo-obstruction have increasingly been described as a complication after total hip replacement and hip arthroplasty. Clinicians caring for these patients should be aware that increases in intra-abdominal pressure can exert a negative impact on renal function long before the overt abdominal compartment syndrome has developed.

Finally, as in all critically ill patients, attention must be paid to the influences of mechanical ventilation on renal function.

Diagnosis of AKI in the Elderly

Serum creatinine (SC) is known to be an unreliable marker of acute renal dysfunction in most patients. Also, renal function must reach a steady state before SC can be of some value. Another major weakness in the value of SC as a diagnostic tool is the variable daily production rate of creatinine, which is highly dependent on muscle mass. This drawback is of greatest importance in the elderly, since muscle mass continuously decreases with age as it is replaced by adipose tissue [3]. SC also varies enormously according to intravascular volume. This becomes an important concern in older patients as they are more susceptible to developing dehydration [4]. As a consequence, for the same level of renal dysfunction, SC will generally tend to be much lower in elderly patients. This often leads to delayed recognition and late treatment initiation. Moreover, contrary to ICU patients, SC is seldom analyzed on a daily basis in most ward patients, and urine output is not reliably measured in the absence of an indwelling catheter. These considerations make the use of conventional criteria for the diagnosis of AKI even less appealing in these patients.

By including slight rises in creatinine in their classifications, the RIFLE and AKIN criteria strive to achieve an earlier diagnosis of AKI. Although they are not currently recommended for use in routine clinical practice, novel biomarkers such as neutrophil gelatinase-associated lipocalin, kidney injury molecule-1, interleukin-18, and cystatin C, have been useful for the early diagnosis of AKI and for prognosis prediction in several studies. With the exception of cystatin C, none of these biological markers have been specifically studied in older patients. However, most adult studies did include elderly patients.

Renal Replacement Therapy in Elderly Patients

Once renal injury is established and the diagnosis has been made, the most beneficial measures are those which avoid further renal function deterioration or additional renal insults [4]. Due to increased autonomic dysfunction, decreased cardiovascular reserve and frequent comorbidities and medications, elderly patients are more prone to hemodynamic complications during dialysis, such as interdialytic hypotension, hypertension, and arrhythmias. They are also more vulnerable to bleeding problems, and to neurologic complications due to rapid changes in serum electrolytes and osmolality. Continuous renal replacement therapies have not been specifically studied in the elderly. It is nevertheless known that they are generally associated with a more stable hemodynamic profile, as well as a lower risk of disequilibrium syndrome, an increased risk of bleeding related to continuous anticoagulation as well as higher costs. Among these uncertainties, it remains clear that clinical management choices should never be based on chronological age alone [8].

Renal Recovery and Mortality after AKI in the Elderly

The available studies on outcomes in elderly patients with AKI reported conflicting results. [9, 10]. In fact, while some studies reported no difference between age groups, several studies showed less renal recovery in older patients. These discrepancies may be due to variations in patient populations between studies, such as the prevalence of underlying chronic kidney disease.

However, it is evident that the chances of recovery are greatly decreased in the presence of chronic renal impairment. As the prevalence of chronic kidney disease is higher in elderly patients, it is quite probable that this population has a lower likelihood for substantial recuperation of renal function.

The available information on mortality is also inconclusive since it suffers from the same limitations as the data on renal recovery, as it is derived from similar studies. Indeed, not all studies showed an increased mortality risk in older patients with AKI, as compared to younger patients.

Conclusion

Age-related changes render the elderly highly susceptible to AKI. A high prevalence of comorbidities also contributes greatly to this increased risk, and brings along frequent iatrogenic complications such as drug toxicity. The most frequent causes are decreased perfusion related to dehydration, hypotension or cardiac insufficiency, medication or radiocontrast-related toxicity, sepsis, and surgery. Prevention is the key in the management of these patients, thus early identification is of utmost importance. Unfortunately SC is a poor diagnostic tool for AKI, even more so in the elderly. Although past studies have yielded conflicting results, older patients probably have lesser chances of renal recovery and a may have a higher mortality than their younger counterparts.

References

1 Lopes JA, Fernandes P, Jorge S, Gonçalves S, Alvarez A, Costa e Silva Z, França C, Prata MM: Acute kidney injury in intensive care unit patients: a comparison between the RIFLE and the Acute Kidney Injury Network classifications. Crit Care 2008;12: R110.

2 Nash K, Hafeez A, Hou S: Hospital-acquired renal insufficiency. Am J Kidney Dis 2002;39:930–936.

3 Martin JE, Sheaff MT: Renal ageing. J Pathol 2007;211:198–205.

4 Cheung CM, Ponnusamy A, Anderton JG: Management of acute renal failure in the elderly patient: a clinician's guide. Drugs Aging 2008;25:455–476.

5 Ronco C, Haapio M, House AA, Anavekar N, Bellomo R: Cardiorenal syndrome. J Am Coll Cardiol 2008;52:1527–1539.

6 McCullough PA, Adam A, Becker CR, et al., CIN Consensus Working Panel: Epidemiology and prognostic implications of contrast-induced nephropathy. Am J Cardiol 2006;98(suppl):5K–13K.

7 Noor S, Usmani A: Postoperative renal failure. Clin Geriatr Med 2008;24:721–729.

8 Michalsen A: Some thoughts on intensive care for elderly patients at the end of their lives. ICU Manage 2009;9:8–10.

9 Schmitt R, Coca S, Kanbay M, Tinetti ME, Cantley LG, Parikh CR: Recovery of kidney function after acute kidney injury in the elderly: a systematic review and meta-analysis. Am J Kidney Dis 2008;52:262–271.

10 Schiffl H: The pathogenesis of acute kidney injury may differ between elderly and younger patients. Am J Kidney Dis 2009;52:1198–1199.

Prof. Claudio Ronco, MD
Department of Nephrology, Dialysis & Transplantation, International Renal Research Institute
San Bortolo Hospital
Viale Rodolfi 37, IT–36100 Vicenza (Italy)
Tel. +39 444 753650, Fax +39 444 753973, E-Mail cronco@goldnet.it

Ronco C, Bellomo R, McCullough PA (eds): Cardiorenal Syndromes in Critical Care.
Contrib Nephrol. Basel, Karger, 2010, vol 165, pp 322–328

Blood Purification in Sepsis: A New Paradigm

Zhiyong Peng · Kai Singbartl · Peter Simon · Thomas Rimmelé · Jeffery Bishop · Gilles Clermont · John A. Kellum

Clinical Research, Investigation, and Systems Modeling of Acute Illness (CRISMA) Laboratory, Department of Critical Care Medicine, University of Pittsburgh Medical Center, Pittsburgh, Pa., USA

Abstract

Sepsis is one of the main causes of death in critically ill patients. The pathophysiology of sepsis is complex and not completely understood. The proinflammatory and anti-inflammatory response leads to cell and organ dysfunction and, in many cases, death. Thus, the goal of the intervention is to restore the homeostasis of circulating mediators rather than to inhibit selectively the proinflammatory or anti-inflammatory mediators. Blood purification has been reported to remove a wide array of inflammatory mediators. The effects are broad-spectrum and auto-regulating. Blood purification has also been demonstrated to restore immune function through improving antigen-presenting capability, adjusting leukocyte recruitment, oxidative burst and phagocytosis, and improving leukocyte responsiveness. A great deal of work has to be done in order to find and optimize the best extracorporeal blood purification therapy for sepsis. New devices specifically target the pathophysiological mechanisms involved in these conditions. High-volume hemofiltration, hemoadsorption, coupled plasma filtration adsorption, and high cutoff membrane are now being tested in septic patients. Preliminary data indicate the feasibility of these modified techniques in sepsis. Their impact on patient prognosis, however, still needs proof by large randomized clinical trials. Finally, the emerging paradigm of sepsis-induced immune suppression provides additional rationale for the development of extracorporeal blood purification therapy for sepsis.

Although substantial advances in the understanding of the pathogenesis of sepsis have been made, the mortality from sepsis remains relatively unchanged. Extracorporeal blood purification as an adjunctive therapy for sepsis has been used for almost a decade, but the efficacy is still disputed. In recent years there has also been considerable progress in our understanding of the technical capabilities and the rationale for its use. Although data from large-scale randomized clinical trials are lacking, some technologies seem promising in animal studies and small-scale clinical observations. However, over the last decade important new insights have been gained as to sepsis pathophysiology. It is time for us to reevaluate blood purification in light of these new findings.

Changing Concepts in the Pathophysiology of Sepsis

The pathophysiology of sepsis is complex and not completely understood. However, it is generally accepted that circulating proinflammatory and anti-inflammatory mediators appear to participate in a complex cascade of events which leads to cell and organ dysfunction and, in many cases, death. In the early course of this syndrome, a proinflammatory phase is recognized which is characterized by a release of proinflammatory mediators in response to such an overwhelming microbial invasion. The release of such mediators (e.g. IL-6, TNF) plays an important role in the defense of microbial invasion. However, prolonged activation or systemic spillover of mediators are potentially injurious to the host in many ways including oxidative stress, compromised tissue oxygen delivery, and diffuse coagulopathy. Clinically, this 'spillover' of such mediators is also associated with hyperdynamic shock, and the development of multisystem organ failure. Nevertheless, overwhelming septic shock with fulminate hemodynamic failure and death within a few hours is seldom observed in the modern intensive care unit. In most cases of sepsis, persisting inflammation promptly yields anti-inflammatory mediators which are interpreted as a protective mechanism to prevent the excessive production of proinflammatory mediators. However, the prolonged release of anti-inflammatory mediators (e.g. IL-10, IL-4, IL-1 receptor antagonist) leads to impaired immunity. Although it remains generally accepted that systemic inflammation in response to severe infection is in large part responsible for the development of shock and the occurrence of organ injury [1], it is becoming increasingly apparent that sepsis also induces severe immune suppression [2] and this defect plays a major role in mortality [3, 4]. Failure to address both the cytotoxic and immune suppressive components of sepsis may help to explain the lack of success in developing effective treatments for sepsis. Unfortunately, the two phenomena produce a conundrum: how can innate immunity be simultaneously curtailed to limit tissue damage and augmented to improve bacterial clearance? The answer, at least in part, appears to come from the multi-compartmental aspects of systemic infection and the role of chemokines and cytokines in directing leukocyte traffic. A useful model of immune response and dysfunction is immunosenescence. Older animals and humans exhibit greater cytokine response yet worse bacterial clearance and increased mortality [3]. Indeed, older individuals are also much more susceptible to sepsis. Recent evidence points to the critical role of chemokine gradients between the nidus of infection and the systemic circulation [5, 6]. Systemic activation of inflammation recruits immune effector cells into the circulation and into tissues away from the pathogen, simultaneously impairing pathogen clearance and increasing remote organ injury. Thus an important goal of treatment would be to reduce systemic inflammation without reducing inflammation at the local site of infection.

What Does Blood Purification Do?

Given the complexity of the host response to infection, an ideal therapy for sepsis must be able to attenuate both pro- and anti-inflammatory arms of the host response, promote pathogen clearance at the site of infection, and influence a variety of cells and mediators given the enormous redundancy of inflammatory signaling networks. In this regard, what does blood purification do?

Broad-Spectrum Removal of Inflammatory Mediators
Early blood purification devices were designed to remove waste products and water from patients with renal failure. It was then observed that renal replacement therapy could remove inflammatory mediators from the plasma of septic patients. Subsequently, an improvement in hemodynamics and, in small case series, improved survival was observed in patients with septic shock. Most of the immune mediators in sepsis are water-soluble and fall into the middle-molecular-weight category and hence can theoretically be removed by a blood purification system designed for renal support. These mediators include eicosanoids, leukotrienes, complement, cytokines, chemokines, and other potentially important small peptides and vasogenic substances. Available blood purification technologies can remove these inflammatory mediators via convection, diffusion, and adsorption. The effects are broad-spectrum and autoregulating. These advantages provide a powerful rationale for blood purification used in sepsis. Existing technology can achieve biologically relevant reductions in lipopolysaccharides (LPS), cytokines (e.g. IL-6, TNF-α, IL-10), chemokines, activated complement factors (e.g. C3a, C5a), coagulation factors, eicosanoids, and leukotrienes.

Altering Chemokine Gradients. We have preliminarily shown that a single four-hour treatment with hemoadsorption attenuated the increased mediator concentrations induced by experimental sepsis, and improved organ function and 1-week survival in the rat [7]. Mean survival time was significantly longer (5.7 vs. 4.5 days, $p < 0.05$), and overall survival to 7 days was significantly better (52 vs. 28%, hazard ratio 0.49, $p < 0.05$) with hemoadsorption compared to sham. In addition, multiple plasma cytokines, HMGB1, alanine aminotransferase and creatinine were significantly lower 72 h after sepsis hemoadsorption compared to sham. Of interest, the effect of treatment on cytokines and markers of organ function were not apparent immediately after the treatment but emerged over the first 48–72 h. Even more importantly hemoadsorption reduced mortality secondary to active infection even in the absence of antibiotics. This raises an important question: how does a treatment that attenuates the cytokine response improve survival in infection? Given the recent advances in the understanding of immune suppression in sepsis, we hypothesize that by removing cytokines and chemokines from the plasma compartment we are enhancing localization of neutrophils to the nidus of infection in the abdomen. Future studies are needed to test this hypothesis.

Restoration of Immune Function

As mentioned, the anti-inflammatory response can result in a pool of hyporesponsive white blood cells unable to perform immune-effector functions. This state has been called 'immunoparalysis'. Markers of this response, such as decreased monocytic cell-surface expression of human leukocyte antigen (HLA)-DR, are strongly associated with secondary nosocomial infection and mortality. Presumably, this is because cells are unable to respond to new infections or even clear the initial insult. Perhaps the best characterized aspect of this response is leukocyte 'reprogramming', as with prolonged exposure these cells become either hyporesponsive to previously proinflammatory stimuli, such as endotoxin and IL-1a, or even actively secrete anti-inflammatory cytokines (e.g. IL-10). In addition, monocytes and dendritic cells have been found to downregulate antigen recognition and display molecules, such as the LPS co-receptor (CD14), and the major histocompatibility complex (MHC) II proteins responsible for display of exogenous antigen to lymphocytes. Decreased antigen presentation by macrophages is only one mechanism through which the anti-inflammatory response encompasses adaptive immunity. Widespread apoptosis of T cells, B cells and dendrocytes occurs in the periphery and lymphoid tissue. Monocytes and neutrophils are important effector cells during this process. Blood purification has been reported to restore the immune function through the regulation of monocytes and neutrophils [8, 9].

Regulating Neutrophil Function. This rapid recruitment of leukocytes is crucial for pathogen clearance. The process of extravasation into inflamed tissue is a multistep and sequential process that requires the regulation of adhesion molecules and chemokines. When the leukocytes come into contact with the pathogen, the pathogen is digested through phagocytosis. Oxidative burst is a crucial reaction that occurs in phagocytosis to degrade internalized particles and pathogens. Oxidative burst and phagocytosis capacity have been shown to be impaired in septic patients. In an experimental porcine sepsis model, high-volume hemofiltration (HVHF; 100 ml/kg/h) improved oxidative burst and phagocytosis in polymorphonuclear leukocytes, which suggested that neutrophil function was stabilized by HVHF [8]. HVHF significantly reduced bacterial translocation and endotoxemia and improved long-term survival (>60 h), 67 vs. 33% with low volume hemofiltration versus 0% in controls.

Improving Monocyte Function. One of the most important functions of monocytes is antigen presentation. MHC II and CD14 expressed in monocytes are the critical effector molecules with regard to the antigen-presenting process. HLA-DR is a component of the MHC II, which is the central element in the process of antigen presentation. Reduction of HLA-DR expression on monocytes is considered to be a marker of deactivation of monocytes. It was reported that the percentage of HLA-DR was less than 30% in immunosuppressive patients. Downregulation of MHC II and CD14 expression on monocytes was also significantly improved by HVHF in the study by Yekebas et al. [8] discussed above. In a clinical study, coupled plasma filtration adsorption (CPFA) increased the HLA-DR expression in monocytes of septic patients [9].

Restoration of Leukocyte Responsiveness. In vitro whole blood LPS stimulus is an important method to evaluate the leukocyte responsiveness. Large amounts of cytokines will be produced after LPS in healthy subjects. However septic patients are in an immunosuppressive state characterized by significantly decreased cytokines after stimulus. HVHF prevented sepsis-induced in vitro endotoxin hyporesponsiveness both in animals and patients. CPFA also improved the leukocyte hyporesponsiveness after LPS stimulus in septic patients [9].

What Has Been Done to Improve These Technologies?

Over the last two decades, extracorporeal therapies used for blood purification have been improved thanks to technologic progress and some new techniques and strategies that have emerged. To augment convective clearance, hemofiltration rates were increased and HVHF was developed. To improve convective or diffusive clearances of middle molecular weight mediators during hemofiltration, membranes with appropriately high permeability, large surface and a nominal cutoff of 60–150 kD have been developed. Several types of high cutoff membranes of different chemical compositions have then been tested and shown to decrease cytokine levels and norepinephrine requirements. However, the cost was seen in significant albumin loss during higher ultrafiltration rates.

In extracorporeal blood purification techniques, the application of adsorbents has been proven to increase remarkably the removal of accumulated plasma mediators [10]. Therefore, an alternative strategy for the reduction of soluble immune mediators and modulation of cellular immunity is provided by adsorption technologies. Such strategies employ adsorptive materials in the extracorporeal circulation in order to achieve biologically relevant binding and thus removal of respective target mediators. The binding of these target molecules to the adsorbents is a result of hydrophobic interactions, electrostatic attraction, hydrogen bonding and van-der-Waals forces. These adsorptive materials include activated charcoal, polymyxin B-immobilized sepharose, amino-acid-immobilized adsorbents, polyamine-immobilized adsorbents, adsorbents having hydrophobic alkali chain ligand, and adsorbents with special antibodies. The present technology for appropriately modifying several intrinsic characteristics of the adsorbents has helped to fine-tune extracorporeal devices for more defined clinical applications. These properties, together with improved biocompatibility, have allowed the development of adsorptive techniques to obtain clearances and total removal rates of target compound that would be unthinkable with conventional hemodialysis or hemofiltration.

In order to remove the mediators with high efficacies, some hybrid techniques have been produced comprising the combination of advantages of different purification methods. CPFA is one of these techniques [9]. It involves plasma separation followed by an adsorptive step-over sorbent allowing non-specific removal of mediators. After return of the cleaned plasma to the circuit, standard hemofiltration is applied. As a low flow of plasma is used, it allows a longer contact time of mediators with the sorbent. Two small studies have shown that CPFA improves hemodynamic status and immune function in patients with severe sepsis and multi-organ dysfunction [9]. Interestingly, changes in monocyte function and hemodynamics were achieved without any measurable effect on plasma levels of TNF and IL-10.

In the future, the optimal blood purification strategy could be to combine physicochemical principles in one single technique. For example, a new hemofiltration membrane having enhanced adsorption properties was recently developed, allowing combination of high convective exchanges with endotoxin adsorption [11].

Conclusion

Severe sepsis and septic shock yield a profound immunologic disequilibrium. The overall concept of blood purification is to attenuate this overwhelming systemic overflow of pro- and anti-inflammatory mediators released at the early phase of septic shock, and restore immunologic equilibrium. Although several new extracorporeal support techniques are now available, they are still at the early stages of clinical testing. Initial studies appear promising and might provide significant advances. A solid proof of their efficacy and an assessment of their cost-effectiveness are needed before wider use is implemented. However, emerging theories of sepsis-induced immune dysfunction fit well with the concept of blood purification in sepsis and will likely spur on the advance of this therapy.

References

1 Bone RC: The pathogenesis of sepsis. Ann Intern Med 1991;115:457–467.
2 Hotchkiss RS, Coopersmith CM, McDunn JE, Ferguson TA: Tilting toward immunosuppression. Nat Med 2009;15:496–497.
3 Turnbull IR, Clark AT, Stromberg PE, et al: Effects of aging on the immunopathologic response to sepsis. Crit Care Med 2009;37:1018–1023.
4 Adib-Conquy M, Cavaillon JM: Compensatory anti-inflammatory response syndrome. Thromb Haemost 2009;101:36–47.
5 Alves-Filho JC, Freitas A, Souto FO, et al: Regulation of chemokine receptor by Toll-like receptor 2 is critical to neutrophil migration and resistance to polymicrobial sepsis. Proc Natl Acad Sci USA 2009;106:4018–4023.
6 Call DR, Nemzek JA, Ebong SJ, Bolgos GL, Newcomb DE, Remick DG: Ratio of local to systemic chemokine concentrations regulates neutrophil recruitment. Am J Pathol 2001;158:715–721.
7 Peng ZY, Wang H, Carter M, DiLeo MV, Kellum JA: Hemoadsorption improves long-term survival after sepsis in the rat (abstract). Crit Care Med 2008;36:A1.
8 Yekebas EF, Eisenberger CF, Ohnesorge H, et al: Attenuation of sepsis-related immunoparalysis by continuous veno-venous hemofiltration in experimental porcine pancreatitis. Crit Care Med 2001;29:1423–1430.
9 Ronco C, Brendolan A, Lonnemann G, et al: A pilot study of coupled plasma filtration with adsorption in septic shock. Crit Care Med 2002;30:1250–1255.
10 Peng ZY, Carter MJ, Kellum JA: Effects of hemoadsorption on cytokine removal and short-term survival in septic rats. Crit Care Med 2008;36:1573–1577.
11 Rimmele T, Assadi A, Cattenoz M, et al: High-volume haemofiltration with a new haemofiltration membrane having enhanced adsorption properties in septic pigs. Nephrol Dial Transplant 2009;24:421–427.

John A. Kellum, MD, FACP, FCCM
604 Scaife Hall, CRISMA Laboratory, Critical Care Medicine, University of Pittsburgh
3550 Terrace Street
Pittsburgh, PA 15261 (USA)
Tel. +1 412 647 7125, Fax +1 412 647 8060, E-Mail Kellumja@upmc.edu

Ronco C, Bellomo R, McCullough PA (eds): Cardiorenal Syndromes in Critical Care.
Contrib Nephrol. Basel, Karger, 2010, vol 165, pp 329–336

Ciné Phase-Contrast Magnetic Resonance Imaging for the Measurement of Renal Blood Flow

John R. Prowle[a] · Maurice P. Molan[b] · Emma Hornsey[b] · Rinaldo Bellomo[a]

Departments of [a]Intensive Care and [b]Radiology, Austin Health, Melbourne, Australia

Abstract

During critical illness, reductions in renal blood flow (RBF) are believed to be a major cause of kidney dysfunction, and therapy is often aimed at restoration of RBF. Despite this, our ability to measure RBF during critical illness has been limited by the invasiveness of the available techniques. Ciné Phase-Contrast Magnetic Resonance Imaging (CPC-MRI) represents an entirely noninvasive, contrast-free method of measuring blood flow with the potential of enabling the measurement of blood flow to major organs including the kidney. We have recently assessed the feasibility of measuring RBF by means of CPC-MRI in 2 critically ill patients with septic acute kidney injury and were able to compare such measurements to those obtained in a normal volunteer.

Acute kidney injury (AKI) is characterized by an abrupt and sustained decline in glomerular filtration rate (GFR). In the intensive care unit (ICU) AKI generally occurs in the setting of systemic illness, in particular sepsis. Under these circumstances alterations in renal blood flow (RBF) are considered central to the pathogenesis of AKI. However, our ability to measure RBF in critically ill patients is limited and surprisingly few measurements of RBF have been carried out in contemporary critically ill patients. This may be explained the lack of accurate, noninvasive and safely applicable techniques for measuring RBF in these patients.

The most widely cited methods for measurement of RBF are clearance techniques first described in the 1940s [1]. Using these methods, clearance of a substance, para-aminohippuric acid (PAH), completely excreted from the circulation into the urine on first pass through the renal circulation is measured; this clearance being theoretically equal to RBF. However, such estimates are only accurate in the presence of intact tubular function and are thus inapplicable in many AKI cases unless renal vein sampling is used to correct for impaired PAH extraction. This requirement greatly increases the invasiveness of the measurement process. Furthermore, since urinary PAH measurements are needed, the accuracy of this technique is difficult to estimate during oliguria. A number of other techniques for measurement of RBF have been investigated; these include trans-renal dye dilution, renal venous or arterial endo-luminal Doppler-ultrasound flow probes, and nuclear medicine washout techniques. Collectively these methods are highly invasive and/or insufficiently accurate for estimation of RBF in critically ill patients with AKI [2] suggesting the need to develop and test the feasibility of new less invasive, more accurate, reproducible and safe techniques. Here, we describe a proof of concept study which aimed to assess the feasibility of measuring RBF in critically ill patients with AKI by means of a technique called Ciné Phase-Contrast Magnetic Resonance Imaging (CPC-MRI), which appears to have the potential to improve our ability to estimate RBF in critically ill patients with AKI.

Ciné Phase-Contrast MRI

MRI techniques detect atomic nuclei with non-zero quantum mechanical spin. The most import of these are hydrogen nuclei, or protons, occurring in water and organic molecules. These nuclei possess magnetic moments that align with strong magnetic fields and interact with applied radiofrequency pulses. The results of these interactions are detected as induced currents, providing spatial and compositional information about the tissue under examination.

In phase-contrast MRI [3] a magnetic field gradient is employed, so that the magnetic moments of protons align with the field at a speed proportional to their proximity to the gradient. If such a magnetic field gradient is applied, reversed and applied for the same duration in the opposite polarity, the net effect on protons in static tissues is zero. However, if a vessel is aligned perpendicular to the field gradient, protons in the bloodstream will be more affected when they are closer to the maximum intensity of the field. Thus, after application and reversal of the field gradient, there will be an alteration in the magnetic moment (a phase change) only in protons moving toward or away from the field, with the magnitude of this phase change being proportional to their velocity. Summation of these phase changes generates a 2D cross-sectional velocity map and allows an estimate of instantaneous blood flow rate within a region of interest. This region of interest can correspond to a vessel cross-section. To calculate blood flow over the cardiac cycle, ECG gating is used to obtain data throughout the RR interval, building up a cinematographic phase-contrast image of blood flow during 1 heartbeat. From these data a flow-time relationship can be plotted. Multiplication of the flow-time integral by the heart rate then provides total blood flow in the given vessel (fig. 1).

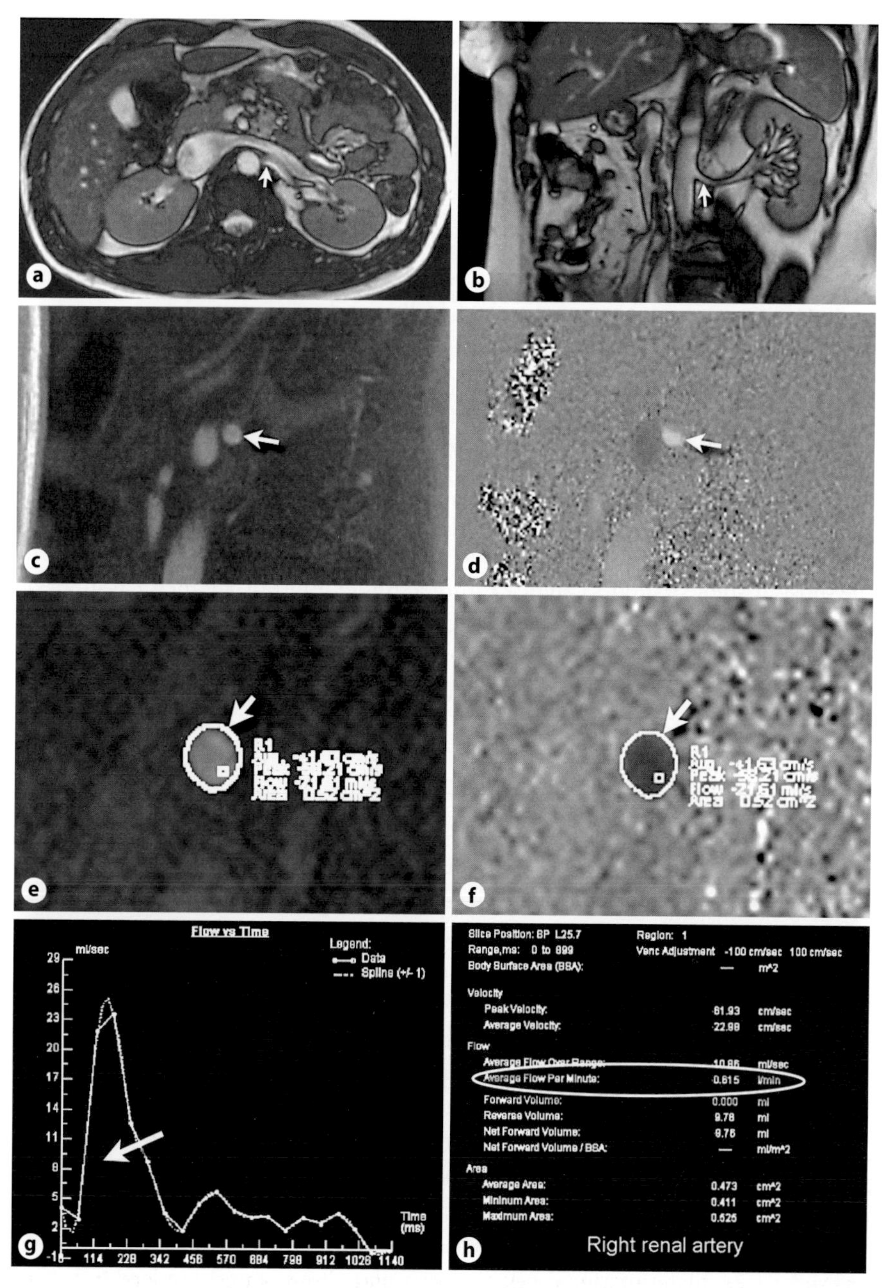

Fig. 1. Measurement of renal blood flow using phase-contrast MRI. **a, b** Scout images allow a cross-sectional view of the renal artery (arrows) to be localized. **c** Sagittal cross-section through renal artery, anatomical view. **d** Sagittal cross-section through renal artery, phase-contrast. **e** Definition of a region of interest (arrow) around the renal artery. **f** Region of interest defines area in which phase changes are analyzed. **g** ECG gating allows an image cycle modeling flow over one RR interval to be modeled and a flow-time curve generated. The area under the curve (arrow) represents the blood flow rate during one heartbeat. **h** Multiplication by heart rate provides flow per minute in the final output.

CPC-MRI is regarded as the most accurate technique available for measurement of flows in large vessels and has been extensively validated in the setting of cardiac MRI [3]. It is completely noninvasive, requiring no injections, intravascular devices or exposure to ionizing radiation. In normal individuals this technique has been used to reproducibly measure RBF [4] and has been validated against PAH clearance techniques [5, 6].

MRI safety requirements preclude the examination of patients possessing implanted electronic devices or some potentially mobile ferro-magnetic foreign bodies. Even non-ferro-magnetic electrically conductive materials can become heated in a strong magnetic field. Thus, some invasive monitoring devices used in the ICU, such as continuous cardiac output thermodilution pulmonary artery catheters and urinary catheter temperature probes, should not be brought into the MRI scanner. Completion of local safety screening requirements is mandatory for all patients and staff prior to entry into the MRI scanner room. Nevertheless, with appropriate use of modern MRI-compatible equipment, including invasive arterial blood pressure monitoring and advanced mechanical ventilators, all but the most unstable patients can be examined under appropriate medical supervision. Given that a method of safely and reliably measuring RBF during critical illness might now be available, it is pertinent to examine why these measurements are important and what mechanistic insights they might provide.

Renal Blood Flow during AKI

Decreased RBF has been described as an essential element in the pathogenesis of AKI during critical illness [7]. Such decreases could be secondary to changes in systemic blood pressure and cardiac output or active renal vasoconstriction. However, the relationship between GFR and alterations in RBF, renal vascular resistance (RVR) or systemic perfusion pressure is complex. GFR is principally dependent on adequate glomerular filtration pressure and is only directly RBF dependent in terms of the speed with which glomerular oncotic pressure rises to oppose the hydrostatic pressure gradient. The systemic perfusion pressure and the distribution of RVR are the main determinants of glomerular hydrostatic pressure. A rise in RVR can raise or lower filtration pressure and thus GFR, depending on its distribution between pre-glomerular and post-glomerular arterioles. Thus, a pure rise in afferent resistance will tend to decrease both RBF and GFR. In contrast, efferent arteriolar vasodilatation decreases in glomerular pressure and can lead to loss of ultrafiltration in the context of normal or elevated RBF. In contrast, vasoconstriction can reduce RBF while simultaneously acting to maintain glomerular capillary pressure and GFR as occurs in the physiological renin-angiotensin responses to hypotension. When measured, RBF during AKI in man is generally lower than normal [2], although the magnitude of decrease in RBF correlates poorly with the severity of AKI [8]. However, reductions in GFR have been reported in the context of normal or elevated RBF in animal models of sepsis [9], after injection of endotoxin into humans [10] and during the therapeutic use of inflammatory cytokines in malignancy [11], situations where vasodilatory influences may predominate.

As well as influencing glomerular hemodynamics, reduced RBF is believed to mediate AKI by causing renal ischemia [7]. However, the true role of ischemia in the causation of AKI is ill understood and may be overstated [12]. Falls in RBF in AKI are generally of smaller magnitude than those in GFR, while, as most renal metabolic activity involves the active reabsorption of filtered electrolytes, if GFR falls so does renal oxygen demand. In keeping with these observations, overt histopathological abnormalities, as seen in ischemia-reperfusion models of AKI, are largely absent from clinical cases of AKI during critical illness [13]. Subtler renal tubular cell injury or metabolic dysfunction are likely to play a role in the pathogenesis of AKI, however. Biomarkers of AKI now allow the detection of early or subclinical tubular injury before loss of GFR is evident in plasma biochemistry. Such damage is likely to be multifactorial in nature. A number of mechanisms may then link tubular injury to loss of GFR. In particular, persistent loss of glomerular ultrafiltration pressure may be important to the initiation and maintenance of AKI [14]. Persistent afferent vasoconstriction, perhaps mediated by aberrant triggering of tubuloglomerular feedback, could play a role in this process [14].

Given the complex and variable relationship between RBF and GFR during AKI and the dearth of clinical data available in critically ill patients [2], better methodology for the measurement of RBF is required. CPC-MRI has great potential to further our knowledge by allowing safe and reliable measurement of this crucial variable in the critically ill.

Early Observations

Using the CPC-MRI technique, we have measured RBF in AKI (table 1). Measurements were made in a healthy volunteer and were consistent with normal values for RBF (table 1, Normal). A wide variety of MRI hardware and software is available. Performance of pilot examinations in normal individuals is recommended when first using MRI-based blood flow measurements in an institution [3].

Examinations have also been performed in patients with AKI complicating critical illness. Over 50% of AKI in the ICU is seen in the context of sepsis. Many such patients have hyperdynamic, vasodilatory shock with elevated cardiac output despite high dose vasopressor therapy. RBF measurement in a typical patient with septic shock demonstrates that severe AKI can occur despite values of RBF in the normal range (table 1, Septic AKI). The maintenance of normal renal oxygen delivery in the context of reduced metabolic activity suggests generalized renal ischemia was unlikely to be occurring when these observations were made. However, despite nearly normal RBF in this individual, RBF

Table 1. Hemodynamic measurements with CPC-MRI in 3 individuals

Setting	Normal	Septic AKI	Established AKI
Age, years	36	54	74
Sex	male	male	male
Body surface area, m^2	1.98	1.97	2.03
Diagnosis	none	pneumonia	urosepsis
Baseline creatinine, µmol/l	78	62	110
Peak creatinine, µmol/l	–	206	441
Day after AKI	–	1	6
Renal replacement therapy?	–	yes	yes
Mean arterial pressure, mm Hg	85	75	100
Vasopressor	–	noradrenaline 33 µg/min	–
Cardiac index, liters/min/m^2	3.5	8.7	3.4
Systemic vascular resistance, dyn/s/cm^5/m^2	1,852	617	2,126
Creatinine clearance, ml/min	120	11	5
Renal blood flow index, ml/min/m^2	687	584	255
Renal fraction, %	19.6	6.7	7.4
Renal vascular resistance index	9,436	9,173	28,567
Filtration fraction, %	17	1.4	1.3

as a fraction of cardiac output was only one third of normal, as cardiac output was significantly elevated in the context of hyperdynamic septic shock. In other words, while RVR was similar to that in healthy individuals, it was elevated in comparison to the very low systemic vascular resistance seen in hyperdynamic septic shock. Additional inferences can be made from these observations; the large decreases in GFR suggest that glomerular capillary pressures were likely to be low. Since systemic blood pressure was maintained nearly normal with vasopressors, this would be consistent with a redistribution of RVR.

In contrast, in a typical patient with established AKI (table 1), low RBF was seen. This was despite resolution of any hypotensive state and systemic vasodilatation. Reduced renal fraction of cardiac output remained the consistent finding in AKI. These data suggest persistence of afferent arteriolar vasoconstriction in established AKI. Recovery of renal function may be associated with a combination of efferent vasoconstriction and afferent relaxation, as restoration of GFR seems to precede any rise in total RBF.

Implications for Future Research

Interpretation of RBF and vascular resistance is most meaningful in the context of cardiac output and systemic resistance. CPC-MRI provides the potential to measure cardiac output during the same examination. Use of this technology enables highly accurate cardiac output measurement in less sick patients who might not otherwise require invasive monitoring. As this is a noninvasive investigation, examinations may be repeated throughout the clinical course of AKI. In the future, the effects of kidney-directed vasoactive therapy could be investigated using this technique, or patients screened as candidates for such therapy based on RBF measurements.

Limitations of PC-MRI Technology

Most ICU patients can be successfully examined. However, some will inevitably be excluded by contraindications to MRI or extreme clinical instability. Awake patients may not tolerate the enclosed scanner or provide movement artifacts. Additionally slow irregular heart rates (wide variation in RR interval) can confound ECG gating and produce unreliable results. More significantly, MRI examinations are expensive, consuming approximately 1 h of scanner time and a similar period of one-to-one medical supervision. Large clinical studies with repeated measurements could be logistically difficult or too costly. However, MRI may provide a gold standard for comparison/calibration of new methods for assessing blood flow or assessing treatment effects in a subgroup of a larger clinical trial. Similarly, baseline measurements of RBF prior to overt AKI may be more practical and could be investigated as risk stratification assessment in, for instance, patients with chronic liver or cardiac disease at risk of acquiring AKI.

Conclusions

CPC-MRI is a promising method of measuring absolute and relative RBF in critically ill patients with AKI. Such measurements could provide unique insights into the pathogenesis of AKI and allow physiological assessment of potential therapies.

References

1 Chasis H, Redish J, Goldring W, Ranges HA, Smith HW: The use of sodium *p*-aminohippurate for the functional evaluation of the human kidney. J Clin Invest 1945;24:583–588.

2 Prowle JR, Ishikawa K, May CN, Bellomo R: Renal blood flow during acute renal failure in man. Blood Purif 2009;28:216–225.

3 Arheden H, Stahlberg F: Blood flow measurements; in de Roos A, Higgins CB (eds): MRI and CT of the Cardiovascular System, ed 2. Philadephila, Lippincott Williams & Wilkins, 2006, pp 71–90.

4 Bax L, Bakker CJ, Klein WM, Blanken N, Beutler JJ, Mali WP: Renal blood flow measurements with use of phase-contrast magnetic resonance imaging: normal values and reproducibility. J Vasc Interv Radiol 2005;16:807–814.
5 Sommer G, Corrigan G, Fredrickson J, Sawyer-Glover A, Liao JR, Myers B, Pelc N: Renal blood flow: measurement in vivo with rapid spiral MR imaging. Radiology 1998;208:729–734.
6 Wolf RL, King BF, Vicente ET, Wilson DM, Ehman RL: Measurement of normal renal artery blood flow: cine phase-contrast MR Imaging vs clearance of p-aminohippurate. AJR Am J Roentgenol 1993;161:995–1002.
7 Schrier RW, Wang W: Acute renal failure and sepsis. N Engl J Med 2004;351:159–169.
8 Prowle JR, Ishikawa K, May CN, Bellomo R: Renal plasma flow and glomerular filtration rate during acute kidney injury in man. Ren Fail 2010;32, in press.
9 Langenberg C, Wan L, Egi M, May CN, Bellomo R: Renal blood flow in experimental septic acute renal failure. Kidney Int 2006;69:1996–2002.
10 Bradley SE, Chasis H, Goldring W, Smith HW: Hemodynamic alterations in normotensive and hypertensive subjects during the pyrogenic reaction. J Clin Invest 1945;24:388–404.
11 Mercatello A, Hadj-Aïssa A, Négrier S, et al: Acute renal failure with preserved renal plasma flow induced by cancer immunotherapy. Kidney Int 1991;40:309–314.
12 Saotome T, Ishikawa K, May CN, Birchall IE, Bellomo R: The impact of experimental hypoperfusion on subsequent kidney function. Intensive Care Med 2010;36:533–540.
13 Langenberg C, Bagshaw SM, May CN, Bellomo R: The histopathology of septic acute kidney injury: a systematic review. Crit Care 2008;36:S198–S203.
14 Alejandro V, Scandling JD, Sibley RK, et al: Mechanisms of filtration failure during postischemic injury of the human kidney. A study of the reperfused renal allograft. J Clin Invest 1995;95:820–831.

Prof. Rinaldo Bellomo
Department of Intensive Care
Austin Health, 145 Studley Road
Heidelberg, Vic. 3084 (Australia)
Tel. +61 3 9496 5992, Fax +61 3 9496 3932, E-Mail rinaldo.bellomo@austin.org.au

Ronco C, Bellomo R, McCullough PA (eds): Cardiorenal Syndromes in Critical Care.
Contrib Nephrol. Basel, Karger, 2010, vol 165, pp 337–344

The Meaning of Transient Azotemia

Shigehiko Uchino

Intensive Care Unit, Department of Anesthesiology, Jikei University School of Medicine, Tokyo, Japan

Abstract

Acute kidney injury (AKI) is common in hospitalized patients and its associated mortality is high. The causes of AKI are commonly divided into 3 groups: pre-renal, intra-renal, and post-renal. According to this paradigm, pre-renal azotemia (PRA) represents a separate entity characterized by a rapidly reversible increase in serum creatinine and urea concentration. This rapid reversibility is believed to reflect a functional reduction in glomerular filtration rate as opposed to established structural kidney injury, which leads to acute tubular necrosis (ATN). This PRA vs. ATN paradigm is well established in the medical and renal literature and widely discussed in textbooks. However, there is no consensus definition for PRA or ATN. The typical description for PRA in the literature is 'reversible increase in serum creatinine and urea concentrations', 'characterized by intact renal parenchymal function but renal hypoperfusion'. Therefore, although the term PRA implies that it is defined histopathologically, it also contains a functional aspect (transient azotemia, TA). Early recognition of PRA or ATN is considered important because PRA can be reversed with fluid resuscitation, but such treatment causes edema in lungs as well as other tissues and therefore can be harmful in ATN. However, evidence suggests that PRA cannot be diagnosed prospectively and is clinically the same as TA, that urinary analysis and biochemistries cannot distinguish PRA and ATN in septic AKI, and that ATN is histologically uncommon in septic AKI. Recent observational studies also found that TA cannot be distinguished from ATN epidemiologically and that the existence of TA is related to high hospital mortality. These findings suggest the need for specific and focused investigations directed at identifying effective treatments to decrease the incidence of TA in hospitalized patients.

Acute kidney injury (AKI) is common in hospitalized patients and its associated mortality is high. The causes of AKI are commonly divided into 3 groups: pre-renal, intra-renal, and post-renal. According to this paradigm, so-called pre-renal azotemia (PRA) represents a separate entity characterized by a rapidly reversible increase in serum creatinine and urea concentration. This rapid reversibility is believed to reflect a functional reduction in glomerular filtration rate as opposed to established structural kidney injury, which leads to more prolonged AKI, so-called acute tubular necrosis (ATN). This PRA vs. ATN paradigm is well established in the medical and renal literature and widely discussed in textbooks [1, 2]. These two causes have been reported to account for approximately 70% of all cases of AKI.

Early recognition of PRA or ATN is considered important because PRA can be reversed with fluid resuscitation, but such treatment causes edema in lungs as well as other tissues and therefore can be harmful in ATN. Multiple studies have been conducted to deal with this issue using urinalysis, serum and urinary biochemistries and ultrasound. These studies have argued that PRA and ATN can be diagnosed with reasonable accuracy using such tests. However, more recently, the usefulness and ability of these tests have been questioned. In addition, a systematic review of the histopathology of septic AKI showed that the majority of studies reported normal histology or only mild, nonspecific changes and that ATN was uncommon. These observations call into question the validity and usefulness of the PRA vs. ATN paradigm in AKI.

Similarity of PRA and TA

Although the term PRA is often used in the literature and textbooks, rather surprisingly there is no consensus definition for PRA. The typical description for PRA in the literature is 'reversible increase in serum creatinine and urea concentrations', 'characterized by intact renal parenchymal function but renal hypoperfusion' [1, 2]. Therefore, although PRA implies that it is defined histopathologically (reduced renal perfusion and intact renal parenchyma), it also contains a functional aspect (transient azotemia, TA). This description is easily understood; however, it is too vague to be used in clinical studies.

There are multiple clinical studies that have attempted to distinguish PRA and ATN using urinalysis, serum and urinary biochemistries and ultrasound. In these studies, PRA was defined in various ways (table 1). PRA was often defined clinically using medical history, diagnoses, physical findings, urinary analysis and biochemistries, and response to treatment. However, the most common definition of PRA in these studies is based on its functional aspect: recovery of renal function (decrease in serum creatinine) after correction of hemodynamic abnormalities. For example, Nickolas et al. [3] tested the sensitivity and specificity of a single measurement of urinary neutrophil gelatinase-associated lipocalin (NGAL) to detect AKI in 635 hospitalized patients. They found that NGAL had a very good sensitivity and specificity to distinguish AKI from PRA or chronic kidney disease. In this study, PRA was defined as 'a new-onset increase in serum creatinine level that satisfied our criteria for altered kidney function and either resolved within 3 days with treatment aimed at restoring

Table 1. Definitions of PRA used in studies differentiating PRA and ATN

Author	Year	Test	Definitions of PRA
Perlmutter [11]	1959	Urine:serum urea nitrogen ratio	Oliguria and azotemia lasting less than 48 h
Espinel [12]	1976	FE-Na	Prompt increase in urinary output and creatinine clearance effected by hemodynamic improvement
Miller [13]	1978	Urinary indices	Return of renal function to normal within 24 to 72 h after correction of hemodynamics
Platt [14]	1991	Doppler ultrasound	Clinical judgment (definitions not mentioned)
Chew [15]	1993	Urinary enzymes	Rapid recovery of renal function after treatment of hypotension or dehydration
Steinhauslin [16]	1994	FE-lithium, FE-UA	Decrease in plasma creatinine toward normal values within 72 h of correction of hemodynamic abnormalities
Izumi [17]	2000	Doppler ultrasound	Not clearly mentioned, but FENa used
Carvounis [18]	2002	FE-urea	Prompt increase in urinary output and creatinine clearance after hemodynamic improvement
Parikh [19]	2004	Urinary IL-18	Multiple definitions but included improvement after treatment
Pepin [20]	2007	FE-Na, FE-urea	Two of 4 criteria (history, physical findings, urine analysis, rapid return to baseline renal function within 7 days)
Perazella [21]	2008	Urine microscopy	Improvement to baseline after fluid resuscitation and/or hemodynamic manipulation within 48 h
Nickolas [22]	2008	Urinary NGAL	Resolved within 3 days or FENa <1%

PRA = Pre-renal azotemia; ATN = acute tubular necrosis; FE = fractional excretion; Na = sodium; UA = uric acid; IL-18 = interleukin-18, NGAL = neutrophil gelatinase-associated lipocalin.

perfusion, such as intravenous volume repletion or discontinuation of diuretics (in the setting of historical and laboratory data suggesting decreased renal perfusion), or was accompanied by fractional excretion of sodium less than 1% at presentation'. They also defined AKI when it 'sustained for at least 3 days despite volume resuscitation'. Therefore, essentially, PRA was distinguished from AKI based on the short duration of AKI (TA). When it was mentioned, this duration was variable among the studies, from 24 h to 7 days (most commonly 72 h).

When a patient is diagnosed as PRA, the only thing certain is that the patient had a short duration of AKI, and it is uncertain that this patient had reduced renal perfusion or intact renal parenchyma. Furthermore, such a diagnosis can only be made retrospectively.

Recent Studies and Systematic Reviews for PRA and ATN

Another issue related to the PRA vs. ATN paradigm in AKI is that some evidence suggests that PRA and ATN cannot be distinguished. Bagshaw et al. [4] conducted a systematic review of studies for urinary biochemistry and microscopy in septic AKI and identified 27 articles including 1,432 patients. Urinary biochemistry or derived indices were reported in 24 articles and microscopy in 7 articles. The majority were small single-center reports and had serious limitations: only 54% of patients had AKI; many studies failed to include a control group; time from diagnosis of sepsis or AKI to measurement of urinary tests was variable, and there were numerous potential confounders. Urinary sodium, fractional excretion of sodium, urinary-plasma creatinine ratio, urinary osmolality, urinary-plasma osmolality ratio, and serum urea-creatinine ratio showed variable and inconsistent results. A few reports of urinary microscopy described muddy brown/epithelial cell casts and renal tubular cells in patients with septic AKI, whereas others described normal urinary sediment.

They extended their systematic review to experimental animal studies as such studies seemed ideal for accurately defining the timing and severity of the renal insult and for correlating any urinary findings with histopathology while minimizing confounding factors [5]. Twenty-seven articles fulfilled all inclusion criteria. The fractional excretion of sodium exhibited a decrease, no change, or an increase from baseline in 11, 5, and 5 studies, respectively. The urine osmolality decreased from baseline in all endotoxin-induced models but showed an early transient increase in 6 studies of cecal-ligation perforation. Proteinuria or urinary enzymuria was reported in only 7 studies and urinary microscopy was described in 1 study. Only 10 studies simultaneously reported on histopathology. In all these studies, histology either was normal or showed minor ultrastructural changes on electron microscopy. Two studies suggest that urinary biochemistry, indices, and microscopy cannot be used reliably to classify the clinical course of AKI in septic patients into ATN and PRA. Logically, given that approximately 50% of AKI in critical illness is associated with sepsis, it seems statistically similarly impossible to prospectively distinguish PRA from ATN on clinical grounds and/or urinary findings in critically ill patients overall. While a specific group of such patients might exist in which this distinction can be achieved prospectively, no studies have yet identified it.

Of relevance to these concerns, the exclusive presence of ATN in patients with delayed recovery has also recently been questioned. Hotchkiss et al. [6]

conducted an autopsy study of 20 critically ill septic patients and found that, despite a high prevalence of AKI (65%), only 1 patient had pathologic evidence of ATN. Langenberg et al. [7] reviewed all studies reporting the histopathology of septic AKI including the study by Hotchkiss et al. [6] and found that only 22% of patients had features suggestive of ATN. They also reviewed septic AKI animal models and found 4 primate studies and 13 rodent studies. In these, only 37 and 23%, respectively, showed features of ATN on histopathological examination. In 2 additional studies performed in dog and sheep models, there was no evidence of ATN. Overall, when ATN was absent, studies reported a wide variety of kidney morphologic changes in septic AKI, ranging from normal (in most cases) to marked cortical tubular necrosis.

Recently, in keeping with such evolving paradigms, the Acute Kidney Injury Network (AKIN) proposed that, given the theoretical and practical difficulties in the usage of the historical terms of PRA and ATN, these terms be discarded and replaced with 'volume-responsive AKI' and 'volume-unresponsive AKI' [8]. We understand that these new terms still require validation and recognize that there are variable and graded responses to volume administration. However, the AKIN approach may help decrease some of the confusion associated with traditional terminology.

The Meaning of TA

If PRA, which is clinically the same as TA characterized by reversibility, is a distinguished syndrome from ATN, it should have two important epidemiological features: a bimodal distribution in the AKI duration (short-lived AKI due to TA and prolonged AKI due to ATN), and lack of or a very small independent association between TA and hospital mortality.

To the best of my knowledge, there are only 2 epidemiological studies that specifically examined the meaning of TA [9, 10]. Tian et al. [9] conducted a retrospective cohort study for patients admitted to their medical wards in a US teaching hospital. Of 6,033 patients, 735 had AKI (defined as an increment in the serum creatinine level of 0.3 mg/dl or greater in 48 h). Of these, 443 (60%) had serum creatinine levels that subsequently decreased by 0.3 mg/dl or greater within 48 h and 197 returned to normal levels within 48 h. These 197 patients (fully reserved AKI) had a substantially higher mortality rate (14.2%) than those patients without AKI (1.3%, $p < 0.01$) but this mortality rate was similar to that of all AKI patients (14.8%). The multivariate logistic regression analysis revealed that 'fully reserved AKI' had an adjusted odds ratio of 4.4 for mortality.

We also conducted a retrospectively cohort study for patients admitted to a university-affiliated hospital in Australia [10]. In this study, patients were divided into 3 groups: pre-renal, ATN and no-AKI. Patients were classified in the no AKI group if they did not satisfy any of the RIFLE criteria for AKI (peak creatinine

Table 2. Odds ratios for hospital mortality in multivariate logistic regression analysis

	All patients		Re-admission	
No AKI	1.00 (reference)			
ATN	6.07 (5.31–6.94)	p < 0.0001	6.49 (5.23–8.01)	p < 0.0001
TA	2.26 (1.86–2.76)	p < 0.0001	2.02 (1.50–2.72)	p < 0.0001
AKI duration				
1 day	1.93 (1.46–2.56)	p < 0.0001	1.85 (1.22–2.81)	p = 0.0036
2 days	2.28 (1.64–3.16)	p < 0.0001	1.55 (0.90–2.65)	p = 0.11
3 days	3.34 (2.28–4.89)	p < 0.0001	3.60 (2.04–6.37)	p < 0.0001
4–7 days	2.71 (1.95–3.77)	p < 0.0001	2.43 (1.45–4.07)	p = 0.0007
8–14 days	4.09 (2.51–6.64)	p < 0.0001	2.36 (0.99–5.60)	p = 0.050
>2 weeks	4.01 (1.59–10.1)	p = 0.0033	5.69 (1.50–21.6)	p = 0.011

AKI = Acute kidney injury; ATN = acute tubular necrosis; TA = transient azotemia.

less than 1.5 times the baseline). Renal recovery was defined when patients with AKI had creatinine reduced to the no AKI range during hospital stay without receiving renal replacement therapy. If renal recovery occurred within 3 days after diagnosing AKI, these patients were classified in the pre-renal group. The other patients (no renal recovery, received renal replacement therapy, renal recovery after more than 3 days) were classified in the ATN group. Because approximately three quarters of patients did not have a known baseline creatinine, sensitivity analyses were conducted by separately studying only patients with more than one admission (measured baseline creatinine available).

Among 20,126 study patients, 3,641 patients (18.1%) had AKI. In 1,600 patients, AKI recovered during their hospital stay. The distribution of AKI duration was not bimodal. Renal function recovered within 3 three days in 1,172 patients(pre-renal group). Patients whose renal function recovered after 3 days (n = 428), or whose renal function did not recover to baseline during their hospital stay (n = 1,817), or who were treated with renal replacement therapy (n = 224) were classified in the ATN group (n = 2,469). Table 2 shows the results of multivariate logistic regression analysis for hospital mortality. Even after excluding confounding factors, pre-renal AKI was an independent predictor of hospital mortality with a high odds ratio (2.26, $p < 0.0001$). Because the threshold of 3 days to distinguish PRA and ATN was arbitrary, multivariate analysis for hospital mortality was repeated for different durations of AKI (1, 2, 3, 4–7, 8–14 and >14 days). There was a gradual increase in the odds ratio for hospital mortality and no obvious threshold to distinguish PRA from ATN. Even 1 day of TA had a significantly increased odds ratio for hospital mortality. The sensitivity analysis for patients with more than 1 admission (for whom baseline kidney function was available) during the study period confirmed the findings for all study patients.

These 2 studies confirmed that patients with TA, although definitions were different between the 2 studies, had significantly higher hospital mortality compared to patients with no AKI. We also found that there was no bimodal distribution in AKI duration to justify the separation of TA from ATN. These findings strengthen the view that AKI is a continuum of injury and that the artificial separation of PRA as an entity is not justified on epidemiological grounds.

Conclusion

Although both terms PRA and ATN imply that there are pathophysiological differences between the two, renal biopsy is rarely conducted in AKI patients. Evidence suggests that: (1) PRA cannot be diagnosed prospectively and is clinically the same as TA; (2) urinary analysis and biochemistries cannot distinguish PRA and ATN in septic AKI, and ATN is histologically uncommon in septic AKI.

Recent epidemiological studies, including ours, have added new information on the issues of the PRA vs. ATN paradigm: TA cannot be distinguished from ATN epidemiologically and the existence of TA is related to high hospital mortality. These findings suggest the need for specific and focused investigations directed at identifying effective treatments to decrease its incidence in hospitalized patients.

References

1 Lameire N, Van Biesen W, Vanholder R: Acute renal failure. Lancet 2005;365:417–430.

2 Blantz RC: Pathophysiology of pre-renal azotemia. Kidney Int 1998;53:512–523.

3 Nickolas TL, O'Rourke MJ, Yang J, Sise ME, Canetta PA, Barasch N, Buchen C, Khan F, Mori K, Giglio J, Devarajan P, Barasch J: Sensitivity and specificity of a single emergency department measurement of urinary neutrophil gelatinase-associated lipocalin for diagnosing acute kidney injury. Ann Intern Med 2008;148:810–819.

4 Bagshaw SM, Langenberg C, Bellomo R: Urinary biochemistry and microscopy in septic acute renal failure: a systematic review. Am J Kidney Dis 2006;48:695–705.

5 Bagshaw SM, Langenberg C, Wan L, May CN, Bellomo R: A systematic review of urinary findings in experimental septic acute renal failure. Crit Care Med 2007;35:1592–1598.

6 Hotchkiss RS, Swanson PE, Freeman BD, Tinsley KW, Cobb JP, Matuschak GM, Buchman TG, Karl IE: Apoptotic cell death in patients with sepsis, shock, and multiple organ dysfunction. Crit Care Med 1999;27:1230–1251.

7 Langenberg C, Bagshaw SM, May CN, Bellomo R: The histopathology of septic acute kidney injury: a systematic review. Crit Care 2008;12:R38.

8 Himmelfarb J, Joannidis M, Molitoris B, Schetz M, Okusa MD, Warnock D, Laghi F, Goldstein SL, Prielipp R, Parikh CR, Pannu N, Lobo SM, Shah S, D'Intini V, Kellum JA: Evaluation and initial management of acute kidney injury. Clin J Am Soc Nephrol 2008;3:962–967.

9 Tian J, Barrantes F, Amoateng-Adjepong Y, Manthous CA: Rapid reversal of acute kidney injury and hospital outcomes: a retrospective cohort study. Am J Kidney Dis 2009;53:974–981.

10 Uchino S, Bellomo R, Bagshaw SM, Goldsmith D: Transient azotaemia is associated with a high risk of death in hospitalized patients. Nephrol Dial Transplant 2010 [Epub ahead of print].

11 Perlmutter M, Grossman SL, Rothenberg S, Dobkin G: Urine serum urea nitrogen ratio; simple test of renal function in acute azotemia and oliguria. JAMA 1959; 170:1533–1537.

12 Espinel CH: The FENa test. Use in the differential diagnosis of acute renal failure. JAMA 1976;236:579–581.

13 Miller TR, Anderson RJ, Linas SL, Henrich WL, Berns AS, Gabow PA, Schrier RW: Urinary diagnostic indices in acute renal failure: a prospective study. Ann Intern Med 1978;89:47–50.

14 Platt JF, Rubin JM, Ellis JH: Acute renal failure: possible role of duplex Doppler US in distinction between acute pre-renal failure and acute tubular necrosis. Radiology 1991; 179:419–423.

15 Chew SL, Lins RL, Daelemans R, Nuyts GD, De Broe ME: Urinary enzymes in acute renal failure. Nephrol Dial Transplant 1993;8: 507–511.

16 Steinhäuslin F, Burnier M, Magnin JL, Munafo A, Buclin T, Diezi J, Biollaz J: Fractional excretion of trace lithium and uric acid in acute renal failure. J Am Soc Nephrol 1994;4:1429–1437.

17 Izumi M, Sugiura T, Nakamura H, Nagatoya K, Imai E, Hori M: Differential diagnosis of prerenal azotemia from acute tubular necrosis and prediction of recovery by Doppler ultrasound. Am J Kidney Dis 2000;35: 713–719.

18 Carvounis CP, Nisar S, Guro-Razuman S: Significance of the fractional excretion of urea in the differential diagnosis of acute renal failure. Kidney Int 2002;62:2223–2229.

19 Parikh CR, Jani A, Melnikov VY, Faubel S, Edelstein CL: Urinary interleukin-18 is a marker of human acute tubular necrosis. Am J Kidney Dis 2004;43:405–414.

20 Pépin MN, Bouchard J, Legault L, Ethier J: Diagnostic performance of fractional excretion of urea and fractional excretion of sodium in the evaluations of patients with acute kidney injury with or without diuretic treatment. Am J Kidney Dis 2007;50:566–573.

21 Perazella MA, Coca SG, Kanbay M, Brewster UC, Parikh CR: Diagnostic value of urine microscopy for differential diagnosis of acute kidney injury in hospitalized patients. Clin J Am Soc Nephrol 2008;3:1615–1619.

22 Nickolas TL, O'Rourke MJ, Yang J, Sise ME, Canetta PA, Barasch N, Buchen C, Khan F, Mori K, Giglio J, Devarajan P, Barasch J: Sensitivity and specificity of a single emergency department measurement of urinary neutrophil gelatinase-associated lipocalin for diagnosing acute kidney injury. Ann Intern Med 2008;148:810–819.

Shigehiko Uchino, Associate Professor
Department of Anesthesiology, Jikei University School of Medicine
3-19-18, Nishi-Shinbashi, Minato-ku
Tokyo 105-8471 (Japan)
Tel. +81 3 3433 1111, Fax +81 3 5401 0454, E-Mail s.uchino@jikei.ac.jp

New Horizons in Critical Care Nephrology
Ronco C, Bellomo R, McCullough PA (eds): Cardiorenal Syndromes in Critical Care.
Contrib Nephrol. Basel, Karger, 2010, vol 165, pp 345–356

Acute Kidney Injury in the Pediatric Population

Cristiana Garisto[a] · Isabella Favia[a] · Zaccaria Ricci[a] · Marco Averardi[a] · Sergio Picardo[a] · Dinna N. Cruz[b]

[a]Department of Pediatric Cardiosurgery, Bambino Gesù Hospital, Rome, [b]Department of Nephrology, Dialysis & Transplantation, International Renal Research Institute, San Bortolo Hospital, Vicenza, Italy

Abstract

The care of acute kidney injury (AKI) in critically ill children shares several features with adult AKI with some critical distinctions: in both settings, however, the exact identification of renal dysfunction, in-depth knowledge of disparate risk factors and patient-specific management are the primary targets in order to provide optimal care. This article will specifically review recent work published on pediatric AKI about definition and epidemiology, the possible etiologies in specific conditions, and the newest laboratory investigations necessary to diagnose AKI severity. A short description of pediatric renal replacement therapies and their potential application to extracorporeal membrane oxygenation will also be described.

Epidemiology, Diagnosis and Outcome

Acute kidney injury (AKI) incidence is generally less than 1% when renal replacement therapy (RRT) requirement is used to define it [1]: requirement for RRT is a most strict AKI definition. Incidence increases when less strict AKI definitions are used, such as doubling of serum creatinine (SCr). Study population characteristics will also influence the estimate of AKI incidence; studies including more severely ill patients will show a higher rate of AKI development. Two recent prospective pediatric AKI studies showed the influence of these factors. In the first study, almost all patients admitted to the pediatric intensive care unit (PICU) were eligible for enrollment and AKI was defined as SCr level doubling [1]. In the second study, only patients receiving invasive mechanical ventilation and vasopressors were included and AKI was defined as a 1.5 times or greater SCr increase [2].

Table 1. pRIFLE criteria

	Estimated creatinine clearance	Urine output
Risk	decrease by 25%	<0.5 ml/kg/h for 8 h
Injury	decrease by 50%	<0.5 ml/kg/h for 16 h
Failure	decrease by 75% or <35 ml/min/1.73 m^2	<0.3 ml/kg/h for 24 h or anuric for 12 h
Loss	persistent failure >4 weeks	
End stage	end-stage renal disease (persistent failure >3 months)	

Both population and definition characteristics in the latter study led to a higher estimate of AKI incidence (82 vs. 4.5%) than in the former. However, the actual occurrence of AKI in each PICU may not have truly differed. AKI ascertainment merely differed in relation to the study population and AKI definition. Currently, AKI generally is defined by changes in SCr level, reflecting changes in glomerular filtration rate (GFR) [3, 4]. For many reasons, SCr level is an inaccurate marker of GFR and a late marker of AKI. SCr concentrations are highly affected by muscle mass, which not only changes with age and height of a child, but also with changes in muscle mass associated with hospitalization-associated malnutrition. Notably, critically ill newborn are the patients at highest risk for AKI with the lowest muscle mass: it frequently occurs to observe oliguric neonates with a pseudo-normal SCr level. As GFR decreases acutely, SCr secretion by the renal tubules will increase, leading to falsely low SCr values that may not capture AKI occurrence. Despite these problems, acute SCr change is the reference standard, albeit imperfect, by which epidemiologic AKI studies are based. Another potential problem in interpreting findings of AKI epidemiologic studies is the issue surrounding baseline renal function. If a lower estimate of baseline SCr is used to detect acute changes in SCr, then the estimated incidence of AKI will be higher; when a more conservative or higher estimate of baseline SCr is used, then AKI incidence will clearly be lower. A report on the different methods for defining baseline renal function found that the incidence and severity distribution of AKI were highly affected by how baseline renal function was defined [5].

The work endorsed in the last 6 years by the Acute Dialysis Quality Initiative for the definition of AKI [3] was recently modified in order to be applicable to the pediatric setting [2]. Such pediatric-modified RIFLE (pRIFLE: pediatric Risk, Injury, Failure, Loss, End-Stage Renal Disease) criteria (table 1) were essentially modified by the utilization of estimated creatinine clearance (calculated using the Schwartz formula) in order to index disparate creatinine values to the different ages and weights of the pediatric population. Minor modifications were made also to the urine output criteria (the 'Risk' and 'Injury' time for oliguria evaluation were slightly prolonged). pRIFLE confirmed that AKI occurred in 82% of the most critically ill children admitted to the PICU. AKI defined by these criteria was an independent risk factor for mortality and increased hospital length of stay. Although the pRIFLE criteria are not currently applicable in the clinical setting for medical decision-making, they provide a multidimensional research tool to assist with AKI descriptive and outcome studies and also allows clinicians and researchers to speak the same language when discussing the presence of AKI.

A great amount of work has been recently done on the utilization of early renal injury biomarkers in pediatric AKI diagnosis. In this respect, neutrophil gelatinase-associated lipocalin (NGAL) is one that has received the most attention in the past year. Human NGAL is a 25-kDa protein covalently bound to gelatinase, and is one of the most upregulated transcripts in the kidney very early after acute injury. The NGAL Meta-Analysis Investigator Group published the results of the analysis of data from 19 studies, 7 of which were pediatric (about 690 children enrolled) [6]. The authors found that in the literature, different definitions of AKI, various settings of AKI, and varying timings of NGAL measurement with regard to a renal insult have been used to assess the predictive value of NGAL level, thus creating effective modifiers of NGAL's usefulness as a biomarker. Furthermore, a clear cutoff NGAL concentration for the detection of AKI has not yet been reported. Overall, NGAL levels clearly appeared to be of diagnostic and prognostic value for AKI: this was improved in cardiac surgery patients compared with critically ill patients, and in children compared with adults. The diagnostic accuracy of plasma/serum NGAL was similar to that of urine NGAL. This biomarker was a useful prognostic tool with regard to the prediction of RRT initiation and in-hospital mortality. It is possible that in the next year this and other biomarkers will routinely enter the routine panel examinations for pediatric AKI diagnosis.

Children with AKI requiring RRT, particularly neonates and those with multiple organ dysfunctions, have a high mortality rate, ranging from 30 to 70% [5, 7]. However, mild pediatric AKI may also be associated with mortality, independent of illness severity [2]. Additional research is needed to determine the strength of the association between milder forms of AKI and mortality. Of particular interest is the long-term renal outcome of patients who have hospital-acquired AKI: up to 60% of children who had hospital-acquired AKI had some form of renal abnormality at the 1- to 5-year follow-up evaluation, defined as hematuria, hypertension, microalbuminuria, low GFR [8].

Distinctive Clinical Features of Pediatric AKI

Infants physiologically show *low GFR*, as a result of low mean arterial blood pressure and high renovascular resistance. This makes the infant kidney vulnerable.

Adequate GFR is maintained by postglomerular, efferent arteriolar vasoconstriction, which is mainly dependent on angiotensin II activity higher than in adults. This results in a higher sensitivity of neonates to the administration of inhibitors of ACE compared with adults [7]. Similarly, the use of prostaglandins synthesis inhibitors, such as indomethacin, used to promote the closure of patent ductus arteriosus, may blunt the vasodilation needed to maintain adequate perfusion of the newborn kidney. *Hypoxemia* reduces renal blood flow and GFR, induces hypovolemia, hypotension and activation of renin-aldosterone-angiotensin system (RAAS). Preservation of adequate oxygenation in this phase is crucial [7]. Vasomotor AKI deriving from *septicemia* is usually part of a multiorgan failure. Induced vasoactive mediators (RAAS, endothelin, thromboxane A_2) may also be present in non-infected patients as a consequence of preexisting congestive heart failure [9]. *Hypothermia* is associated with renal vasoconstriction and fall in GFR. Preservation of body temperature is essential for the infant's renal function [7]. *Positive pressure ventilation* decreases venous return and cardiac output and increases renal sympathetic nervous activity and vasopressin serum levels. In this case, fluid administration that might potentially overcome the problem is hampered by the need for fluid restriction [7]. *Nephrotoxicity* of drugs or of intravenous contrast media in the preoperative period may induce tubular injury. *Cardiopulmonary bypass* (CPB) per se may affect renal function both because of hemodynamic changes and because of the activation of immune responses. Hemodynamic changes are determined by the balance between minute oxygen consumption (VO_2) and perfusion pressure during CPB. VO_2 is a function of perfusion pressure and, in turn, the non-pulsatile flow of CPB alters the arterial resistance occurring during normal pulsatile flow [7]. Although the exact influence of this on the kidney function of infants is not known, it is likely that CPB itself induces impairment of organ perfusion. This is in agreement with the finding of the association between CPB duration and the risk of AKI in infants [10]. Surgical trauma, blood contact with CPB surface, endotoxemia, ischemia and high levels of free hemoglobin contribute to the initiation or the maintenance of the *systemic inflammatory response syndrome* in infants. Miniaturization of CPB circuit and the use of intraoperative ultrafiltration are improvements that could have a significant impact on the prevention of AKI in infants during CPB [7]. The *degree of cardiac performance* plays a central role in the preservation of kidney function. Both persistence of cardiac failure and/or residual heart defect or failed correction expose the infant to the risk of AKI.

Other specific patient populations who are at higher risk of developing AKI have been studied in some detail. Patients receiving stem cell transplants are at substantial risk of developing AKI for several reasons, including the extensive use of nephrotoxic medications, veno-occlusive disease in association with hepatorenal syndrome, the high incidence of sepsis, and tumor lysis syndrome [8]. Because of the large amounts of fluid received during their treatment, these patients are also at particularly high risk of developing substantial fluid overload.

AKI Therapy in Infants

Similarly to the adult setting, a 'magic bullet' does not exist to prevent, protect or heal pediatric AKI. As a general rule, optimization of hemodynamics and cardiac function is the priority in all children with a congenital heart defect or unstable hemodynamics. Infants suffering from low cardiac output with significantly impacted kidney perfusion are usually given with inotropes (dopamine, milrinone), vasopressors (high-dose dopamine, epinephrine), splanchnic vasodilators (milrinone, fenoldopam) and diuretics (furosemide) that must be titrated in order to achieve the best equilibrium between cardiac function (contractility), mean arterial pressure (systemic flow), blood oxygenation (pulmonary flow) and vital organ perfusion (cerebral, renal, hepatic, gastrointestinal function).

Loop diuretics are the most used in the oliguric critically ill children and furosemide is by far the most popular one [11]. Dosage and administration modality vary from boluses (1 mg/kg every 12 or 8 h) to continuous infusion (up to 10–20 mg/kg/day). The administration strategy in these patients has been recently re-evaluated in the light of the results obtained in post-heart surgery infants treated with continuous vs. intermittent furosemide. Continuous furosemide infusion has been demonstrated to be generally more advantageous if compared to bolus administration in that it yields an almost comparable urinary output with a much lower dose, less hourly fluctuations [12] and less urinary wasting in sodium and chloride [13]. Doses of continuous furosemide infusion range from 0.1 to 0.2 mg/kg/h. The increase of the initial dose until the desired urinary output is reached is the most used strategy. However, in a 3-day trial on 13 post-heart surgery infants and children, van der Vorst et al. [14] suggested a starting dose of 0.2 mg/kg/h and eventual tapering. The major side effects of diuretic use are electrolyte disturbance (hypokalemia and hyponatremia), metabolic alkalosis with hypocloremia and diuretic resistance. This last effect consists in an absolute or relative inefficiency of diuretic standard dosing and may depend on heart failure per se (inability to reach the optimal peak intraluminal levels of drug; hypoalbuminemia that causes less intravascular binding of the diuretics and less delivery to the proximal tubular cells), hyponatremia (hyperaldosteronism, vasopressin production and less free water excretion), and the so-called 'braking effect' (decreased clinical responsiveness to diuretics with time due to possible sodium retention secondary to volume changes and activation of the RAAS and adrenergic mechanisms). Strategies to overcome diuretic resistance include use of continuous infusions and increased dosages of loop diuretics, use of combined therapy to block sodium resorption at multiple sites of action, correction of electrolyte imbalance, metabolic derangements and excessive intravascular depletion [11].

Recently, a great amount of interest on the use of fenoldopam, a dopamine-receptor agonist, has been raised. Fenoldopam selectively binds to dopamine-1 receptors on smooth muscle cells of renal and splanchnic vascular beds. Activation of these receptors increases intracellular cyclic adenosine monophosphate-dependent protein kinase A activity, enhancing relaxation of vascular smooth muscle [15]. In the kidneys an increased concentration of cyclic adenosine monophosphate in the proximal tubules and medullary portion of the ascending loop of Henle inhibits the sodium-potassium adenosine triphosphatase pump and the sodium-hydrogen exchanger. Fenoldopam infusion blunts aldosterone production and results in increased renal blood flow, urinary sodium excretion, and urine output. Fenoldopam is rapidly titratable, with an elimination half-life of about 10 min [16]. In a retrospective series of 25 post-CPB neonates, Costello et al. [17] reported a significant improvement of diuresis with the use of fenoldopam as compared with chlorothiazide and furosemide. In a recent prospective controlled trial in post-CPB neonates, low-dose fenoldopam infusion did not show beneficial effects in urine output compared with a control population [18]. A slight trend to a higher urine output and a more negative fluid balance early after surgery could be observed. Fenoldopam did not cause any side effect and drug-related tachycardia or hypotension. In this light, it is possible that neonatal kidneys were relatively resistant to stimulation of dopamine-1 receptors related to ontogenic differences in receptor density, affinity, coupling to intracellular second messengers or more distal mechanisms [19]: these receptors might require higher fenoldopam doses to achieve significant clinical effects. Further studies are required to determine if larger doses are beneficial in neonates undergoing cardiac surgery.

Renal Replacement Therapy in Infants

The indication to RRT in children has changed through the years and the present tendency is that of a wider application of this kind of treatment [1]. This may be due to two main causes: although no clear recommendation is made for the application of RRT in children without AKI, it is now widely accepted that RRT is able to positively affect the clinical course of fluid overload, multiple organ disease syndrome (MODS) and the correlated capillary leak syndrome [7]. Up to 20% of all pediatric MODS are represented by children who underwent cardiac surgery [20], then the prevention of fluid accumulation has been clearly associated with an improved survival [21, 22].

The two dialysis modalities most frequently used in infants are peritoneal dialysis (PD) and continuous RRT (CRRT).

Peritoneal Dialysis

PD is relatively easy to perform, it does not require heparinization or vascular access (often complicated in infants) and it is also generally well tolerated in hemodynamically unstable patients [7]. Nonetheless, one of the main disadvantages of PD in these patients is a relative lack of efficiency, especially in water removal with direct consequences on fluid balance and frequent limitation of parenteral nutrition in particular when the treatment of a highly catabolic patient is required [23]. Given these limitations, the early application of PD in order to achieve the prevention and treatment of fluid overload is presently accepted [24]. In particular, infants and children with specific risk factors for AKI should be considered for the preventive use of PD: prevention of volume overload prompted some authors to deliver postoperative prophylactic PD in neonates and infants after complex congenital cardiac surgery [15].

Continuous Renal Replacement Therapies
Both ultrafiltration and solute clearance occur rather slowly in patients undergoing PD. Consequently, PD may not be the optimal modality for patients with severe volume overload who require rapid ultrafiltration, or for patients with severe life-threatening hyperkalemia who require rapid reduction of serum potassium. Moreover, the amount of ultrafiltration is often unpredictable due to the impaired peritoneal perfusion in hemodynamically unstable patients like post-heart surgery infants. In such conditions, the ultrafiltration capacity of PD may not cope with the desired amount of fluid removal and, if scarce, it is obligatorily dedicated to patient's weight loss rather than to the balance of nutrition intake. These limitations of PD explain the ongoing diffusion of extracorporeal dialysis in critical pediatric patients and in post-heart surgery infants [7]. Extracorporeal dialysis can be managed with a variety of modalities, including intermittent hemodialysis, and continuous hemofiltration or hemodiafiltration. The choice of dialysis modality to be used is influenced by several factors, including the goals of dialysis, the unique advantages and disadvantages of each modality, and institutional resources. Intermittent dialysis may not be well tolerated in infants because of its rapid rate of solute clearance [25] and in particular in hemodynamically unstable pediatric cardiac surgery patients [26]. These children are generally treated by CRRT that provides both fluid and solute re-equilibration and removal of proinflammatory mediators.

Commercially available circuits generally have a huge priming volume (in relation to smaller patients) and coupled with monitors that provide an accurate fluid balance have rendered CRRT feasible in infants. Catheter size ranges from 6.5 Fr (7.5 cm long) for less than 10 kg patients to 8 Fr (15–20 cm long) for 11–15 kg patients. Blood priming may be indicated if >8 ml/kg are necessary to fill the CRRT circuit. A full anticoagulation must be always be maintained in order to avoid excessive blood loss in case of circuit clotting [27]. Nevertheless, clinicians are currently forced to apply technical solutions that have been developed for critically ill adults, and the issue of CRRT in infants and neonates still presents many concerns (see below).

Post-heart surgery infants represent a peculiar population in relation to the RRT requirement: firstly, they are most exposed to the risk of water accumulation,

AKI and inflammation due to nature of hemodilutional, hypothermic CPB and to the massive exposure of blood to the artificial surface of CPB [7]. Second, being the exact moment of renal-inflammatory injury (i.e. CPB) is predictable, post-heart surgery patients receive ultrafiltration already during CPB in a preventive way with the filter placed in parallel with the CPB circuit in order to remove inflammatory mediators from the beginning of their generation [7]. There is evidence that this approach is able to exert positive effects on hemodynamics, metabolism and inflammation in the postoperative period [28]. In the setting of MODS and sepsis, several studies showed a statistical difference in the percentage of fluid overload of children with severe renal dysfunction requiring RRT. At the time of dialysis initiation, survivors tended to have less fluid overload than non-survivors [12, 13]. For this reason, in children, priority is now given to the correction of water overload. In fact, different to adult patients where the dialysis dose may play a key role, restoring an adequate water content in small children is the main independent variable for outcome prediction. This concept is much more important in critically ill small children, where capillary leak syndrome is a dramatic manifestation of MODS and a relatively large amount of fluid must be administered in order to deliver an adequate amount of drug infusion, parenteral/enteral nutrition, and blood derivates. With regard to the CRRT modality, the low blood flow rates used in pediatric patients raise concerns about high ultrafiltration during postdilution hemofiltration because of excessive hemoconcentration and filter clotting; in this light, hemodialysis generally appears to be the optimal modality for maximizing clearance of small solutes during CRRT. Nevertheless, if we consider the advantages of hemofiltration with respect to hemodialysis on the clearance of medium and higher molecular weight solutes together with the increased risk of filter clotting, predilution hemofiltration might be the preferred modality in small patients. There are no randomized trials guiding the prescription of CRRT in children: a small solute clearance of 2 l/h · 1.73 m^2 has been recommended in children [cf. www.pcrrt.com].

A peculiar patient with respect to this general description is the infant with AKI and an extracorporeal membrane oxygenation treatment (ECMO) [29]: the CRRT circuit may be placed in series or in parallel and countercurrent to the ECMO circuit. In this last case, the blood into the CRRT circuit runs in the opposite direction with the one in the ECMO circuit with its arterial side connected to the circuit after the pump (generally after the oxygenator) [30]. The venous side is connected to the bladder that collects patient venous return. With this set up the CRRT circuit receives the blood after the ECMO pump, a positive pressure segment, and it returns it to the ECMO bladder where the pressure is close to zero. If this set up causes a minimal recirculation effect, it warrants the best safety features for the ECMO circuit. Shaheen et al. [31] recently presented their experience with two different subgroups of children: one group that required hemofiltration alone and the second group that required hemofiltration and ECMO. Not surprisingly, the authors identified a higher mortality rate in those patients requiring CRRT and ECMO compared with those patients requiring hemofiltration alone. The authors promoted the concept that certain therapies should be reserved for experienced teams. Performing CRRT in a heterogeneous population with large age and weight ranges poses significant clinical and technical challenges. The low frequency of CRRT use, as well as the use of other extracorporeal therapies, also raises problems with maintaining nursing skills. Objective clinical and biochemical markers for commencing CRRT alone or in combination with ECMO remain to be defined. In a recent article by Hoover et al. [32] on fluid management with continuous venovenous hemofiltration (CVVH) in pediatric patients receiving ECMO support, the authors showed that the use of CVVH in ECMO was associated with improved fluid balance, increased caloric intake and less diuretic administration compared with case-matched ECMO controls. The authors concluded that routine application of CVVH to ECMO circuits should be considered and they called for a randomized trial to evaluate such strategy. Similarly, Blijdorp et al. [33] enrolled 15 ECMO newborns with hemofiltration and compared them with 46 patients without hemofiltration (control group). According to these authors, time on ECMO from decannulation until extubation was significantly shorter in the hemofiltration group. No significant differences in mortality were evident. However, the potentially detrimental effects of RRT added to ECMO should not be underestimated.

The CARPEDIEM Project

The Cardio-Renal, Pediatric Dialysis Emergency Machine (CARPEDIEM) project was designed in order to create the basis for the conception of an RRT equipment specifically dedicated to newborns and small infants with a weight range of 2.5–9.9 kg and with a body surface area from 0.15 to 0.5. In these patients the total bloodstream amount ranges from 200 ml to <1 liter meaning that total body water content varies from 1 to 5 liters. Vascular access is also an important issue, since sometimes the smallest 6.5-Fr catheter totally occludes femoral venous drainage or obligates to extremely low extracorporeal blood flow rates. In current clinical practice, application of dialysis equipment to pediatric patients weighing <10 kg is administered 'off label': this means that technology is now substantially 'adapted' to smaller patients with great concerns about outcomes and side effects of such extracorporeal therapy. On the other side, it remains to be verified if a monitor dedicated to neonates and small infants, able to interact with 'problematic' vascular accesses (elevated negative or positive pressures), designed in order to provide exact blood flow rates between 10 and 50 ml/min and with an hourly ultrafiltration error <5 ml/h, would be able to optimize and finally improve the care of pediatric AKI. Finally, since industries do not perform specific tests for treatments in smaller patients and safety features are not specifically projected, legal concerns may arise when operators decide to prescribe these therapies.

In this light, the CARPEDIEM project is aimed to search for funding and financial partnerships in order to achieve five short- and medium-term goals: (1) identify optimal prescriptions and technical requirements for neonatal RRT; (2) design a dedicated equipment; (3) manufacture such a machine and make possible the large-scale production of its disposable material; (4) validate its use in clinical practice, and (5) develop a multicenter trial to define ideal prescription and application of neonatal RRT. The CARPEDIEM project has been conceived in Vicenza, Italy, with the collaboration of two Italian pediatric cardiosurgical centers (Milan and Rome). At the moment, technical support is entrusted to an Italian dialysis company. Such a network is expected to develop to a multinational study group when the clinical phase has been reached.

Conclusion

The development of recent standardized definitions of AKI and improvement of the knowledge about pediatric AKI epidemiology have allowed to stratify outcome by AKI severity and by several risk factors: neonates, children undergoing CPB surgery, stem cell transplantation, or with multiple organ dysfunction syndrome are at highest risk for AKI. This might lead to optimize and tailor AKI therapy and prevention in critically ill children. However, despite advances in acute pediatric dialysis therapy and in overall care of critically ill children, all forms of pediatric AKI are still associated with a high mortality rate, necessitating more research in early AKI identification and therapeutic trials. Furthermore, strong evidence (i.e. coming from a prospective trial) about the impact of fluid balance and its eventual control on outcome in pediatric oliguric patients is still lacking. Finally, dedicated RRT machines should be applied and validated to the specific setting of pediatric renal replacement in order to improve safety and efficacy, and to extend the application of this fundamental approach to pediatric renal dysfunction.

References

1 Bailey D, Phan V, Litalien C, Ducruet T, Merouani A, Lacroix J, et al: Risk factors of acute renal failure in critically ill children: a prospective descriptive epidemiological study. Pediatr Crit Care Med 2007;8:29–35.

2 Akcan-Arikan A, Zappitelli M, Loftis LL, Washburn KK, Jefferson LS, Goldstein SL: Modified RIFLE criteria in critically ill children with acute kidney injury. Kidney Int 2007;71:1028–1035.

3 Bellomo R, Ronco C, Kellum JA, Mehta RL, Palevsky P: Acute renal failure – definition, outcome measures, animal models, fluid therapy and information technology needs. Second International Consensus Conference of the Acute Dialysis Quality Initiative (ADQI) Group. Crit Care 2004;8: R204–R212.

4 Schneider J, Khemani R, Grushkin C, Bart R: Serum creatinine as stratified in the RIFLE score for acute kidney injury is associated with mortality and length of stay for children in the pediatric intensive care unit. Crit Care Med 2010;38:933–939.

5 Zappitelli M, Parikh CR, Akcan-Arikan A, Washburn KK, Moffett BS, Goldstein SL: Ascertainment and epidemiology of acute kidney injury varies with definition interpretation. Clin J Am Soc Nephrol 2008;3:948–954.

6 Haase M, Bellomo R, Devarajan P, Schlattmann P, Haase-Fielitz A, NGAL Meta-Analysis Investigator Group: Accuracy of neutrophil gelatinase-associated lipocalin in diagnosis and prognosis in acute kidney injury: a systematic review and meta-analysis. Am J Kidney Dis 2009; 54:1012–1024.

7 Picca S, Ricci Z, Picardo S: Acute kidney injury in an infant after cardiopulmonary bypass. Semin Nephrol 2008;28:470–476.

8 Zappitelli M: Epidemiology and diagnosis of acute kidney injury. Semin Nephrol 2008;28: 436–446.

9 Torre-Amione G, Kapadia S, Benedict C, Oral H, Young JB, Mann DL: Proinflammatory cytokine levels in patients with depressed left ventricular ejection fraction: a report from the Studies of Left Ventricular Dysfunction (SOLVD). J Am Coll Cardiol 1996;27:1201–1206.

10 Picca S, Principato F, Mazzera E, Corona R, Ferrigno L, Marcelletti C, Rizzoni G: Risks of acute renal failure after cardiopulmonary bypass surgery in children: a retrospective 10-year case-control study. Nephrol Dial Transplant 1995;10:630–636.

11 Dickerson HA, Chang AC: Diuretics; in Chang AC, Towbin JA (eds): Heart Failure in Children and Young Adults. Philadelphia, Saunders, 2006, pp 453–467.

12 Luciani GB, Nichani S, Chang AC, Wells WJ, Newth CJ, Starnes VA: Continuous versus intermittent furosemide infusion in critically ill infants after open heart operations. Ann Thorac Surg 1997;64:1133–1139.

13 Singh NC, Kissoon N, al Mofada S, Bennett M, Bohn DJ: Comparison of continuous versus intermittent furosemide administration in postoperative pediatric cardiac patients. Crit Care Med 1992;20:17–21.

14 Van der Vorst MM, Ruys-Dudok van Heel I, Kist-van Holthe JE, den Hartigh J, Schoemaker RC, Cohen AF, Burggraaf J: Continuous intravenous furosemide in haemodynamically unstable children after cardiac surgery. Intensive Care Med 2001;27: 711–715.

15 Lokhandwala MF, Barrett RJ: Cardiovascular dopamine receptors: physiological, pharmacological and therapeutic implications. J Auton Pharmacol 1982;2:189–215.

16 Weber RR, McCoy CE, Ziemniak JA, Frederickson ED, Goldberg LI, Murphy MB: Pharmacokinetic and pharmacodynamic properties of intravenous fenoldopam, a dopamine-1 receptor agonist, in hypertensive patients. Br J Clin Pharmacol 1988; 25:17–21.

17 Costello JM, Thiagarajan RR, Dionne RE, Allan CK, Booth KL, Burmester M, Wessel DL, Laussen PC: Initial experience with fenoldopam after cardiac surgery in neonates with an insufficient response to conventional diuretics. Pediatr Crit Care Med 2006;7:28–33.

18 Ricci Z, Stazi GV, Di Chiara L, Morelli S, Vitale V, Giorni C, Ronco C, Picardo S: Fenoldopam in newborn patients undergoing cardiopulmonary bypass: controlled clinical trial. Interact Cardiovasc Thorac Surg 2008;7:1049–1053.

19 Cheung PY, Barrington KJ: Renal dopamine receptors: mechanisms of action and developmental aspects. Cardiovasc Res 1996;31:2–6.

20 Tantaleán JA, León RJ, Santos AA, Sánchez E: Multiple organ dysfunction syndrome in children. Pediatr Crit Care Med 2003;4:181–185.

21 Goldstein SL, Somers MJ, Baum MA, Symons JM, Brophy PD, Blowey D, Bunchman TE, Baker C, Mottes T, McAfee N, Barnett J, Morrison G, Rogers K, Fortenberry JD: Pediatric patients with multi-organ dysfunction syndrome receiving continuous renal replacement therapy. Kidney Int 2005;67:653–658.

22 Sutherland SM, Zappitelli M, Alexander SR, Chua AN, Brophy PD, Bunchman TE, Hackbarth R, Somers MJ, Baum M, Symons JM, Flores FX, Benfield M, Askenazi D, Chand D, Fortenberry JD, Mahan JD, McBryde K, Blowey D, Goldstein SL: Fluid overload and mortality in children receiving continuous renal replacement therapy: the prospective pediatric continuous renal replacement therapy registry. Am J Kidney Dis 2010;55:316–325.

23 Ricci Z, Morelli S, Ronco C, Polito A, Stazi GV, Giorni C, Di Chiara L, Picardo S: Inotropic support and peritoneal dialysis adequacy in neonates after cardiac surgery. Interact Cardiovasc Thorac Surg 2008;7:116–120.
24 Alkan T, Akçevin A, Türkoglu H, Paker T, Sasmazel A, Bayer V, Ersoy C, Askn D, Aytaç A: Postoperative prophylactic peritoneal dialysis in neonates and infants after complex congenital cardiac surgery. ASAIO J 2006;52: 693–697.
25 Ronco C, Ricci Z: Renal replacement therapies: physiological review. Intensive Care Med 2008;34:2139–2146.
26 Flynn JT: Choice of dialysis modality for management of pediatric acute renal failure. Pediatr Nephrol 2002;17:61–69.
27 Wilkins B, Morrison A: Pediatric CRRT; in Atlas of Hemofiltration. Philadelphia, Saunders, 2002, pp 59–62.
28 Elliott MJ: Ultrafiltration and modified ultrafiltration in pediatric open heart operations. Ann Thorac Surg 1993;56:1518–1522.
29 Santiago MJ, Sánchez A, López-Herce J, Pérez R, del Castillo J, Urbano J, Carrillo A: The use of continuous renal replacement therapy in series with extracorporeal membrane oxygenation. Kidney Int 2009;76:1289–1292.
30 Ricci Z, Polito A, Giorni C, Di Chiara L, Ronco C, Picardo S: Continuous hemofiltration dose calculation in a newborn patient with congenital heart disease and preoperative renal failure. Int J Artif Organs 2007;30: 258–261.
31 Shaheen IS, Harvey B, Watson AR, Pandya HC, Mayer A, Thomas D: Continuous venovenous hemofiltration with or without extracorporeal membrane oxygenation in children. Pediatr Crit Care Med 2007;8:362–365.
32 Hoover NG, Heard M, Reid C, Wagoner S, Rogers K, Foland J, Paden ML, Fortenberry JD: Enhanced fluid management with continuous venovenous hemofiltration in pediatric respiratory failure patients receiving extracorporeal membrane oxygenation support. Intensive Care Med 2008;34:2241–2247.
33 Blijdorp K, Cransberg K, Wildschut ED, Gischler SJ, Jan Houmes R, Wolff ED, Tibboel D: Haemofiltration in newborns treated with extracorporeal membrane oxygenation: a case-comparison study. Crit Care 2009;13:R48.

Dr. Zaccaria Ricci
Piazza S. Onofrio 4
IT–00100 Rome (Italy)
Tel. +39 0 64456115, Fax +39 0 444993949
E-Mail z.ricci@libero.it

Author Index

Subject Index